DESIGN OF HAZARDOUS MECHANICAL STRUCTURES, SYSTEMS AND COMPONENTS FOR EXTREME LOADS

by

JOHN D. STEVENSON

and

OVIDIU COMAN

ASME PRESS

Library of Congress Cataloging-in-Publication Data

Stevenson, J. D. (John David), 1933–
Design of hazardous mechanical structures, systems and components for extreme loads / John D. Stevenson, Ovidiu Coman.
p. cm.
Includes bibliographical references and index.
ISBN 0-7918-0242-6
1. Structural stability—Standards. I. Coman, Ovidiu. II. Title.
TA656.S765 2006
624.1'72—dc22 2005031812

ACKNOWLEDGMENT

The authors wish to thank Dr. A.N. Bierbraer, St. Petersburg, Russia, author of Seismic Analysis of Structures, Saint Petersburg, NAUKA 1998 and his earlier publications (in Russian) who was an inspiration for this book and who made extremely valuable review comments. We also wish to thank Ms. Margaret Pifer for her assistance in preparing figures and Mrs. Lori Bodziony for her preparation of the book text.

FOREWORD

This book is meant to identify and summarize the extreme loads to include loads resulting from external natural phenomena and man-induced, as well as internal accident hazards phenomena applicable to design of hazardous facility mechanical structures, systems and components. It includes first order definition of the loads and approximate analysis procedures used to design such mechanical structures, systems and components developed over the past 30 years for hazardous facilities to include nuclear, petrochemical and biomedical facilities.

Chapter 1 identifies the extreme loads to be considered and industries to which the extreme loads apply. It also identifies the design classifications and safety goals applied to mechanical Structures, Systems and Components, SSC.

Chapter 2 discusses risk-based and risk-informed SSC Design Basis.

Chapter 3 identifies and describes generalized load phenomena.

Chapter 4 discusses design acceptance criteria as defined by the American society of mechanical Engineers and related organization standards.

Chapter 5 addresses earthquake related design and analysis and Chapter 6 discusses in detail natural hazard phenomena design loads other than seismic loads.

Chapter 7 addresses external blast load design and analysis and Chapter 8 high energy system (accident) SSC rupture loads.

Chapter 9 discusses vibration response of fluid distribution systems.

Chapter 10 covers extreme load design of buried tunnels, vaults and piping.

Chapter 11 discusses Quality Assurance and Control in Engineering, Construction and Procurement of Safety Related Structures, Systems and Components in Hazardous Facilities.

TABLE OF CONTENTS

CHAPTER

1

INTRODUCTION

1.1 PURPOSE

This book has been prepared as a guide to the structural and leak-tight integrity design and analyses of structures, distribution systems and components (SSC), which are addressed by construction[1] standards prepared by the American Society of Mechanical Engineers, ASME and other closely related Standards. Particular application is to SSC, which are subject to extreme load phenomena, whether of a design basis external natural, man-induced or accident phenomena. The book is intended for use in assisting structural-mechanical engineers in their supply of engineering services to the nuclear, petrochemical and biomedical industries.

1.2 SCOPE

1.2.1 Structures

The structures addressed by this guide include metal:

- stacks and towers guyed and unguyed
- vertical tanks

or any other structure addressed by ASME and related standards such as those of the American Petroleum Institute (API), American Institute of Steel Construction (AISC), American Iron and Steel Institute (AISI), American Water Works Association (AWWA), Manufacturer Standardization Society of the Valve and Fitting Industry Standard Practice (MSS-SP), the International Building Code (IBC) and the American Society of Civil Engineers (ASCE).

Not included in this scope are building and other SSC that are supplied as products for, which there are no widely accepted industry construction or procurement standards, and where the manufacturer, supplier or vendor provides a warranty of the product.

1.2.2 Distribution Systems

The distribution systems addressed by this guide include metal:

- piping
- duct

[1] Construction as used in the context of this book includes the activities of administrative control, documentation, material selection, material qualification, design, fabrication, erection, examination, testing and quality assurance.

- tubing
- distribution system supports

or any other distribution system addressed by ASME and related standards, such as those developed by AISC, AISI, MSS-SP, IBC, AWWA and ASCE, as well as the Sheet Metal and Air Conditioning Contractors National Association (SMACNA).

Not included in the scope are distribution systems, which are supplied as part of a product for which there are no widely accepted industry construction standards and where the manufacturer, supplier or vendor provides an acceptable warranty of the product.

1.2.3 Components

The components addressed by this guide include metal:

- pressure vessels
- boilers
- heat exchangers
- chillers
- condensers
- horizontal tanks
- glove boxes
- pumps
- circulators
- valves
- dampers
- cranes and hoists
- compressors
- fans
- filters
- component supports

addressed by ASME and related standards developed by API, AISC, AISI, MSS-SP, IBC, SMACNA and ASCE, as well as the Crane Manufacturers Association of America (CMAA).

Not included in the scope of this book are mechanical components that are supplied as part of a product for which there are no widely accepted industry construction standards and specific electrical components where the manufacturer, supplier or vendor provides an acceptable warranty of the product.

1.2.4 Standards

Standards without a publishing organization listed are from ASME.

The specific latest edition standards addressed or referenced in this guide include:

- General
  - NQA-1, "Quality Assurance Program Requirements for Nuclear Facilities"
  - OM-S/G Standards and Guides for the Operation and Maintenance of Nuclear Power Plants
  - IBC-2000, "International Building Code," International Code Council
  - ASCE–4, "Seismic Analysis of Safety Related Nuclear Structures and Commentary," American Society of Civil Engineers
  - ASCE–7, "Minimum Design Loads for Buildings and Other Structures," American Society of Civil Engineers

- Boilers & Pressure Vessels
  - B&PVC Section I, "Power Boilers"
  - B&PVC Section III, "Rules for Construction of Nuclear Facility Components"
  - B&PVC Section IV, "Heating Boilers"
  - B&PVC Section VIII, "Pressure Vessels"
  - BPE Bio-Processing Equipment
  - B19.1, "Safety Standard for Air Compressor Systems"
- Pumps
  - QME-1, "Qualification of Active Mechanical Equipment Used in Nuclear Power Plants"
  - B73.1, "Specification for Horizontal End Suction Centrifugal Pumps for Chemical Process"
  - B73.2, "Specification for Vertical In-Line Centrifugal Pumps for Chemical Process"
  - API-610, "Centrifugal Pumps for Petroleum, Heavy Duty Chemical and Gas Industry Services," American Petrol Institute
- Valves
  - QME-1, "Qualification of Active Mechanical Equipment Used in Nuclear Power Plants"
  - B16.34, "Valves-Flanged Threaded and Welding End"
  - B16.41, "Functional Qualification Requirements for Power Operated Valve Assemblies for Nuclear Power Plants"
- Piping
  - B31.1, "Power Piping"
  - B31.3, "Process Piping"
- Air Handling Equipment
  - AG-1, "Code on Nuclear Air and Gas Treatment"
  - N509, "Nuclear Power Plants Air-Cleaning Units and Components"
  - SMACNA, "HVAC Systems Duct Design," Sheet Metal and Air Conditioning Contractors National Association
- Cranes and Hoists
  - NOG-1, "Rules for Construction of Overhead and Gantry Cranes"
  - NUM-1, "Rules for Construction of Cranes and Monorails and Hoists"
  - B30.2, "Overhead and Gantry Cranes"
  - B30.9, "Slings"
  - B30.10, "Hoists"
  - CMAA-70, "Specification for Electric Overhead Traveling Cranes," Crane Manufacturer Association of America
- Tanks
  - B&PVC Section III, Subsections NC and ND
  - API-620, "Design and Construction of Large, Welded, Low-Pressure Storage Tanks," American Petroleum Institute
  - API-650, "Atmospheric Welded Steel Tanks for Oil Storage," American Petroleum Institute
  - AWWA-100, "Welded Steel Tanks for Water Storage," American Water Works Association
- Steel Stacks
  - STS-1, "Steel Stacks"
- Steel Supports
  - AISC S328, "Load and Resistance Factor Design Specification for Structural Steel Buildings," American Institute for Steel Construction
  - AISC N690, "Nuclear Facility – Steel Safety Related Structures for Design Fabrication and Erection," American Institute for Steel Construction
  - B&PVC Section III – NF, "Supports"
  - MSS-SP-58, "Pipe Hangers and Supports – Materials, Design and Manufacture," Manufacturers Standardization Society of the Valve and Fitting Industry
  - AISI, "Specification for Design of Cold Formed Steel Structural Members," American Iron and Steel Institute

- Anchors
  - ACI 349 Appendix B, "Steel Embedments-Code Requirements for Nuclear Safety Related Concrete Structures," American Concrete Institute

1.3 FACILITIES REQUIRING CONSIDERATION OF EXTREME LOADS

Table 1.1 lists external event hazard, which should be considered as potential design basis loads on SSC in hazardous facilities.

TABLE 1.1. INDIVIDUAL EXTERNAL, NATURAL AND MAN-INDUCED ACCIDENT AND MALEVOLENT HAZARDS THAT MIGHT BE CONSIDERED A DESIGN BASIS FOR EXTREME LOADS

Hazards	Status	Remarks
Aircraft Impact (Accidental)	3	—
Airports within 5 miles of plant:		—
Impact (large aircraft)		—
Impact (small aircraft)		—
Fire (large aircraft)		—
Fire (small aircraft)		—
Airports more than 5 miles from plant (d):		—
Impact (large aircraft)		—
Impact (small aircraft)		—
Fire (large aircraft)		—
Fire (small aircraft)		—
Avalanche	4	—
Coastal Erosion	1	See Note 1.
Drought[1]	1	Assumes multiple source of ultimate heat sink or ultimate heat sink if not affected by drought, e.g., cooling tower with adequately sized basin.
Earthquake	0	—
Fire, External	0 ,2, 3	—
Flood	4	—
Fog	2	Could, however, increase probability of man-made hazard involving surface vehicles or aircraft.
Forest Fire	2	Site cleared for such fire. Control-room habitability required for smoke.
Frost	2, 5	Snow and ice govern.
Hail	2, 5	Other missiles govern.
High Summer Temperature[1]	1	Ultimate heat sink to be conservatively deigned for extended drought conditions.
High Tide, High Lake Level or High River Stage	0, 4, 5	Included under flood.
Ice Buildup	2, 5	See remark under snow for roof loading. Ice effects on intake structures on northern lakes may require design consideration. Ice blockage of rivers causing flooding is included under river flooding.
Landslide	4	—
Lightning	2	Plant lightning protected by use of industry protection standard or a grounding mat.
Low Lake or River Water Level[1]	1	Ultimate heat sink should be conservatively designed for 30 days of evaporation.
Low Winter Temperature[1]	2	Thermal stresses and embrittlement not significant or they are covered by design codes or standards.
Malevolent Vehicle Intrusion Explosion and Fire (Road, Rail, Ship Vehicle or Air)	0, 6	Generally not considered a "design basis" prior to mid-1990s.

TABLE 1.1. (Continued)

Hazards	Status	Remarks
Meteorite	3	Less than 10^{-7} probability of occurrence per year.
Military, Industrial or Transportation Vehicle Accident (Blast, Missiles, Toxic Substances)	0, 3	—
Pipeline Accident	0, 3	—
(Blast, Missiles, Toxic Substances)		—
Precipitation-Rain and Snow	0, 4, 5	Included under flood.
Retaining Structure Failure (Flood Overtapping or Sudden Drawdown)	3, 4	—
Sandstorm	4	Extreme wind should include these phenomena. Blockage of air intakes with particulate matter and ?? roof loads.
Seiche	0, 4, 5	Included under flood.
Soil Shrink-swell Consolidation	1,	See Note 2.
Storm Surge	0, 4, 5	Included under flood.
Surface Vehicle, Pipeline and Military or Industrial Facility Accident	4	—
Tornado	0, 1	In high- and moderate-tornado-activity regions.
Toxic Gas	5	Control-room habitability required for toxic gas accident. This assumes no operator action outside the control room is required to render the consequences of the event acceptable.
Tsunami	4	—
Volcanic Activity	3, 4	Currently there is no criteria for anti-volcanic design other than IAEA.
Waterspout	4	Considered in conjunction with tornado. Tornado governs. Loading due to water in spout not governing.
Wave Action	0, 4, 5	Included under flood.
Wind	0	—

1. These natural hazards provide some of the design basis for the ultimate heat sink.
2. Site-related characteristics, such as subsidence due to subsurface pumping, mining, cavities or alteration of groundwater regions; active surface faulting, liquefaction potential, sensitive soils and rocks that have expansive, heave and shrinkage characteristics; flood-plane levels are natural phenomena, which should be considered and evaluated during the site suitability evaluation process. Such characteristics either result in (1) the site being considered unsuitable; or (2) proper design consideration and construction techniques are employed to mitigate the hazard.

Note: The following exclusion criteria can be used to eliminate postulated hazards from being included as a design basis.

0. Normally considered in design of safety-related structures, systems and components in hazardous facilities.
1. A phenomenon that occurs slowly or with adequate warning with respect to the necessary time required for taking appropriate protective or mitigating action.
2. A phenomenon that in itself has no significant impact on the operation of a hazardous facility design basis.
3. A phenomenon that usually has a probability of occurrence less than the upper limit for design-basis probability of exceedence.
4. Locate the hazardous facility sufficiently distant from the postulated phenomenon to mitigate its effects (siting).
5. A phenomenon that is included or enveloped by design for another phenomenon. For example, tornado loads may envelope the affects of a small airplane crash.
6. Design consideration is by the Safeguards Group when it is responsible for construction to resist malevolent acts.

Comments:

a. Duration of flood governed by duration of wind.
b. Wind includes hurricane, cyclone or typhoon wind effects.
c. For sites farther than five miles from an airport, estimates of probability of impact are required if the site is near a heavily traveled commercial air traffic lane, or if there are unusual aircraft operations in the vicinity.

1.3.1 Introduction

There are two basic categories of industrial facilities: conventional and potentially hazardous. In the hazardous category are three sub-categories: nuclear, petrochemical and biomedical. It is the hazardous sub-category of facilities that requires consideration of extreme loads for safety-related structures, systems and components contained in those facilities. At the owner's or user's option, extreme load design requirements may be extended to mission-dependent structures, systems and components as a matter of economic necessity.

Conventional industrial facilities typically divide siting criteria and postulated design loads into two categories. Those loads that have a probability of exceedence per year on the facility greater than 10^{-1} are termed service or normal loads. In the second category, termed upset, are severe or occasional loads, which typically have a mean probability of exceedence between 10^{-1} and 10^{-2} per year. If these loads have the ability to (1) significantly impair the facility's mission; (2) cause significant damage to the physical plant; or (3) lead to serious injury or loss of human life, these postulated loads become design basis loads.

Hazardous facility SSC that have a safety-related function typically consider a third category of loads, termed extreme faulted or abnormal loads, whose combined probability of exceedence and resultant probability of failure would result in consequences to the public in excess of acceptable limits as defined by National Industry Consensus Standards or Regulatory Authorities. Where large or undefined releases of toxins, explosive products or radiological materials are postulated, a safety goal range of 10^{-5}/year to 10^{-7}/year or less probability of release typically has been defined. This safety goal typically is set at two orders of magnitude below what would be established as safety goals associated with structural or leak-tight integrity of conventional industrial facilities. As a result of this definition, extreme loads and extreme load combinations considered in design have a practical range of occurrence per year between 10^{-2} to about 10^{-6}, depending on the consequence and the credit taken for design features to prevent or mitigate the consequences of the extreme event occurrence.

1.3.2 Nuclear Facilities

Nuclear facilities are those that store or process significant amounts of radioactive material and waste in product form such that a significant unmitigated risk exists and that nuclear material or waste will be released beyond the facility boundary and affect public health and the environment. Included in this category are:

- Nuclear power plants
- Nuclear spent nuclear fuel storage facilities
- Radioactive waste storage facilities
- Nuclear research reactors
- Isotope production reactors
- Radioactive material processing plants
- Radioactive material storage facilities
- Shipping of radioactive material and waste

Guidance as to what constitutes significant amounts of radioactive material and waste and high radiological levels can be found in Federal Regulation 10CFR71, Appendix A [1].

A relatively complete and detailed set of safety-related guidelines for construction, operation, maintenance and decommissioning of civilian-type nuclear facilities have been issued in the form of Federal Regulations, Standard Review Plans and Regulatory Guides issued by the United States Nuclear Regulatory Commission (U.S. NRC). Noncivilian nuclear facilities requirements and guidance as to construction, operation, maintenance and decommissioning have been developed by the United States Department of Energy (U.S. DOE), with oversight from the Defense Nuclear Facility Safety Board (DNFSB).

For mechanical systems and components, these government regulatory requirements and guidelines are supplemented by voluntary industry consensus standards prepared primarily by the American Nuclear Society and the American Society of Mechanical Engineers.

TABLE 1.2. TYPICAL PRELIMINARY HAZARD CATEGORIZATION OF NUCLEAR FACILITIES BASED ON RADIOLOGICAL INVENTORY

Hazard category of the facility		Inventory GBq[2] Beta-gamma	Alpha
Cat. 1S[1]	Reactor System	>100,000	>10,000
Cat. 1	High	>400,000	>4,000
Cat. 2	Medium	<400,000	<4,000
Cat. 3	Low	<40,000	<400

1. Cat. 1S nuclear reactors where core melt is credible with loss of reactor cooling, or control requires a containment system.
2. 1MBq = 27 curves.

Preliminary hazard classification of nuclear facilities is based on radiological inventory as shown in Table 1.2. Final facility hazard classification on unmitigated release of radioactivity is shown in Table 1.3.

1.3.3 Petrochemical and Biomass Facilities

Petrochemical facilities are those that store or process quantities and associated product forms of toxic, flammable, combustible or explosive materials or waste such that a significant unmitigated risk exists that these materials or waste could cause serious public health and environmental damage beyond the facility boundary. Included in this category are:

- Refinery plants
  - Alkylation units
  - Cat cracking units
  - Cat reforming units
  - Crude units
  - Hydrocracking units
  - Hydrotreation units
- Munitions process and storage plants
- Fertilizer process and storage plants
- Grain process and storage plants
- Transportation of hazardous materials facilities

TABLE 1.3. TYPICAL FINAL HAZARD CATEGORIZATION OF NUCLEAR FACILITIES BASED ON UNMITIGATED RADIOLOGICAL EXPOSURE TO THE PUBLIC OR COLLOCATED WORKER

Hazard Category of the Facility		Unmitigation Radiological Release Dose, S_v (REM) Collocated Worker	Public
Cat. 1S[1]	Reactor System	>1.0 (100)	>0.25 (25)
Cat. 1	High	>1.0 (100)	>0.25 (25)
Cat. 2	Medium	>0.25 (25)	>0.05 (5)
Cat. 3	Low	>0.05 (5)	>0.01 (1)

1. A robustly designed containment system is required so mitigation of release by containment function is assumed when performing normally unmitigated release computations. Robustness is normally defined when design-basis accident conditions are required to be considered as a normal loading condition (design and service level A), and seismic stresses are defined by elastic analysis with inelastic energy absorption factor (F_u) taken as 1.0 as defined in DOE Standard 1020, or R_p is taken as 1.0 as defined in the ASCE 7-2005 Standard.

A series of safety-related guidelines associated with construction, operation and maintenance of these facilities have been developed by the Center for Chemical Process Safety of the American Institute of Chemical Engineers. The most comprehensive guideline to petrochemical facility safety can be found in "Guidelines for Engineering Design for Process Safety" [2]. For mechanical SSC, these guidelines are supplemented primarily by standards of the American Petroleum Institute and the American Society of Mechanical Engineers.

Loads and acceptance criteria typically considered in the design of petrochemical facilities are typically defined in the ASCE-7 and ASME B31.3, as well as ASME B&PVC Section VIII Standards. With the exception of external blast loads, the loads and acceptance criteria are those defined in the hazardous facility category associated with severe load condition. The external blast load is defined in the extreme load category. Additional discussion of hazard design of petrochemical facilities can be found in Section 1.6.2.

1.3.4 Bio-Hazard Facilities

Currently the design and construction of Biomedical facilities are considered hazardous or essential facilities as defined in the ASCE-7 Standard. Additional discussion of the hazard design of biomedical facilities can be found in Section 1.6.3.

1.4 DESCRIPTION OF POTENTIAL DESIGN BASIS HAZARDS

In Table 1 is a list of external events that might be considered in the design of a hazardous facility SSC. Many of the hazards listed there do not become "design basis hazards" or "Loads" because of the exclusion criteria that are also identified in Table 1.

1.5 LOADS TYPICALLY CONSIDERED IN DESIGN OF SAFETY-RELATED STRUCTURES, SYSTEMS AND COMPONENTS (SSC) IN HIGH-HAZARD FACILITIES

1.5.1 Service Load

Design Loads

Normal or Design loads, such as dead (D) and live (L) loads, on systems and components are usually considered with predetermined safety margins in the form of load factors or allowable stresses and are applied as defined in applicable National Building [3, 4] and ASME Construction [5, 6] Codes.

Service or normal loads on mechanical distribution systems and components, in addition to D and L loads, would include loads defined by the facility owner, which typically include design pressure (P) and normal reaction (R_O) loads as well as temperature effect loads (T_o), if differential temperature loads exceed about 100F° (55C°).

1.5.2 Severe Loads

Human Occupancy

The severe loads considered in design are dependent on human occupancy to a significant degree. In general, human occupancy is defined as 72 man hours of occupancy over a 24-hour period.

Natural Hazards

a. Seismic Load, E_O

The severe seismic load is typically defined in National Building Codes as a general requirement for all buildings or other facilities with human occupancy. It is often defined as the equivalent to a median 2×10^{-3}/yr or 10 percent probability of exceedence in 50 years for ordinary buildings, and 1×10^{-3}/yr for essential or commercially hazardous buildings by use of a 1.25 importance multiplication factor applied to ordinary building equipment seismic loads. In recent years this earthquake has been redefined as having a median 4×10^{-4}/yr or 2 percent probability of exceedence in 50 years. However, the earthquake accelerations thus defined are multiplied by a 0.67 factor for structural design purposes, thereby effectively reducing earthquake accelerations for structural design purposes to the earlier 2×10^{-3}/yr criterion.

b. Wind and Precipitation (Rain, Snow or Ice) Loads, W, S
   The design basis wind velocity is also defined on a probabilistic basis. Its selection is typically based on either a several-minute average fastest wind or more recently, since 1995, on a maximum 3-second gust wind speed for a 50-year recurrence interval or 2×10^{-2} mean probability of exceedence for the site exposure level based on the ASCE-7 Standard. The basic wind load pressure on SSC is determined as a function of wind velocity squared, air density, importance factor, height above ground and exposure coefficients as defined in Chapter 6, "References". Note that severe precipitation (rain, snow and ice) loads for ordinary or commercial structures, systems or components are also usually defined for design at the 50-year recurrence interval or 2×10^{-2} probability of exceedence level. For hazardous facilities in the severe load category, the National Building Codes typically decrease this to 10^{-2}/yr for wind and precipitation by use of a multiplication term called an Importance Factor.

   Storm sewers intended to carry off-site precipitation are typically designed for a 25-year recurrence interval (4×10^{-2}/yr probability of exceedence). The excess precipitation not carried by the sewer system is by surface runoff, which may or may not be a design consideration.

Operating Transients Occasional and Upset Loads

Plant-specific anticipated operational steady-state or transients pressure, P_o, component or distribution system support reaction load, R_o, temperature, T_o, and temperature-induced restraint of free end displacement, P_E, loads are included in this category. These typically have 10^{-1} to 10^{-2}/yr probability of exceedence levels.

1.5.3 Extreme Loads

Consequences of hazardous facility safety-related SSC failures differ significantly from those in conventional facilities. In conventional facilities the life safety of human occupants is usually the primary concern. Hazardous facilities potentially release significant amounts of toxic, flammable, explosive or radioactive materials to the environment beyond the site boundaries; hence they have both a short-term and long-term effect on public health and safety and the environment as well as the facility worker. The prevention of such failures in hazardous facilities and the mitigation of their potential consequences are the primary objective of hazardous facility safety design. The types of hazards considered in hazardous facility designs include:

a. Natural phenomena hazards
b. Man-induced external hazards, both accidental and malevolent
c. Postulated design basis accidents
d. Fire

Natural Phenomena Hazards

a. Design Basis Earthquake Loads
   The Design Basis Earthquake (E_s) ground motion has a very low probability of being exceeded during the design life of the plant. It represents the maximum level of ground motion to be used for design purposes for safety-related SSC in the extreme load category.

It should not be confused with the ASCE 7-05 earthquake level, E_o. The E_S evaluation is based on the seismotectonic model and/or detailed knowledge of the seismology, geology and engineering foundation parameters of the site, region and area.

The mean probability of exceedence established for the highest earthquake category hazardous facility SSC is typically 10^{-4}/yr or median probability of exceedence of 10^{-5}/yr [7]. Median rather than mean statistics are typically used for earthquakes because earthquake levels at a particular probability level tend to have highly skewed probability density functions, which tend to make mean values rather unstable.

Regardless of the seismic hazard potential, a design basis earthquake is recommended for every hazardous facility. The recommended minimum level regardless of the site location is a peak-free field surface ground acceleration of 0.1g (zero period of the design median shaped ground response spectrum). This corresponds to a peak spectral ground motion acceleration of 0.25g for 5 percent critically damped systems in the resonant region of response in the frequency range typically between 2 hz and 10 hz.

b. Extreme Straight Winds

Straight winds are observed at a fixed weather station and generally follow an (Extremal) Type I or other skew-positive probability density function. Historically, such wind design parameters have been averaged over a relatively long period of time (i.e., 10 minute to 1.0 hour or fastest one-quarter mile or one-mile wind) and require a gust factor greater than 1.0 to be applied to the design of SSC. Maximum design basis winds for this category are typically developed from yearly maximum wind speed gathered over a 20- to 100-year period. Since 1995 the fastest 3-second duration gust wind is defined for SSC design purposes by ASCE 7-05 Code at the 2×10^{-2}/yr probability of exceedence level. A wind speed having a probability of occurrence equal to or less than 10^{-3}/yr to 10^{-4}/yr has been used in the past to define these wind loads in the extreme straight wind category. At probability levels lower than about 10^{-4}/yr the rare winds associated with cyclonic tornado winds tend to govern design. Note that the hazardous material dispersion potential of straight winds and cyclonic winds are similar and typically exceed those for earthquakes flooding and precipitation. The probability of exceedence levels for wind loads are generally set equal to or lower than those for earthquake and flooding (precipitation) for design purposes.

The applications of rare hurricane, typhoon and cyclone storm effects are usually limited to the coastal regions of large land masses and islands of the Atlantic East and Gulf of Mexico coasts of North America, Central America, Eastern Pacific Coast as well as the Western Pacific and the Indian Ocean adjacent to the Asian main load.

c. Rare Cyclonic Winds

Rare cyclonic winds are distinguished from extreme straight winds in that they affect a particular weather station only once during several or (10's of year) intervals. As a result, statistics for this type of wind are collected on an area basis. Typically, a 400-mile (600-km) length of coastline is used to record hurricane or typhoon statistics and a 5° latitude-longitude square is used to predict tornado wind design velocities [8].

d. Tornado

Tornadoes, which are defined as very high-intensity, small-radius cyclonic winds, occur over large continental land masses that do not have significant natural barriers to the mixing of warm moist air from tropical seas with cold air from the temperate zone. They can also occur within the general circulation of a hurricane, typhoon or cyclone. Historically worldwide they have been considered for design purposes at only relatively hazardous facility sites. When considered they have typically been defined at the 10^{-5} to 10^{-7} per year probability of exceedence level based on generic or large region tornado statistics. The very low probability of exceedence levels (10^{-5} to 10^{-7}) typically established for tornadoes as compared to earthquakes (10^{-3} to 10^{-5}) can be explained at least in part by the potentially wide and rapid dispersion of toxic materials by wind action when this is generally not the case with earthquakes, unless they are accompanied by fire or the potential rupture of a high-energy system.

Unfortunately, one of the regions of the world, east of the Rocky Mountains, with the highest probability of damaging tornadoes is the U.S. and to a lesser extent Canada; hence, tornadoes often have been considered in the design of safety-related SSC in hazardous facilities in this Continental North America geographical area. Such a high tornado hazard zone typically has tornado wind speed exceeding straight wind speed below the 10^{-4}/yr probability of exceedence level.

e. Tsunamis and Seiche

Tsunamis (ocean waves) are the long water waves (with wave periods from 5 to 60 minutes, or in a longer range) impulsively generated by mechanisms such as exploding islands (volcanic eruptions), submerged landslides, rockfalls into bays or the ocean, tectonic displacements associated with earthquakes and underwater explosions. Seiches are similar to tsunamis but typically occur in smaller enclosed bodies of water such as bays and lakes. When tsunami or seiche waves reach a coast, initially a drawdown occurs and has the appearance of a rapidly occurring low tide after which water surges back in the form of a "tidal wave." The elevation above the tide level (at the time of the tsunami or seiche) reached by this water is called the run-up elevation. Run-up elevations vary considerably from point to point along the coast and are very sensitive to the shape of the coastline as it relates to the direction of the source mechanism for the wave.

f. Floods and Storm Surge

All NPP located on bodies of water subject to flooding or storm surge should be designed to survive a design basis flood or surge typically called the probable maximum flood (PMF). The most expedient design solution is to locate plant-grade and safety-related SSC above the maximum flood or surge level. Storm surge levels are functions of wind duration as well as the shape of the local coastline.

g. Precipitation (Rain, Ice and Snow)

It is often current design practice not to identify precipitation (rain, ice or snow) as design loads except in the severe load category. However, it is recommended that they also be included in the extreme load category as well as in roof-design-associated ponding and that the effects of blockage of the primary roof drain systems are included. The probabilities of exceedence for the precipitation load should be the same as for extreme winds. It should also be understood that storm sewer systems for industrial facilities are typically designed for only a 25-year return period flood or precipitation. Excess flood or precipitation beyond the 25-year return period should be capable of surface runoff and the SSC should be able to withstand any loads associated with the surface runoff.

h. Wind-Borne Missiles

Design basis wind-borne missiles typically are associated with two generic types of missiles: (1) penetrating type with a relatively high velocity, rigidity, small mass and impact area; and (2) impact type with somewhat reduced velocity, crushable, large mass and impact area as discussed in Chapter 5.

Man-Induced External Hazards

a. Turbine Missiles

Failures of turbines or other types of large rotating equipment such as flywheels, resulting in throwing parts of the turbines as missiles over several kilometers distance, have been known to occur. The failures generally fall into two categories: failure at operating speed due to design, material, or environmental factors; or failure at a high overspeed due to a failure of the speed and load control system.

In recent years, turbines with 100-percent inspection and quality control and advanced manufacturing procedures, as well as turbine control systems with double or triple overspeed control redundancy, have become standard. Turbine manufacturers have done reliability analyses of the potential failure modes of the steam turbines or the control systems and have predicted rates of failure below those considered credible for hazardous facility design. However, to reinforce this conclusion, turbines are generally orientated so that no line- of-sight missile could strike the safety-related portions of the hazardous facility.

b. Aircraft Crash

The statistical probability of accidental aircraft impacts on a hazardous facility depends on its location with respect to flight paths and proximity to commercial or military airfields and upon the accident (crash) rates for different types of aircraft in the site vicinity. The size (mass) and velocity of the various aircraft assumed to crash upon a hazardous facility should be defined. The effect of postulated burning fuel penetration and impact shock load effects on internal SSC should also be taken into account when assessing the probability of damage or failure if the incident probability is considered high enough to warrant its design basis consideration.

In general, accidental aircraft crash should be considered a design basis event if the probability of the event exceeds 10^{-5}/yr.

c. Industrial Facility (External) Accident

A fixed industrial facility may present several types of potential hazards to a nearby hazardous facility, including fire (and resulting smoke and combustion gases), explosion (with attendant pressure wave, ground shock and missiles) and release of toxic or flammable vapors or gases.

Design probability assessment has played a direct role in missile protection. In general, all postulated missiles are evaluated as to their potential for becoming design basis missiles on the basis of probability, as discussed in Chapter 2.

The principal hazards posed to a hazardous facility from an external fire are the effects on control room habitability and emergency power diesel generator air takes from smoke and combustion gases.

Design of the safety-related SSC of a hazardous facility from external explosion effects is necessary when the potential pressure effects on the hazardous facility SSC exceed the effects of a design basis wind taking due consideration of the fact that the wind load effect tends to be statically applied and the blast load dynamically applied.

Hazard facility siting should avoid sites very near other hazard facilities such as munitions, hazardous chemicals, or petrochemical production or storage facilities.

d. Pipeline Accident

A pipeline, transporting materials that are in a liquid state under normal temperature and atmospheric pressure, will not usually cause a substantial hazard condition even if leakage occurs. However, accidents to pipelines transporting vapors under high pressure or reduced temperature so as to liquefy the substance can lead to leaks of natural gas, propane and other flammable explosive or toxic vapors that may have potentially unacceptable consequences on a nearby hazardous facility. The applicable hazards are overpressure due to air blast, thermal load resulting from deflagration, missile hazard and vapor concentration within the plant. Most of the parameters involved in determining the effect of a pipeline accident are site-rèlated variables and evaluation must be made on a case-by-case basis. The evaluation involves quantity-distance relationships, site topography, site meteorology, prevailing wind direction, wind speed and atmospheric stability class, as well as the characteristics of the potentially hazardous material.

For the purpose of estimating the potential of a pipeline accident, the following assumptions are typically made:

1. Ignition, deflagration or detonation occur.
2. Meteorology (wind direction, wind speed and stability) is such that a gas cloud with the proper mixture to detonate is formed. (Very conservative assumption since the necessary meteorological conditions for such cloud formation are rare in most instances.)
3. About 1.0 mile (1.6 kilometers) of pipeline is assumed to rupture for application to a typical site.

e. Retaining Structure Failure
Retaining structure (dam) failures can result in either abnormally high or low water levels, either of which may be a hazard to a facility.

When dam design, construction, maintenance and inspection are performed according to acceptable modern standards and practices, the probability of failure should be very small in the absence of overtopping due to flood or landslide or extreme earthquake. Historical statistics show an average frequency of dam failure of one failure per 10,000 dam-years worldwide (for dams greater than 15 meters in height). Such a large probability reflects many instances of poor design, construction, maintenance or intervening environmental hazard. A frequency of one retaining structure failure in 10,000 years is recommended for use in identifying those hazard combinations, which are sufficiently large to require consideration by the hazardous facility designer. It is up to the designer to further investigate the failure probabilities for dams and other water-retaining structures in the facility area to determine whether lower hazard probabilities are justifiable.

In performing these studies, it is recommended that the designer investigate three categories of excessive failure risk: earthfill dams, dams with high seismic risk and dams located on poor foundation material. If it can be demonstrated that the structure is in none of these three categories, and that it is subjected to regular inspection and maintenance according to accepted professional practice, then a probability of 10^{-6} per dam-year (100 times lower than that indicated by the historical data) is considered representative. Dams with probabilities of failure exceeding 10^{-5}/yr should be considered design basis events.

f. Accidental Surface Vehicle Explosion
Surface vehicle explosions (truck, train, barge in river or canal, or ocean ship) present a potential hazard to a hazardous facility from the standpoint of both blast overpressure and explosion-generated missiles.

To determine whether a surface vehicle explosion is likely to occur, the probability of occurrence and the duration of the effects should be estimated.

For some hazardous facilities in the U.S. (nuclear power plants) surface vehicle explosions, including those on highway, train, ship or river barge, are defined as design bases independent of probabilities [9]. An investigation of stand-off distance is then made to determine if the postulated blast warrants its consideration as a design basis for the plant. For other types of nuclear facilities a probability of exceedence in excess of 10^{-5}/yr is typically considered a design basis event.

g. Malevolent Vehicle Intrusion and Explosion
Malevolent vehicle intrusion and explosion involves the potential of a land, sea or air vehicle intruding on the plat vital areas and detonating. Prior to 1994 it was the responsibility of facility security forces to preclude the occurrence of this potential. Since 1994, with respect to land vehicles, engineered features have typically been added to prevent intrusion within a stand-off distance for a given quantity of explosives.

Since the destruction of the World Trade Center in 2001 the potential for malevolent vehicle intrusion and explosion has been given much greater attention at all types of hazardous facilities. The actual steps taken have varied on a plant-specific basis and have generally been considered safeguard information, not available publicly. In Chapter 6 the various engineering principles necessary to protect against a malevolent vehicle intrusion and explosion are discussed.

Design-Basis Accidents

The following four phenomena describe the potential accident loads to be considered as a design basis.

1. Rupture of high pressure systems resulting in release of toxic, flammable or explosive substances or missiles;
2. Rupture of rotating systems, such as turbine, rotor or fly wheel rupture;

3. Deflegration or detonation of flammable or explosive materials or waste within the facility and
4. Heavy load drops associated with crane or hoist failure.

Fire

As a function of the quantity and type of combustible materials contained in a facility, postulated fire hazard should be considered as a design basis event.

1.6 CATEGORIZATION AND CLASSIFICATION OF FUNCTIONAL SAFETY-RELATED MISSION ESSENTIAL AND PERFORMANCE CATEGORIZATION OF SSC AND PROCEDURES IN HAZARDOUS FACILITIES

Classifications and categorizations of SSC relative to hazardous facilities are discussed in this section. A discussion of classification relative to quality assurance can be found in Section 11.2.

1.6.1 Nuclear Industry

The U.S. Department of Energy (DOE) nuclear industry has developed a number of well-defined hazard categories for facilities [10] as well as safety-related functional classification [11] for reference category [12] SSC. In Reference 10 three categories of hazardous nuclear facilities are identified. Within these facilities there are two safety-related functional classifications developed and with respect to performance associated with external natural phenomenon hazards, there are five reference categories identified, with three primarily related to the external natural phenomena hazard loading on safety-related SSC.

The U.S. Nuclear Regulatory Commission (NRC) has also developed a somewhat different construction classification and categorization system using a single safety (seismic category) [13] and four quality classes [14]. Recently, the U.S. NRC has been developing more detailed construction classification systems consisting of RISC categories 1 through 4 as defined in Reference 17.

Neither the petrochemical nor biochemical industries appear to have developed the detailed classification and categorization systems typically used in the construction of safety-related facilities and SSC in the detail developed by industry standards, organizations and governmental regulators for the nuclear industry. The goals, classification and categorization system adopted in this book are patterned after those currently in use by the U.S. NRC, the U.S. DOE and the International Atomic Energy Agency for nuclear facilities and safety-related SSC contained within those facilities.

Hazard Categorization

The preliminary hazard categorization of facilities, which takes place during the conceptual design phase, is initially a function of the inventory of radioactive material contained within the facility as shown in Table 2 [16]. It is necessary to perform this evaluation to obtain an accurate picture of the construction standards likely to be used and the costs associated therewith.

After the design of the facility is completed, and the unmitigated radiological release from safety-related SSC are quantified with respect to their release to the public and collocated facility workers outside the facility, the hazard category is finalized as shown in Table 1.3 [16].

Hazard Category 3 Facilities

a. *Definition*

   Hazard Analysis shows the potential for only significant localized (in facility) consequences.

b. *Applicability*
Facilities that initially have quantities of hazardous material that exceed limiting values as shown in Table 1.2. Facilities with final hazard categorizations where unmitigated doses exceed values are shown in Table 1.3.

By definition, this category of facilities and hazards cannot release the quantities of materials that could seriously threaten the health and safety of workers (i.e., unmitigated REM dose in excess of 100 REM) more than 100 meters from the release point, with less than 5.0 REM unmitigated release to the public or the environment outside the exclusion area.

c. *Design Basis Events*
The design basis events to be considered in the design of SSC in hazard category 3 facilities are typically the same as those considered for essential facility SSC as defined in National Building Codes [3, 4].

d. *Engineering Analysis*
Limited engineering analysis is needed to determine the preventive and mitigating features relied upon for the specific event sequences identified for design purposes to determine their effectiveness for facility worker protection.

Nuclear Hazard Category 2 Facilities

a. *Definition*
Hazard analysis shows the potential for significant on-site consequences to collocated workers.

b. *Interpretation*
Facilities with the potential for nuclear criticality events, or with sufficient quantities of hazardous material and energy that would require on-site emergency planning responses.

c. Types of Hazard Category 2 Facilities
Generally there are two types of nuclear facilities: storage and process facilities. Storage facilities with respect to mechanical components are typically limited to storage systems, component systems, air-handling systems and components, plus crane hoists and glove boxes. Process facilities typically have large quantities of active mechanical equipment such as pumps and valves and passive mechanical equipment such as tanks, vessels and heat exchangers, as well as distribution systems such as piping, tubing and duct.

d. Accident Sequence Selection
As general guidance for category 2 facilities, it is necessary to include a range of accident conditions to adequately characterize the safety basis for the facility. Accident sequences should be selected to provide insight into the hazards associated with the facility. Because these accidents are used in establishing the technical safety requirements for the facility, their selection is very important. The search should include higher-probability unique events, which pose hazards only to workers, as well as unusual, lower-probability events, which include a reasonable maximum release from the facility. Design basis natural phenomena should also be included in the range of design events.

1. *Operational Accidents:* These result from processes and activities involved in operating the facility and generally cover many diverse routine or nonroutine events with potentially adverse consequences to the workers or public. Fires, explosions, spills, process disturbances and criticality events are included as operational events. A reasonable set of operational events that represents the accident release mechanisms identified should be selected. Explicit consideration should be given to nonroutine or unique events that present significant risks to facility workers. These events should include process explosions or criticalities, which have the potential for serious worker injury or death but would not necessarily cause significant releases outside of the facility.
2. *Design Basis Natural Phenomena Events:* As currently defined by DOE, these events include earthquakes, high winds, tornados, floods, etc., for which the facility has been (or

TABLE 1.4. TYPICAL EXAMPLES OF HAZARD CATEGORY 2[2] POTENTIALLY HAZARDOUS EVENTS CONSIDERED IN DESIGN

No.	(a) Internal Events
1	Equipment fire
2	Room fire
3	Room fire involving radioactive or toxic materials
4	Uncontrolled chemical reaction
5	Chemical exposure
6	Inhalation, ingestion or dermal exposure to toxic, radioactive or carcinogenic materials
7	Compressed gas explosion
8	Gas explosion (oxygen, acetylene, LP gas)
9	High-intensity laser-light exposure
10	Ionizing radiation exposure due to ICF target implosions
11	Ionizing radiation exposure due to contaminated components
12	Non-ionizing radiation exposure
13	High-energy system rupture (internal)
	(b) External Events
1	Earthquake
2	Wind
3	Flood
4	Precipitation
5	Lighting
6	Volcanic action[1]

1. Depending on site proximity to volcanic activity.
2. These events are also used as the basis for safety and performance goals used to define the design requirements for safety-related SSC contined within a hazard category 2 facility.

should have been) designed. Explicit consideration should be given to such sequences and a representative set of accidents described.

3. *External Events:* The effect of facility- or site-specific events such as airplane crashes, transportation accidents or collocated facility accidents in the category 2 facility should also be addressed as identified in Table 1.4.

Hazard Category 1 Facilities

a. *Definition*
   Hazard analysis shows the potential for significant off-site (public and environment) consequences.
b. *Release Mechanisms*
   Refer to the discussion in category 2 for release mechanism analysis techniques, but consider a more stringent performance goal. It should be noted that large reactors would probably be considered highly complex facilities.
c. *Sequence Selection*
   For highly complex facilities, an extensive set of accident sequences need to be categorized so that a reasonable spectrum of sequences is analyzed.
d. *Engineering Analysis*
   Extensive engineering analysis is required for category 1 facilities.
e. *Consequence Analysis*
   The requirements are the same as those given for category 2 facilities except that performance goals are more stringent.

Hazard Category 1S Facilities

a. *Definition*
   Hazard analysis shows the need for robustly designed containment or confinement systems to mitigate off-site doses. Confinements are usually applied to the reactor coolant and control systems within the reactor building, as well as to systems contained within the reactor auxiliary and related structures.
b. *Interpretation*
   Reactors capable of core melt following a loss of coolant or control.
c. *Release Mechanism*
   Refers to discussion in category 2 and category 1 for release mechanisms, but with more stringent performance goals and hazards.

Table 1.5 summarizes the hazardous events considered in the design of this category of facilities.

Functional and Performance Goals for Safety-Related SSC

Functional goals are typically divided into two subsets: safety and mission essential or critical goals. Only safety goals have been quantified numerically.

TABLE 1.5. TYPICAL EXAMPLES OF HAZARD CATEGORY 1S AND 1[3] POTENTIALLY HAZARDOUS EVENTS CONSIDERED IN DESIGN

No.	(a) Internal Events
1	Equipment fire
2	Room fire
3	Room fire involving radioactive or toxic materials
4	Uncontrolled chemical reaction
5	Chemical exposure
6	Inhalation, ingestion or dermal exposure to toxic, radioactive or bio-hazard materials
7	Compressed gas explosion (2)
8	Industrial gas explosion (oxygen, acetylene, LP gas) (2)
9	High-intensity Laser-light exposure
10	Radiation and toxic exposure due to contaminated components
11	Non-ionizing radiation exposure
12	High-energy system rupture (internal)
13	Heavy load drop other industrial accidents above 10^{-3}/yr probability level
	(b) External Events
1	Earthquake
2	Wind (tornado)
3	Wind missile
4	Flood
5	Precipitation (rain and snow)
6	Lighting
7	Volcanic action[1]
8	External blast
9	Vehicle impact: (a) car or truck; (b) ship or boat; (c) aircraft
10	High-energy system rupture (external)

1. Depending on site proximity to volcanic activity.
2. Unless flammable gas, safety-related detection and mitigation systems are installed.
3. These events are also used as the basis for safety and performance goals used to define the design requirements for safety-related SSC contained within a hazard category 2 facility.

Safety Goals

The U.S. NRC has established two safety goals, which are supported by two quantitative objectives. These two supporting objectives are based on the principle that nuclear risks should not be a significant addition to other societal risks. The Commission wants to make clear that no death attributable to nuclear power plant operation will ever be "acceptable" in the sense that the Commission would regard it as a routine or permissible event.

In the vicinity of a large nuclear power plant (category 1S facility) the risk of prompt fatalities to an average individual that might result from reactor accidents should not exceed one-tenth of one percent (0.1 percent) of the sum of prompt fatality risks resulting from other accidents to which members of the U.S. population are generally exposed.

In the area near a large nuclear power plant, the risk of cancer fatalities that might result from the operation should not exceed one-tenth of one percent (0.1 percent) of the sum of cancer fatality risks resulting from all other causes.

The Commission believes that this ration of 0.1 percent appropriately reflects both of the qualitative goals so that individuals and society bear no significant additional risk. However, this does not necessarily mean that an additional risk that exceeds 0.1 percent would by itself constitute a significant additional risk. The 0.1 percent ratio to other risks is low enough to support an expectation that people living or working near large nuclear power plants would have no special concern due to the plant's proximity.

The average individual in the vicinity of the plant is defined as the average individual biologically (in terms of age and other risk factors) and locationally who resides within a mile (1.6 km) from the plant site boundary. This means that the average individual is found by accumulating the estimated individual risks and dividing by the number of individuals residing in the vicinity of the plant.

The U.S. DOE has also adopted two quantitative safety goals to limit the risks of fatalities associated with its nuclear operations. These goals are the same as those established for large nuclear power plants by the NRC and, like the NRC goals, should be viewed as aiming points for performance. These goals are:

a. The risk to an average individual in the vicinity of a DOE nuclear facility for prompt fatalities that might result from accidents should not exceed one-tenth of one percent (0.1 percent) of the sum of prompt fatalities resulting from other accidents to which members of the population are generally exposed. For (1.6 km) evaluation purposes, individuals are assumed to be located within one mile of the site boundary.
b. The risk of cancer fatalities in the area of a DOE nuclear facility that might result from operations should not exceed one-tenth of one percent (0.1 percent) of the sum of all cancer fatality risks resulting from all other causes. For evaluation purposes, individuals are assumed to be located within 10 miles (16 km) of the site boundary.

In striving to reach these goals, DOE nuclear facilities and activities are constructed, operated and decommissioned with: (a) appropriate barriers to prevent or minimize potential radioactive releases; (b) engineered safety features to minimize potential releases; and (c) procedural controls to mitigate the effects of potential releases. These goals shall be addressed for both new and existing facilities. Proposed modifications to existing facilities to achieve these goals shall be prioritized along with other proposed modifications based on their safety significance. DOE shall pursue the evolution of additional potential safety goals for plant and collocated workers to support enhanced safe operations of its facilities.

DOE recognizes there are large uncertainties in the data and available methods for assessing risk levels, especially with respect to potential health effects from nuclear facility operations. DOE safety goals are aiming points and are not substitutes for compliance with DOE directives and nuclear safety-related rules.

The safety goals provide a rational basis and assurance to the public that the limits on cancer deaths do not pose a significant health and environmental threat to the public. However, they provide little guidance to engineers as to how to achieve these goals. Both the U.S. NRC and DOE have provided numerical guidance for the design of nuclear safety-related SSC in the form of performance goals.

Performance Goals

In NRC space consistent with the traditional defense-in-depth approach and the accident mitigation philosophy requiring reliable performance of containment or accident mitigation systems, the overall mean frequency of a large release of radioactive materials to the environment from a reactor accident should be less than 1 in 1,000,000 per year, 10^{-6} of reactor operation for all hazards. This is typically extended to 1 in 10,000,000 per year, 10^{-7} from any single hazard source. Similarly, numerical probabilistic safety also has been established for DOE facilities as 10^{-5}/yr for a safety class and performance category 4 and 10^{-4}/yr for safety significant performance category 3 SSC; 10^{-4}/yr for a hazard category 2 facility.

The potential release of a large quantity of radioactive material to the environment and the public in general requires the failure of engineered confinement or containment systems. Design of confinement or containment systems is generally based on the postulated unmitigated release of available radioactive materials. An exception to this design assumption is large nuclear reactor containment, because of its robust design that treats accident loads as normal design loads with the associated higher safety margins. These loads are combined with dynamic external natural phenomena hazard loads to be carried elastically without reduction due to inelastic energy absorption. As a result they are not assumed to fail when considering unmitigated safety analysis. Other containment or confinement components and systems may or may not be assumed to fail in unmitigated safety analyses as a function of the robustness of their design.

In U.S. DOE nuclear facility confinements inelastic response of safety-related SSC is permitted in response to design-basis accident plus external man-induced and natural phenomena hazards. Hence, safety analyses for DOE facilities typically assume that the confinement systems are assumed to fail when making their unmitigated safety assessments.

Functional (Safety) Classifications of SSC

In NRC space, quality classifications as defined in *R.G. 1.26* [14] indicate that construction standards be used. However, the load requirements are essentially the same for all quality classes A through C. Risk informed categorizations are also being developed [1-15].

In the DOE space industry there are typically four functional classifications of SSC and procedures. The first two are safety classifications, and the third is an essential to the mission classification, and the fourth comprises all those SSC and procedures not included in the first three classifications.

DOE has safety-classified SSC in DOE Std. 3009 as safety and safety significant. Existing DOE Orders, Standards and Guidelines indicate use of specific industry codes and standards as defined in DOE Order 420.1 [16], which provide different levels of design criteria and quality control.

Unlike the NRC, DOE has also mandated different natural hazard load levels as a function of Performance Categorization and Safety and Safety Significant classifications, as shown in Table 1.6.

Functional and Performance Classification of Safety-Related SSC

a. *NRC Safety Class 1*

   There has been very little application of the functional safety-related NRC Safety Class 1 for SSC outside the reactor cooling control and protection systems of large nuclear reactors capable of core melt in the case of loss of coolant or control. This designation carries with it the most stringent requirements (e.g., most rigorous design enhanced inspection, testing and maintenance and special instrumentation and control systems). This NRC Safety Class designation is typically associated with ASME B&PVC Section NB Class 1 Construction.

b. *DOE Safety Class, Performance Category (PC-4)*

   This SSC designation has been traditionally reserved for SSCs needed for public and environmental protection. Accordingly, safety class 1 SSCs are those that accident analysis indicates are needed to prevent accident consequences from exceeding unmitigated release to the public in excess of 25 REM. This is the next most stringent safety classification. It is typically associated with ASME B&PVC Section III Class 2 Construction in NRC facilities and ASME B&PVC Section VIII in DOE facilities.

   It is recommended that this safety classification be applicable to safety-related SSC in DOE Hazard Category 1 facilities and be extended to Hazard Category 2 facility SSCs that

TABLE 1.6. MEAN OR MEDIAN PROBABILITY OF EXCEEDENCE LIMITS TYPICALLY USED IN DESIGN OR PERFORMANCE CATEGORIZED NUCLEAR SSC[1,2,5]

	EXTERNAL								INTERNAL		
Performance or Design Category	Median Earthquake[6]	Mean Straight Wind	Mean Tornado	Mean Precipitation[3]	Mean Flood	Blast Determinate	Mean Accidental Aircraft Crash	High-Energy System Rupture	Design[7] Basis Accident 1	Design[7] Basis Accident 2	Design[7] Basis Accident ... N
5 (NRC)	$10^{-5(8)}$	1×10^{-2}	1×10^{-7}	10^{-2}	1×10^{-4}	1×10^{-6}	1×10^{-6}	1×10^{-6}	As	As	As
4	10^{-4}	2×10^{-4}	2×10^{-6}	10^{-2}	1×10^{-3}	1×10^{-5}	1×10^{-5}	1×10^{-5}	per	per	per
3	1×10^{-3}	2×10^{-3}	2×10^{-5}	10^{-2}	2×10^{-3}	1×10^{-4}	1×10^{-4}	1×10^{-4}	SAR	SAR	SAR
2	1×10^{-3}	1×10^{-2}	—	2×10^{-2}	1×10^{-2}	—	—	—	—	—	—
1	2×10^{-3}	2×10^{-2}	—	2×10^{-2}	2×10^{-2}	—	—	—	—	—	—
0	—	—	—	—	—	—	—	—	—	—	—

Note: Performance Categories Assigned to SSC:

5 Safety-Related[1] large water cooled nuclear power plants > 1,000 MW_t associated with reactor safety, LNG storage and processing > 200 cu. meters.

4 New spent fuel less than three years old[2] requiring forced cooling and large quantities of volatile and gaseous high level radioactive waste, storage or processing facility, safety-related items nuclear research or power reactors larger than 100 MW_t.

3 Spent fuel dry storage not requiring forced cooling and type B packaging solid and liquid nuclear waste and material storage, facility safety significant items and research nuclear reactors larger than 1 MW_t safety items[3].

2 Essential or normal hazardous facilities, police stations, hospitals, petrochemical plants, nuclear reactors equal to or less than 1 MW_t and secondary support safety significant items for U.S. DOE hazard category 2 use National Building Code Requirements.

1 Ordinary residential, commercial or industrial facilities use national building code requirements.

0 Low cost structures with no permanent human occupancy—iocal building code requirements if any

1. Safety-related items are often divided into two classes: safety and safety significant. The design requirements for safety significant items are usually one performance category less than safety class items in the same facility.
2. In U.S. NRC licensed facilities safety-related SSC are usually categories as PC-5.
3. Precipitation probability levels shown have often been used in nuclear facility design treated as a severe load. It is recommended that they be taken equal to flood probabilities of exceedence in future designs.
4. In U.S. DOE facilities 10^{-4}/yr, and 10^{-3}/yr, respectively, straight wind exceedence probability are used for PC 4 and PC-3 SSC, respectively.
5. Safety-related item functions are serviced by primary and secondary support systems. A primary support system is one that needs to operate within 72 hours of the event and should be classified the same as the safety-related item it supports. A secondary support system needed after 72 hours can be one performance class less than the safety-related item it supports.
6. Median peak ground accelerations and spectral shapes are defined for design purposes. Mean pga at the 10^{-4}/yr probability of exceedence may be substituted for median probability of exceedence.
7. Design-basis accidents typically are defined as loss of coolant, high-energy system rupture and heavy load drops within the nuclear facilities.
8. Mean earthquake probability of exceedence level taken as one order of magnitude higher.

provide a confinement or radiological release barrier function. It is not anticipated that Safety Class SSC would be found in Hazard Category 3 facilities.

c. *DOE Safety Significant Class Performance Category (PC-3)*
This category of SSCs is provided to ensure that safety-related SSCs will be given adequate attention in the safety analysis report (SAR) and facility operations programs. Safety significant SSCs are those of importance to defense in depth or collocated worker safety as determined in hazard analysis. Control of such SSCs does not require meeting the level of stringency associated with safety classes. They are typically associated with ASME B&PVC Section VIII in DOE facilities. The PC-3 categorization is associated with safety-significant Class barrier design. The PC-3 performance designation carries with it more stringent loading requirements.

d. *DOE Safety Significant Class, Performance Category (PC-2)*
This functional classification and performance categorization are typically found in Hazard Category 2 facilities, which are safety-related but do not perform a barrier function.

A very high percentage of safety-related SSCs fall within the safety significant PC-2 classification as compared to the DOE safety class PC-4 or safety significant PC-3 designation.

Safety significant PC-2 SSC are be expected to meet the requirements of the International Building Code Table 1604.5 Category III for SSC, which perform a safety-related support function. In Table 6 is a summary of probabilistic related loading conditions applicable to the various design classes or performance categories.

Quality Classification

Related to the safety classifications of SSC is a quality classification requirement. The quality classification system used by the NRC consists of four Quality Groupings A through D, which are defined by the U.S. NRC in *R.G. 1.26* [14]. While not formerly recognized in DOE standards, there are typically 3 quality levels defined for DOE facilities. QL-1 is taken equivalent to NRC quality group A and B, QL-2 is equivalent to NRC quality group C and QL-3 is equivalent to NRC Quality Group D. In the NRC Regulatory Guide, quality group A applies to SSCs that would be classified as ASME B&PVC Section III Class1 (Subsection NB). It generally applies to SSCs that make up the safety-related SSC of the cooling system of a large power reactor plant.

Quality group B in NRC space applies to engineered safety systems necessary to shut down a large power reactor and maintains it in a safe shutdown state. It would also be applied to containment or confinement SSC for a large power reactor plant. In DOE space it generally applies to SSCs that are in safety class and performance category PC-4.

Quality group C in NRC space generally applies to ancillaries for quality group A or B SSC and in DOE space applies to safety significant class and performance category PC-2 SSC.

Quality Group D in NRC space quality class D applies to non-nuclear safety components. In DOE space the ASME B&PVC Section VIII and B31.3 Codes are permitted in lieu of ASME B&PVC Section III Code requirements.

For DOE safety classification, DOE Order 420.1 and DOE Std. 3009 are used as a means to select a proper construction or procurement code or standard as suggested in Table 1.7. The loading to be considered in conjunction with these standards is defined by the performance category. In NRC space quality classification in R.G. 1.26 defines the loadings as being constant for quality levels A through C. Recommended NRC Standards are also found in Table 1.7.

Design (Limit State) Classification

Within a particular standard, there are typically several limit (damage) states permitted. For example, the ASME B&PVC Section III Division 1 has six such design classifications: design, service levels A, B, C, D and test. These classifications are characterized by different acceptance criteria associated with allowable stress, strain, deflection, deformation, etc., limits.

TABLE 1.7. CONSTRUCTION,[1] PROCUREMENT AND QUALITY STANDARD RECOMMENDATIONS FOR SAFETY-RELATED SSC

ITEMS	Natural Phenomena Hazard[a] Design-Basis Events			
	Safety Class		Safety Significant	
	Q-1 (k) PC-4 (b) SDC-5 (i) CLASS A (j) Sesmic class 1 (m)	Q-2 (k) PC-3 (c) SDC-4 (i) CLASS B (j) Seismic class 1 (m)	Q-3 (k) PC-3 (b,d,n) SDC-3 (i) Class C (j) Seismic class 2 (m)	S (k) PC-2 (c) SDC-2 (i) CLASS D (j) UBC-2000
	(A) Geotechnical (i)1			
Site-specific sampled borings with SPT tests and undisturbed tests	ASTM D 1586 and ASTM D 1587	ASTM D 1586 and ASTM D 1587	ASTM D 1586 and ASTM D 1587	ASTM D 1586 and ASTM D 1587
Cone penetration tests	ASTM D 3441	Can be used to support borings ASTM D 3441	Can be used to support borings ASTM D 3441	Optional ASTM D 3441
Field shear wave velocity measurements	ASTM D 4428/4428M	ASTM D 4428/4428M	ASTM D 4428/4428M	Not required
Dynamic laboratory testing for damping and modulus degration	ASTM D 4015 or D3999	Option (j) ASTM D 4015 or D3999	Optional (j) ASTM D 4015 or D3999	Not required
Static strength tests	ASTM D 3080	ASTM D 3080	ASTM D 3080	Optional (j) ASTM D 3080
Consolidation testing	ASTM D 2435	ASTM D 2435	ASTM D 2435	Optional (j) ASTM D 2435
Index testing	ASTM D4318	ASTM D4318	ASTM D4318	ASTM D4318
Liquefaction testing	ASTM D 5311	Optional (j) ASTM D 5311	Optional (j) ASTM D 5311	Optional (j) ASTM D 5311
Amplification studies	(Example SHAKE91)	(Example SHAKE91)	(Example SHAKE91)	Not Required
Liquefaction analysis	(Example: seed approach with SPT blowcounts CPT data)	(Example: seed approach with SPT blowcounts CPT data)	(Example: seed approach with SPT blowcounts CPT data)	Optional (j)
Soil structure interaction analysis	(Example: SASSI Code)	(Example: SASSI Code)	(Example: SASSI Code)	Optional (j) (Example: SASSI Code)
Dynamic settlement analysis	No standard specified	No standard specified	No standard specified	Not required
Bearing capacity analysis and settlement analysis	No standard specified	No standard specified	No standard specified	No standard specified
Slope stability analysis	Necessity depends on site topography	Necessity depends on site topography	Necessity depends on site topography	Necessity depends on site topography
	(B) Structures (g)			
Reinforced concrete	ACI-349	ACI-349	ACI-349	ACI-318
Structural steel	AISC N690	AISC N690	AISC N690	AISC Bldg.
Reinforced masonry	—	ACI-530 (e)	ACI-530 (e)	ACI-530 (e)
Unreinforced masonry	—	—	—	ACI-530

TABLE 1.7. (Continued)

ITEMS	Natural Phenomena Hazard[a] Design-Basis Events			
	Safety Class		Safety Significant	
	Q-1 (k) PC-4 (b) SDC-5 (i) CLASS A (j) Sesmic class 1 (m)	Q-2 (k) PC-3 (c) SDC-4 (i) CLASS B (j) Seismic class 1 (m)	Q-3 (k) PC-3 (b,d,n) SDC-3 (i) Class C (j) Seismic class 2 (m)	S (k) PC-2 (c) SDC-2 (i) CLASS D (j) UBC-2000
(C) Mechanical Items (h) (1) Distribution Systems				
Piping (process – vital)	ASME BPVC Sect. III, Cl. 1	ASME BPVC Sect. III, Cl. 2	ASME BPVC Sect. III, Cl. 2	ASME B31.1 or B31.3
Piping (process – other)	ASME BPVC Sect. III, Cl. 2	ASME BPVC Sect. III, Cl. 3	ASME BPVC Sect. III, Cl. 3 or B31(o)	ASME B31.1 or B31.3
Duct (HVAC) – round	ASME BPVC Sect. III, Cl. 2 or AG-1, SA	ASME B&PVC Sect. III, Cl. 3 or AG-1, SA	SMACNA	SMACNA
Duct (HVAC) – rectangular	AG-1, SA	AG-1, SA	SMACNA	SMACNA
Duct (process) – round	ASME BPVC Sect. III, Cl. 2 or AG-1, SA	ASME BPVC Sect. III, Cl. 3 or AG-1, SA or B31(o)	ASME BPVC Sect. III, Cl. 3 or B31.1 or B31.3	SMACNA
Duct (process) – rectangular	AG-1, SA	AG-1, SA	SMACNA	SMACNA
Tubing	ASME BPVC Sect. III, Cl. 3	ASME B31.1 or B31.3	ASME B31.1 or B31.3	Manufacturer's Specification
(2) Passive Components				
Pressure vessels (vital)	ASME BPVC Sect. III, Cl. 1	ASME BPVC Sect. III, Cl. 2	ASME BPVC Sect. III, Cl. 3 or Sect. VIII Div. 2(q)	ASME BPVC Sect. VIII-Div. 1(s)
Pressure vessels (other)	ASME BPVC Sect. III, Cl. 2	Sect. VIII-Div. 2	ASME BPVC Sect. III, Cl. 3 or Sect. VIII Div. 2(q)	ASME BPVC Sect. VIII-Div. 1(s)
Heat exchangers (water – vital)	ASME BPVC Sect. III, Cl. 1	ASME BPVC Sect. III, Cl. 2	ASME BPVC Sect. III, Cl. 3 or Sect. VIII Div. 2	ASME BPVC Sect. VIII-Div. 1(s)
Heat exchanger (water – other)	ASME BPVC Sect. III, Cl. 2	Sect. VIII-Div. 2	Sect. VIII-Div. 2	ASME BPVC Sect. VIII-Div. 1(s)
Tanks (atmospheric – other)	ASME BPVC Sect. III Cl. 2	ASME BPVC Sect. III – Cl. 3	API-650	API-650
Tanks (water – vital)	ASME BPVC Sect. III, Cl. 3	API-650	API-650	AWWA 100
Tanks (water – other)	API-650	AWWA 100	AWWA 100	AWWA 100
Glove boxes	ANSI/ASTM C852, ANS 11.16, AISC N690	ANSI/ASTM C852, ANS 11.16, AISC N690	ANSI/ASTM C852, ANS 11.16, AISC N690	ANSI/ASTM C852, ANS 11.16, AISC N690

(continues)

TABLE 1.7. (Continued)

ITEMS	Natural Phenomena Hazard[a] Design-Basis Events			
	Safety Class		Safety Significant	
	Q-1 (k) PC-4 (b) SDC-5 (i) CLASS A (j) Sesmic class 1 (m)	Q-2 (k) PC-3 (c) SDC-4 (i) CLASS B (j) Seismic class 1 (m)	Q-3 (k) PC-3 (b,d,n) SDC-3 (i) Class C (j) Seismic class 2 (m)	S (k) PC-2 (c) SDC-2 (i) CLASS D (j) UBC-2000
(3) Active Components				
Pumps (vital)	ASME BPVC Sect III, Cl. 1	ASME BPVC Sect. III, Cl. 2	ASME BPVC Sect. III, Cl. 3	API-610
Pumps (other)	ASME BPVC Sect. III, Cl. 2	ASME BPVC Sect. III – Cl. 3	API-610	API-610
Circulators	ASME AG-1 Sect. BA	ASME AG-1 Sect. BA	ASME AG-1 Sect. BA	SMACNA
Fans	ASME AG-1 Sect. BA	ASME AG-1 Sect. BA	ASME AG-1 Sect. BA	SMACNA
Valves (vital)	ASME BPVC Sect. III, Cl. 1	ASME BPVC Sect. III, Cl. 2	ASME BPVC Sect. III, Cl. 2	B31.1 or B31.3
Valves (other)	ASME BPVC Sect. III, Cl. 2	B31.1 or B31.3	B31.1 or B31.3	B31.1 or B31.3
Dampers	ASME AG-1 Sect. BA	ASME AG-1 Sect. BA	ASME AG-1 Sect. BA or SMACNA(n)	SMACNA
Cranes and hoists (vital)	ASME N0G-1 Type 1	ASME N0G-1 Type II	ASME N0G-1 Type II or CMAA-70(n)	CMAA-70
Cranes and hoists (other)	ASME N0G-1 Type II	ASME N0G-1 Type II	ASME N0G-1 Type II or CMAA-70(n)	CMAA-70
Engines	IEEE 387	Manufacturer's Specification	Manufacturer's Specification	Manufacturer's Specification
Compressors	ASME BPVC Sect. III-Class 3	Manufacturer's Specification	Manufacturer's Specification	Manufacturer's Specification
(D) Electrical Items (g) (1) Distribution Systems				
Cable trays	IEEE-628 NEMA VE-1	IEEE-628 NEMA VE-1	IEEE-628 NEMA VE-1	Manufacturer's Specification
Conduit	ASME BPVC Sect. III, Cl. 3	ASME B31.1	ASME B31.1	Manufacturer's Specification

TABLE 1.7. (Continued)

ITEMS	Natural Phenomena Hazard[a] Design-Basis Events			
	Safety Class		Safety Significant	
	Q-1 (k) PC-4 (b) SDC-5 (i) CLASS A (j) Sesmic class 1 (m)	Q-2 (k) PC-3 (c) SDC-4 (i) CLASS B (j) Seismic class 1 (m)	Q-3 (k) PC-3 (b,d,n) SDC-3 (i) Class C (j) Seismic class 2 (m)	S (k) PC-2 (c) SDC-2 (i) CLASS D (j) UBC-2000
	(2) Passive Components			
Transformers	IEEE-638	IEEE-638	IEEE-638	IEEEC-57.12 series
Batteries	IEEE-450 IEEE-323 IEEE-344 IEEE-535	IEEE-535 IEEE-450 IEEE-344	IEEE-535 IEEE-450	Manufacturer's Specification
Battery racks	IEEE-450 AISC-N690	IEEE-450 AISC Bldg. Spec.	IEEE-450 AISC Bldg. Spec.	Manufacturer's Specification
Electrical panels and racks	IEEE-420 NEMA-IC-6	IEEE-420 NEMA-IC-6	IEEE-420 NEMA-IC-6	Manufacturer's Specification
	(3) Active Components			
Motor control centers	IEEE-649	IEEE-649	IEEE-649	IEEE-649
Switch gear	IEEE-C-37.82	IEEE-C-37.82	IEEE-C-37.82	Manufacturer's Specification
Motors	IEEE-334 IEEE-308 IEEE-328 IEEE-344	IEEE-334 IEEE-308	IEEE-334 IEEE-308	Manufacturer's Specification
Valve actuators	IEEE-382	IEEE-382	IEEE-382	Manufacturer's Specification
Generators	IEEE-387 IEEE-308 IEEE-323 IEEE-344	IEEE-387 IEEE-308	IEEE-387 IEEE-308	Manufacturer's Specification
Battery chargers and inverters	IEEE-382	IEEE-382	IEEE-382	Manufacturer's Specification
	(E) Instrumentation and Control (g)			
Relays	IEEE-37.99	IEEE-37.99	IEEE-37.99	IEEE-37.99
Switches	IEEE-344	IEEE-344	IEEE-344	IEEE-344
Pressure sensors	IEEE-344	IEEE-344	IEEE-344	IEEE-344
Temperature sensors	IEEE-344	IEEE-344	IEEE-344	IEEE-344
Radiation Sensor	IEEE-344	IEEE-344	IEEE-344	IEEE-344
Radiation alarms	ANSI 13.1 ANSI 42.18	ANSI 13.1 ANSI 42.18	ANSI 13.1 ANSI 42.18	ANSI 13.1 ANSI 42.18

(continues)

TABLE 1.7. (Continued)

ITEMS	Natural Phenomena Hazard[a] Design-Basis Events			
	Safety Class		Safety Significant	
	Q-1 (k) PC-4 (b) SDC-5 (i) CLASS A (j) Sesmic class 1 (m)	Q-2 (k) PC-3 (c) SDC-4 (i) CLASS B (j) Seismic class 1 (m)	Q-3 (k) PC-3 (b,d,n) SDC-3 (i) Class C (j) Seismic class 2 (m)	S (k) PC-2 (c) SDC-2 (i) CLASS D (j) UBC-2000
(F) Fire Protection (g)				
Piping – aboveground	— —	ASME B31.1 or B31.3 NFPA-13 or 14	NFPA-13 or 14	
Piping – buried	— —	ASME B31.1 or B31.3 NFPA-24	NFPA-24	NFPA-24
Valves	—	NFPA-26	NFPA-26	NFPA-26
Pumps	—	NFPA-20	NFPA-20	NFPA-20
Tanks – aboveground	—	NFPA-22	NFPA-22	NFPA-22
Tanks – buried	—	UL-58	UL-58	UL-58
		(G) Supports (1) Mechanical		
Linear	ASME-III-NF-Cl. 2	AISC-N690	AISC-N690	AISC Bldg. Spec.
Plate and shell	ASME-III-NF-Cl. 2	ASME-III-NF-Cl. 3	ASME-III-NF-Cl. 3	AISC Bldg. Spec.
Standard	MSSSP-58	MSSSP-58	MSSSP-58	MSSSP-58
(2) Electrical Item				
Linear	AISC-N690	AISC Bldg. Spec.(f)	AISC Bldg. Spec.(f)	AISC Bldg. Spec.(f)
Standard	MSSSP-58	MSSSP-58	MSSSP-58	Manu. Rec.

[1] Includes the activities of administrative controls, documentation, material section, material qualification, design, fabrication, installation, examination and quality control.

a. Other design-basis hazards, as identified in DOE Order 420.1 for the safety class designation, shall use PC-4 and safety significant class designation PC-3 standards.
b. As defined in DOE Standards 1020 and 1021 located in hazard category 1 facilities as defined in DOE Std. 1027.
c. As defined in DOE Standards 1020 and 1021 located in hazard category 2 and 3 facilities as defined in DOE Std. 1027.
d. For natural phenomena hazard design it is permissible to use PC-2 designated standards provided the high hazard categorization of the IBC-2000 Code is used.
e. Can be used when confinement or containment functions are not required.
f. Alternatively AISI specification may be used for cold-formed steel supports (Unitstrut).
g. This task is "optional" unless there is inadequate correlation or data in the literature for the site-specific soils, in which case the task will be "required."
h. Where an example is indicated rather than an ASTM Standard, the example is one of several approaches available to the profession. When no example is given, the profession has several approaches available and is encouraged to use an approach consistent with the specific site conditions.
i. As defined in Draft *ANS 2.26 Standard,* "Categorization of Nuclear Facility Structures, Systems and Components for Seismic Design and Evaluation."
j. As defined in *U.S. NRC Regulatory Guide 1.26,* "Quality Group Classifications and Standards for Water, Steam and Radioactive Waste Containing Components of Nuclear Power Plants," Rev. 3, February 1976.

TABLE 1.7. (Continued)

k. Safety class as defined in *ANS Standard 58.14,* "Safety and Pressure Integrity Classification Criteria for Light Water Reactors," September 1993.
l. Safety class as defined in *ANS 58.14,* "Safety and Pressure Integrity Classification Criteria for Light Water Reactors," American Nuclear Society, 1993.
m. Seismic Category 1 is required to assume inelastic energy absorption factor $F\mu$=1.0 (i.e., elastic design). For seismic category 2, the F_μ specified in *DOE Std. 1020* may be used.
n. Commercial construction standards listed in this column are required to use enhanced administrative and documentation requirements (see ASME B&PVC Section III, Subsection NCA).
o. Where B31 is indicated it may be either B31.1 or B31.3.

Note: Vital SSC are those that normally contain or distribute significant quantities of radioactive or other toxic substances. Such cases are typically limited to systems with design temperatures greater than 300 EF, design pressures greater than 500 psi, piping larger than 4.0 inches, and vessels and tanks larger than 150 cubic feet. It also applies to systems and components within a given safety-related class or performance category that are considered particularly important to safety.

1. *ACI-318,* "Building Code Requirements for Reinforced Concrete," American Concrete Institute.
2. *ACI-349,* "Code Requirements for Nuclear Safety Related Concrete Structures," American Concrete Institute.
3. *ACI-530,* "Building Code Requirements for Masonry," American Concrete Institute.
4. *AISC-S-326,* "Specification for the Design, Fabrication and Erection of Structural Steel for Buildings," American Institute of Steel Construction.
5. *AISC-N690,* " Specification for Safety Related Structures for Design, Fabrication and Erection of Structural Steel for Buildings," American Institute of Steel Construction.
6. *AISI,* " Specification for Design of Cold Formed Steel Structural Members," American Iron and Steel Institute.
7. *ANS 11.16,* "Design Guide for Radioactive Material Handling Facilities and Equipment," American Nuclear Society, 1988.
8. *ANS 13.1,* "Guide to Sampling Airborne Radioactive Materials in Nuclear Facilities", American Nuclear Society, 1999.
9. *ANSI N42.18,* "Specification and Performance of On-Site Instrumentation for Continuously Monitoring Radioactive Effluents," American Nuclear Society, 1991.
10. *API-610,* "Centrifugal Pump for General Refinery Service," American Petroleum Institute.
11. *API-620,* " Design and Construction of Large, Welded, Low-Pressure Storage Tanks," American Petroleum Institute.
12. *API-650,* "Atmospheric Welded Steel Tanks for Oil Storage," American Petroleum Institute.
13. *ASME AG-1, Div. II-BA,* "Ventilation Air Cleaning and Air Conditioning Fans and Blowers," American Society of Mechanical Engineers.
14. *ASME AG-1, DIV. II-SA,* "Ventilation Air Cleaning and Air Conditioning Dampers and Louvers," American Society of Mechanical Engineers.
15. *ASME B31.1,* " "Power Piping," American Society of Mechanical Engineers.
16. *ASME BPVC Sect. III, Cl. 1,* "Nuclear Components Subsection NB," American Society of Mechanical Engineers.
17. *ASME BPVC Sect. III, Cl. 2,* "Nuclear Components Subsection NC," American Society of Mechanical Engineers.
18. *ASME BPVC Sect. III, Cl. 3,* "Nuclear Components Subsection ND," American Society of Mechanical Engineers.
19. *ASME BPVC Sect. III,* "Supports for Nuclear Components Subsection NF," American Society of Mechanical Engineers.
20. *ASME BPVC Sect. VIII, Div. 1,* "Unfired Pressure Vessels," American Society of Mechanical Engineers.
21. *ASME BPVC Sect. VIII, Div. 2,* "Unfired Pressure Vessels Design by Analysis," American Society of Mechanical Engineers.
22. *ASME NOG-1, Type I,* " Standards on Nuclear Type I Overhead, Underhead, Gantry and Monorail Cranes," American Society of Mechanical Engineers.
23. *ASME NOG-1, Type II,* "Standards on Nuclear Type II Overhead, Underhead, Gantry and Monorail Cranes," American Society of Mechanical Engineers.
24. *ASME QME-1,* "Qualification of Active Mechanical Equipment used in Nuclear Power Plants," American Society of Mechanical Engineers.
25. *ASTM C 852-93 (R1997),* "Standard Guide for Design Criteria for Plutonium Glove Boxes,' American Society for Testing and Materials, 1997.
26. *ASTM D-1586,* "Standard Test Method for Penetration Test and Split-Barrel Sampling of Soils," American Society for Testing Materials.
27. *ASTM D-1587,* "Standard Practice for Thin Walled Tube Sampling of Soils for Geotechnical Purposes," American Society for Testing Materials.
28. *ASTM D-2435,* "Standard Method for One-Dimensional Consolidation Properties of Soils," American Society for Testing Materials.
29. *ASTM D-3080,* "Standard Test Method for Direct Shear Test of Soil Under Consolidated Drained Conditions," American Society for Testing Materials.
30. *ASTM D-3441,* "Standard Test Method for Mechanical Core Penetration Test of Soil," American Society for Testing Materials.

(continues)

TABLE 1.7. (Continued)

31. *ASTM D-3999,* "Standard Test Method for Determination of the Modulus and Damping Properties of Soils Using the Cyclic Tri-axial Apparatus," American Society for Testing Materials.
32. *ASTM D-4015,* "Standard Test Methods for Modulus and Damping of Soils by the Resonant Column Method," American Society for Testing Materials.
33. *ASTM-4318,* "Test Method for Liquid Limit, Plastic Limit and Plasticity Index of Soils," American Society for Testing Materials.
34. *ASTM D-4428,* "Standard Methods for Crushable Seismic Testing," American Society for Testing Materials.
35. *ASTM D-5311,* "Standard Test Method for Load Controlled Cyclic Tri-axial Strength of Soil," American Society for Testing Materials.
36. *AWWA-100,* "Welded Steel Tanks for Water Storage," American Water Works Association.
37. *AWWA-Man. 11,* "Steel Pipe – A Guide for Design and Installation," American Water Works Association.
38. *CMAA-70,* "Specification for Electric Overhead Traveling Cranes," Crane Manufacturer Association of America.
39. *IEEE-308,* "Standard Criteria for Class 1E Power Systems for Nuclear Power Generating Stations," Institute of Electrical and Electronic Engineers.
40. *IEEE-323,* "Standard for Qualifying Class 1E Equipment for Nuclear Power Generating Stations," Institute of Electrical and Electronic Engineers.
41. *IEEE-334,* "Type Test of Continuous Duty Class 1E Equipment for Nuclear Power Generating Stations," Institute of Electrical and Electronic Engineers.
42. *IEEE-344,* "Recommended Practice for Seismic Qualification of Class 1 Equipment," Institute of Electrical and Electronic Engineers.
43. *IEEE-382,* " Standard for Qualification of Actuators for Power Operated Valve Assemblies with Safety Related Functions for Nuclear Power Plants," Institute of Electrical and Electronic Engineers.
44. *IEEE-387,* "Criteria for Diesel-Generator Units Applied as Standby Power Supplies for Nuclear Power Generating Structures," Institute of Electrical and Electronic Engineers.
45. *IEEE-420,* "Standard Design and Qualification of Class 1E Control Boards, Panels and Racks used in Nuclear Power Generating Stations", Institute of Electrical and Electronic Engineers.
46. *IEEE-450,* "Recommended Practice Maintenance, Testing and Replacement of Large Led Storage Batteries," Institute of Electrical and Electronic Engineers.
47. *IEEE-535,* "Standard Qualifications of Class 1E Lead Storage Batteries for Nuclear Power Generating Stations," Institute of Electrical and Electronic Engineers.
48. *IEEE-628,* "Standard for the Design, Installation and Qualifications of Raceway Systems for Class 1E Circuits for Nuclear Power Generating Stations," Institute of Electrical and Electronic Engineers.
49. *IEEE-638,* "Standard for Qualifications of Class 1E Transformers for Nuclear Generating Station," Institute of Electrical and Electronic Engineers.
50. *IEEE-649,* "Standard for Qualifying Class 1E Motor Control Centers for Nuclear Power Generators Station," Institute of Electrical and Electronic Engineers.
51. *IEEE-C-37.82,* "Standard for the Qualification of Switchgear Assemblies for Class 1E Applications in Nuclear Power Generating Stations," Institute of Electrical and Electronic Engineers.
52. *IEEE-C-37.98,* "Standard for Seismic Testing of Relays," Institute of Electrical and Electronic Engineers.
53. *IEEE-C-57.12 Series,* "Standards for Transformers," Institute of Electrical and Electronic Engineers.
54. *MSS_SP-58,* "Pipe Hangers and Supports – Materials, Design and Manufacture," Manufacturers Standardization Society of the Valve and Fitting Industry.
55. *MSS-SP-90,* "Guidelines on Terminology for Pipe Hangers and Supports," Manufacturers Standardization Society of the Valve and Fitting Industry.
56. *NEMA IC-6,* "Industrial Control and Systems Enclosures," National Electric Manufacturer Association.
57. *NEMA VE-1,* "Metallic Cable Tray Systems," National Electric Manufacturer Association.
58. *NEMA SM-23,* "Steam Turbines for Mechanical Drive Service," National Electric Manufacturer Association.
59. *NEMA SM-24,* "Land-Based Steam Turbine Generator Sets 0 to 33,000 KW," National Electric Manufacturer Association.
60. *NFPA-13,* "Installation of Sprinkler Systems," National Fire Protection Association.
61. *NFPA-14,* "Installation of Stand Pipe and Hose Systems," National Fire Protection Association.
62. *NFPA-20,* "Installation of Centrifugal Fire Pumps," National Fire Protection Association.
63. *NFPA-22,* "Water Tanks for Private Fire Protection," National Fire Protection Association.
64. *NFPA-24,* "Installation of Private Fire Service Mains and Their Apparatuses," National Fire Protection Association.
65. *NFPA-26,* "Recommended Practice for the Supervision of Valves Controlling Water Supplies for Fire Protection," National Fire Protection Association.
66. *SMACNA,* "HVAC Systems Duct Design," Sheet Metal and Air Conditioning Contractors National Association.
67. *SMACNA,* "Rectangular Industrial Duct Construction Standard," Sheet Metal and Air Conditioning Contractors National Association.
68. *SMACNA,* "Round Industrial Duct Construction Standard," Sheet Metal and Air Conditioning Contractors National Association.
69. *UL-58,* "Steel Underground Tanks for Flammable and Combustible Liquids," Underwriters Laboratories.

TABLE 1.8. DAMAGE (LIMIT STATE) CLASSIFICATION CHARACTERISTICS AND ASME B&PVC SECTION III AND VIII EQUIVALENCY

SSC Type	Limit (Damage) State A B&PVC Section III, Service Level D F_u,[1] R_p[2] > 1	Limit (Damage) State B B&PVC Section III, Service Level D F_u, $R_p = 1$	Limit (Damage) State C B&PVC Section III, Service Levels B & C Section VIII Div. 2 AD150, $K = 1.2$	Limit (Damage) State D B&PVC Section III, Service Level A Section VIII Div. 2 AD150, $K = 1.0$
Mechanical distribution systems and components	Substantial loss of SSC stiffness, structural and leak-tight integrity, but retaining some margin against total failure. Generally would require replacement.	Some loss of SSC stiffness but retaining structural and leak-tight integrity. Generally will require extensive repairs to be put back in operation.	The SSC remains essentially elastic and retains nearly full stiffness and structural and leak-tight integrity. Service level B limit for continual operability of active components.	SSC undamaged.

[1] As defined in DOE Std. 1020-02.
[2] As defined in ANS 2.26-04 and ASCE-7-02.

In general, there are four limit states identified for mechanical SSC. They are typically identified as classes D through A [17] as shown in Table 1.8.

The different design states D, C, B and A applied to mechanical distribution systems and equipment are accommodated generally by changing the allowable stresses permitted for specific load combinations. For example, for passive equipment and distribution systems subjected to the design-basis earthquake, it is permissible to apply limit state B, which is equivalent to ASME BPVC Section III service level D with total primary membrane stresses calculated on an elastic basis to have an allowable 1.6 to 2.0 times yield stress with no evaluation of secondary stresses.

For ASME active equipment and distribution systems (must move or otherwise change state) service categories A or B of *ASME Boiler and Pressure Vessel Code Section III* are often specified. For ASME service level B allowable primary membrane stresses are typically limited to 0.8 to 1.0 times the specified minimum yield at temperature, with primary and secondary or secondary stresses limited to twice yield. Primary membrane plus bending stresses are limited to 1.5 times the primary membrane stress. In both instances, the computed stresses or strains input are the same and are based on assumed elastic response to the design-basis earthquake motion or other extreme environmental loading. The design and corresponding ASME B&PVC Section III service conditions as well as ASME B&PVC Section VIII Div. 2 requirements are summarized in Table 1.8.

In the design of DOE facility SSC, an inelastic energy absorption factor [11] is typically applied as a divisor for the elastically computed seismic load on the SSCs prior to computing resultant stresses and strains. Elastic computed displacements or deformations and allowable stresses applied to NRC safety-related SSC associated with seismic response are usually compared to acceptance criteria without modification by an inelastic energy absorption factor greater than 1.0.

Recent Changes to Safety Classification of SSC used in NRC-Regulated Facilities

The safety-related distribution systems and components defined in SRP section 3.2 [13] for nuclear power plants typically consists of more than 200,000 feet (65,000 meter) (two-thirds of a nuclear power plant distribution systems and mechanical components) and several thousand mechanical components (pumps, valves, vessels, heat exchangers, tanks, etc.). While different ASME B&PVC classes were used for these various safety-related items (quality class A, B & C) [14], the design-basis load for natural and man-induced hazards and accident loads were taken at the same level regardless of whether they were defined probabilistically or deterministically. The safety-related classification of items was made on the basis of the function of the system of which they were a part. For example, all parts of the reactor coolant system, even small drain lines, were classified as quality class A; hence, ASME B&PVC Section III Class 1, even though failure would have little or no impact on the behavior of the system nor on core melt or the early release of radioactivity.

With the widespread use of probabilistic risk assessment (PRA) over the past 20 years, it has become possible to evaluate and quantify the safety significance and failure probability of individual components and sub-systems. As a result, the NRC has developed a new Federal Regulatory, "Risk-Informed Categorization and Treatment of Structures, Systems and Components for Nuclear Power Reactors" [17].

As a result of this new regulation, it is anticipated that what previously was one safety class of items will be divided into two safety classes, RISC 1 and RISC 3. It is anticipated that the construction requirements for RISC 3 SSC will be considerably less stringent than what had been used before and more consistent with a particular SSC's effects on nuclear safety.

It should also be noted that two levels of safety as part of the RISC 1 and RISC 3 classification is quite similar to the DOE practice of dividing its safety-related classes into safety and safety significant classes.

1.6.2 Petrochemical Industry

Introduction

The petrochemical industry is generally significantly less regulated than the nuclear industry and has much less formal design loading and acceptance criteria. Criteria for design is developed by the owner of petrochemical facilities on a case-by-case basis. For external natural phenomena hazard loads the petrochemical industry in the mechanical SSC arena generally follows the International Building Code Standard, supplemented by ASME, which includes piping system B31.3 [6] and ASME Section VIII [5] for lethal service. The lethal service designation generally affects material selection, material qualification, fabrication, inspection and examination requirements and does not have any direct impact on design.

The American Institute for Chemical Engineers Center for Chemical Process Safety has developed a large number of industry guidelines relative to the safety of petrochemical facilities. Of particular interest to the overall design of such facilities is G-23, "Guidelines for Engineering Design for Process Safety," [2] and G-42, "Guidelines for Chemical Process Quantitative Risk Analysis," [18]

Because of the particularly high hazard related to the explosive potential associated with bulk storage of liquefied natural gas, it is recommended that containers larger than 1,000 cu. ft. (35 cu. meters) at least meet the requirements of DOE [16] safety class, performance category PC-4 requirements [11] or the NRC quality class B standards [14]. It is also recommended that nuclear-type standards be applied to bulk storage or processing of toxic materials where postulated release is capable of causing death or serious injury to the public or collocated worker.

Natural Hazards

For external natural hazards such as earthquakes, the procedures of static or equivalent static analysis, as defined in Chapter 5, are usually followed. Wind, flood and precipitation design is summarized in Chapter 6, which follows the requirements of ASCE-7-05. In regions of high tornado probability, for which design loads are not considered in ASCE 7-05, it is not unusual for individual facility owners to require a tornado resistant design as well when such design is specified: it typically would be at the F-3 scale level or 200 mph.

For impact- and impulse-type dynamic loads in general, it is permissable to consider ductile response locally for both NRC and DOE SSC design.

External Blast Load

Extreme external accident explosion blast loads receive the most attention in petrochemical facility design. Of particular interest in this regard is the Center or Chemical Process Safety *Guideline for Evaluating Process Plant Buildings for External Explosion and Fires* [19]. A companion to this is *Guidelines Evaluating the Characteristics of Vapor Cloud Explosives, Flash Fires and BLEVE*'s [20]. In Chapter 7 can be found procedures for evaluating the effect of external explosions on SSC.

1.6.3 Bio-Medical Industry

Mechanical construction requirements for potentially hazardous biomedical facilities are contained in the ASME-BPVC Biomedical Equipment Standard [21]. In general, this standard references B31.3 for piping and ASME B&PVC Section VIII Div. 1for vessel construction for all classes of components.

The natural hazard design is the same as defined for petrochemical facilities in section 1.6.2. In general, there is no requirement to evaluate accidental external explosion for mechanical SSC involved in the storage or processing of pathogens capable of producing epidemics that are capable of resulting in death or serious injury to the public or collocated worker.

REFERENCES

1. 10CFR 71, 1996, "Packaging and Transport of Radioactive Material-Appendix A," *U.S. Code of Federal Regulations,* June 6.
2. Center for Chemical Process Safety, 1993, "Guidelines for Engineering Design for Process Safety," American Institute of Chemical Engineers, New York, NY.
3. International Code Council, 2003, *International Building Code,* International Code Council, Falls Church, Virginia.
4. *ASCE Standard SEI/ASCE 7-02,* 2003, "Minimum Design Loads for Buildings and Other Structures," American Society of Civil Engineers, New York, NY.
5. *Boiler and Pressure Vessel Code,* 2001, "Rules for Construction of Pressure Vessels-Section VIII," American Society of Mechanical Engineers, New York, NY.
6. *ASME Code for Pressure Piping,* 2002, "ASME B31.3-2002," American Society of Mechanical Engineers, New York, NY.
7. *Regulatory Guide 1.165,* 1997, "Identification and Characterization of Seismic Sources and Determination of Safety Shutdown Earthquake Ground Motion," U.S. Nuclear Regulatory Commission, March.
8. Liu, H., 1991, *Wind Engineering, A Handbook for Structural Engineers,* Prentice Hall, TK.
9. *Regulatory Guide 1.91,* 1978, "Evaluation of Explosions Postulated to Occur on Transportation Routes Near Nuclear Power Plants," U.S. Nuclear Regulatory Commission, Rev. 1, Feb.,
10. *DOE Standard 1027,* 1992, "Hazard Categorization and Accident Analysis Technique for Compliance with DOE Order 5480.23 Nuclear Safety Analysis Reports," U.S. Department of Energy.
11. *DOE Standard 1020,* 2002, "Natural Phenomena Hazards Design and Evaluation Criteria for Department of Energy Facilities," U.S. Department of Energy.
12. DOE Standard 1021, 1993, "Natural Phenomena Hazards Performance Categorization Guidelines for Structures, Systems and Components," U.S. Department of Energy.
13. *SRP Section 3.2.1.,* 1981, "Seismic Classification, Rev. 1," U.S. Nuclear Regulatory Commission.
14. *R.G. 1.26,* 1976, "Quality Group Classification Standards, Rev. 3," U.S. Nuclear Regulatory Commission.
15. *10CFR50.69,* 2003, "Risk-Informed Categorization and Treatment of Structures, Systems and Components for Nuclear Power Reactors," Federal Regulater, Washington, D.C. Vol. 68, No. 95,
16. *IAEA Safety Report,* 2003, "Safety of New and Existing Research Reactor Facilities in Relation to External Events," International Atomic Energy Agency, Vienna, Austria.
17. *Doe Order 420.1,* 1995, "Facility Safety," U.S. Department of Energy, Washington, D.C.
18. DOE Standard 3009, 1994, "Preparation Guide for U.S. Department of Energy Non-Reactor Nuclear Facility Safety Analysis Reports," U.S. Department of Energy,
19. *Guidelines for Chemical Process Quantitive Risk Analysis, 2nd* Ed., 2000, Center for Chemical Process Safety of the AICE, New York.
20. *Guidelines for Evaluating the Characteristics of Vapors, Closed Explosions, Flash Fires and BLEVES,* 1994, Center for Chemical Process Safety of the AICE, New York.
21. *ASME BPE-2002,* 2002, "Bio-Processing Equipment," American Society of Mechanical Engineers, New York, NY.

CHAPTER

2

RISK-BASED AND RISK-INFORMED DESIGN BASES

2.1 INTRODUCTION

Probability of failure or risk-informed assessments are playing an increasingly significant role in defining design load for hazardous facilities [1, 2]. One important role has been in the area of the decision process regarding extreme load design bases. Prior to its employment, hazardous facility design requirements were based, to a considerable degree, on the "minimax" decision rule, which says the worst possible future is sure to be found, so action should be taken to minimize the maximum possible loss. This is popularly referred to as the "what if" basis for design, where loads and particularly their combination, which could be postulated regardless of their probability or conditional probability of occurrence, became design bases.

The minimax rule was tempered somewhat by the reasonable decision rule, which says that any reasonable decision maker under the same circumstances and with the same background would take the same action; hence these decisions tended to be made taking historical precedents into consideration. Finally, there is the acceptable risk decision rule, which say: Take the action where the sum of the products of the probability of exceedence and the consequences of exceedence are equal to or less than a risk associated with some natural phenomena over which society apparently has no control, or is less than some man-made activity that historically has been accepted by society, considering the actual or perceived cost/benefit to society.

The current tendency in hazardous facility design is a continuing shift away from the minimax rule or deterministic-based siting and design toward an acceptable risk rule tempered with the reasonableness rule. This is typically referred to as risk-informed siting and design. However, it should be understood that there is often an inability to develop an accurate probability assessment in many areas because of the lack of the necessary statistical data or ignorance as to the governing physical relationships and variability in the phenomena being investigated. For this reason deterministic bounds are often placed on the applicable design criteria that are developed based on the acceptable risk rules. Use of probabilities, as a basis of design commonly referred to as risk-based design, must await the development of probability estimates with less uncertainty than is now the case.

2.2 PROBABILISTIC-DEFINED SAFETY GOALS AND HAZARD DEFINITION

Rational design of safety-related structures, systems and components (SSC) for extreme loads that put the public and environment at risk, or mission critical SSC, must consider three basic factors: (1) safety goals; (2) hazard or extreme load definition; and (3) SSC failure probabilities given the occurrence of a design-basis phenomena through its full range of probabilities. These factors are affected by risk of death or serious injury society is willing to accept, often perceived on a voluntary versus involuntary basis. Table 2.1 and Figures

TABLE 2.1. CURRENT PROBABILITIES OF DEATH TO THE U.S. POPULATION FROM TYPICAL NATURAL AND MAN-MADE SOURCES

Source	Average No. Deaths/Year	Average No. of People at Risk	Probability
Earthquake	13	50×10^6	2.6×10^{-6}
Tornado	120	100×10^6	1.2×10^{-5}
Automobile accident	45,000	250×10^6	1.8×10^{-4}

Note: Probability for earthquake and tornado deaths based on statistics over the last 100 years. Auto deaths are based on statistics over the past five years.

2.1 and [3] 2.2 present the yearly risks of death probabilities in the U.S. from major man-induced and natural hazards. From Table 2.1 and Figures 2.1 and 2.2, it can be deduced that a 10^{-4}/yr probability of death or serious injury is reasonable and societal acceptable safety goal if the perceived benefit is large enough. Associated with the 10^{-4}/yr safety goal is a performance goal, which defines the probability of SSC failure, which could lead to serious injury or death. Implicitly conventional industrial facilities designed to meet design code standards in common usage meet a probability of failure performance goal of about 10^{-3}/yr. These performance goals apply to service and severe load category probabilities of failure. The service or severe load categories limits range from one to 2×10^{-3}/yr as shown in Table 2.2.

Safety goals established for high-hazard facilities subjected to extreme hazards, such as nuclear power plant reactors, have been established by the Nuclear Regulatory Commission (NRC) at 10^{-6}/yr probability of exceedence for the release of undefined (large) quantities of radioactive materials from all sources from a large nuclear power plant [3]. It is further assumed that the conditional probability of release caused by any one source is 10^{-1}/yr less than the probability of the event given commonly used acceptance or design criteria for extreme events. For SSC in other types of nuclear facilities with lesser radiological consequences, the U.S. Department of Energy (DOE) has established performance goals of 10^{-5}/yr for performance category 4 and 10^{-4}/yr for performance category 3 probability of failure of a particular SSC, such that it cannot perform its safety function [4]. These relatively lower performance goals have led to the development of the extreme category of hazard with probability limits that are also shown in Table 1.6 of Chapter 1.

TABLE 2.2. NORMAL AND SEVERE LOAD PROBABILITIES OF EXCEEDENCE CONSIDERED BY DESIGN CODES

Load	Category	Mean Probability of Occurrence
Dead	Normal	1.0
Pressure[4]	Normal	1.0
Live[1]	Normal	2×10^{-2}
Thermal[3]	Normal	1.0
Earthquake	Severe	2×10^{-3}
Wind	Severe	2×10^{-2}
Snow	Severe	2×10^{-2}
Rain[2]	Severe	2×10^{-2}
Transient	Severe	—

1. While this load is considered in the normal category, it generally would not be exceeded more than once in the design life of the facility.
2. Drainage system usually designed for 4×10^{-2} probability of exceedence.
3. Not usually considered in design unless Δt exceeds 100F°.
4. Design pressures are typically taken as 10 percent above maximum operating pressure.

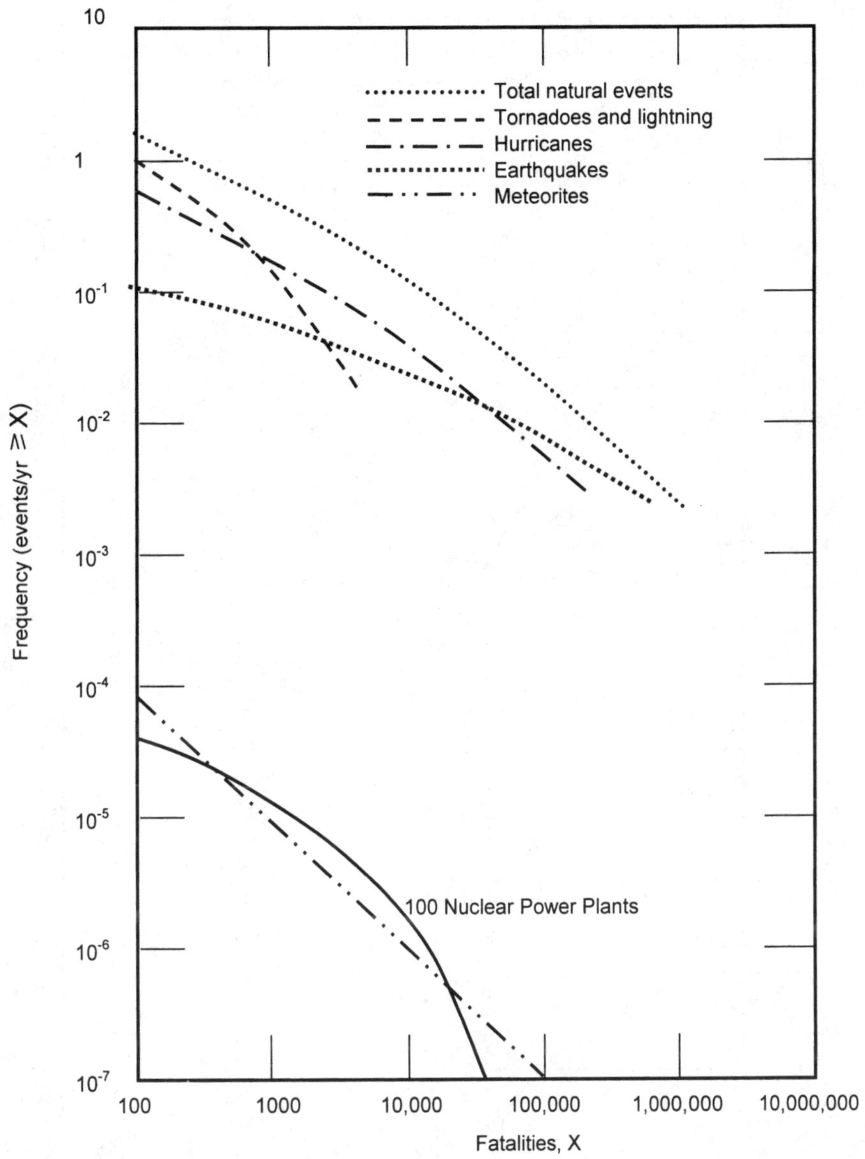

FIGURE 2.1. FREQUENCY OF NATURAL EVENTS INVOLVING FATALITIES

2.3 APPLICATION OF PROBABILITY CONCEPTS FOR EVALUATION OF SERVICE AND SEVERE LOADS

2.3.1 Service or Normal Loads

Service or normal loads on a SSC, such as dead, pressure and live load, are usually considered on a non-probabilistic basis, since their probability of occurrence per year is considered 1.0, and for building and other structures they are defined in applicable industry standards. For mechanical systems and components, they are typically defined by the owner in a design specification.

Elevated temperature effect loads in structures are typically considered only if they exceed about 120° F (50°C) and are treated in the same manner as dead and live loads. For mechanical systems and components, generally they are only considered if they exceed a differential temperature of about 100F° (55C°), and in such cases they are required to meet secondary stress limits independent of primary stresses such as dead and pressure stresses.

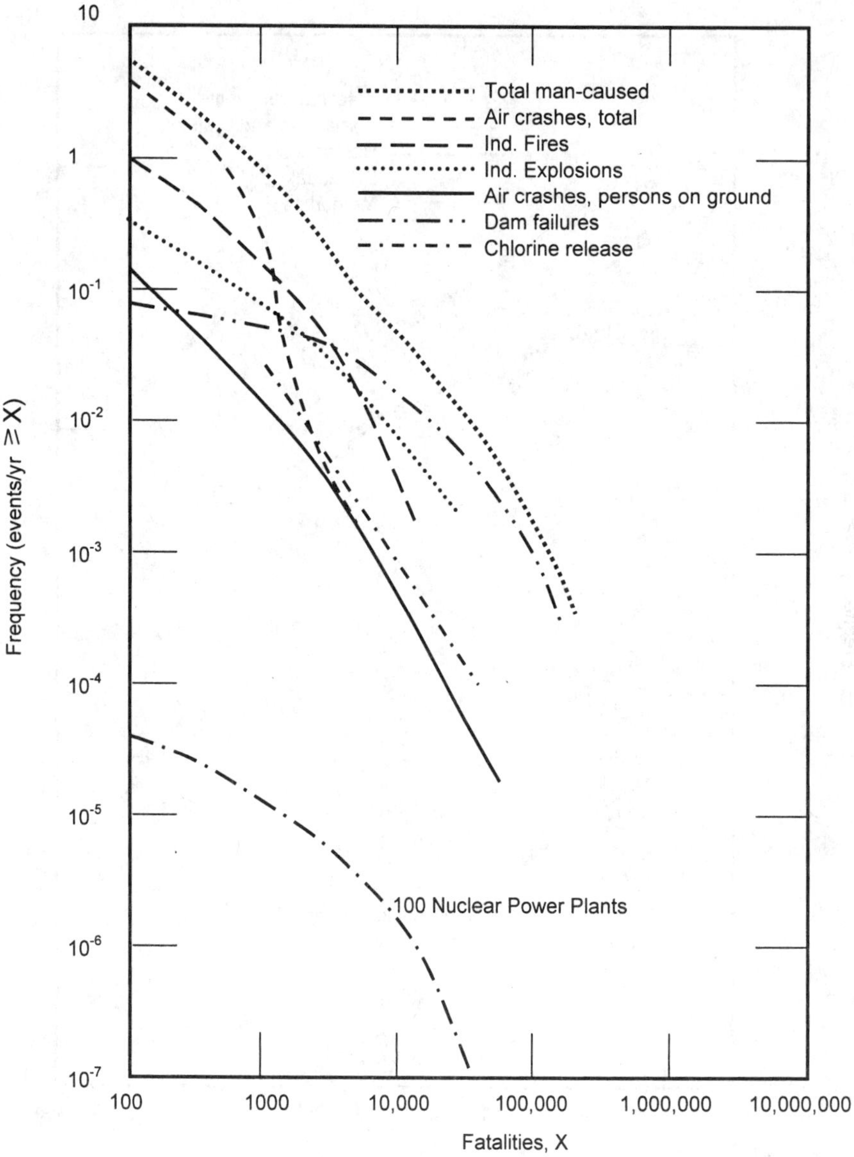

FIGURE 2.2. FREQUENCY OF MAN-INDUCED EVENTS INVOLVING FATALITIES

2.3.2 Anticipated Loads

Natural Hazards

a. Seismic load: The anticipated seismic load is defined in the International Building Code-2003 as that earthquake that has a return period of 2,500 years or probability of exceedence of 4×10^{-4}/yr [5]. As a practical manner this level of earthquake is reduced by a 2/3 factor before being applied to the design of SSC, which actually results in approximately a 500-year return period or 2×10^{-3}/yr probability of exceedence.

b. Wind load: The design-basis wind load is defined on a probabilistic basis and currently is based on the three-second maximum gust for a 50-year mean recurrence interval [6].

Transient Operation Loads

These are occasional loads that are anticipated during the operating life of the facility, such as safety relief valve discharge with no well-defined probability of exceedence.

2.4 APPLICATION OF PROBABILITY CONCEPTS FOR EVALUATION OF EXTREME NATURAL HAZARD LOADS

Hazardous facility failures differ from those in conventional facilities because hazardous SSC in these facilities can potentially cause serious injury and death to collocated workers and the general public, as well as serious injury to the environment. The mitigation or prevention of such failures has been the primary objectives of hazardous facility safety design. An essential and difficult part of any design basis for such facilities is the determination of the probability of life-threatening accidents or the probability distributions of the extreme external environmental or man-induced events. A probabilistic description and load determination for the various extreme or rare events is summarized here.

2.4.1 Design-Basis Earthquake Ground Motion

In general, probabilistic seismic hazard assessment (PSHA) associated with basic development involves four steps: identification of sources to include size, shape and focal depth; establishment of recurrence relationships, magnitude distribution and average rate of occurrence for each source; selection of attenuation relationship; and finally, computation of the site seismic hazard curve [7, 8]. Uncertainty is inherent in each of these steps. For example, the length or size used to model an identified source is only a best estimate, therefore the above steps may be repeated several times using a number of different source models, attenuation expressions, etc.

Step 1: Identification of Sources

When a prospective site for a facility is being reviewed, all seismic sources that can produce damaging ground motion at the site need to be defined. The identification of seismic source zones is based upon the interpretation of geological geophysical and seismological data. A brief discussion of some of the considerations involved in the identification process is presented here.

Although sources can be represented by points, lines and areas, in most regions they are defined as areas (or zones) because the preciseness of a point or line may not properly depict the knowledge (or lack of knowledge) of a potential earthquake location and mechanism. The length of line sources and the size of zones may vary, but each designated source is typically assumed to have uniform earthquake potential. This means that a given magnitude event is equally likely to occur anywhere on or within a source. If a line source is used to represent a well-defined fault, but a portion of the fault is considerably more active than the rest, then enough source lines (within reason) should be used to properly define the fault.

Once the sources are defined, the site-to-source distance distributions can be established. The distance distribution for a given source does not have to be of closed mathematical form, but can be an array of values representing the distance from the site to the segments or sub-areas of a discretized source.

Step 2: Recurrence Relationships, Magnitude, Distributions and Average Rates

The second step involved in a PSHA is the establishment of earthquake recurrence relationships, magnitude distribution and average occurrence rates. A recurrence model specifies the relative number of earthquakes of different magnitude levels. In most cases, earthquake recurrence is expressed by the Gutenberg-Richter *b*-line: $\log N = a - bM$, where M = magnitude; N = expected (or average) number of earthquakes of magnitude greater than or equal to M; a and b are constants for a given source. This relationship plots as a straight line with a y- intercept of "*a*" and slope of "*b*," hence the name *b*-line. Multiple values of a and b can be used to represent different portions of the magnitude scale of a given source. The line(s) can be derived from regression analysis of either recorded data or a combination of recorded and geologic data, with the latter usually resulting in multisloped *b*-lines. In most cases the recurrence expression is presented in its normalized form (i.e., per unit time, per unit distance/area). It should be noted that even though the *b*-line can be used to compute the average occurrence rate of earthquakes, it does not address the time distribution of the events. Forecasting models can be developed and then used to correlate events in space and time.

Use of the Gutenberg-Richter relation implies that magnitudes are exponentially distributed. The magnitude probability density function is typically expressed as:

$$f(M) = c\beta e^{-\beta(M-m_o)},\ where\ c = \frac{1}{1-e^{-B(m_{max}-m_o)}} \tag{2.1}$$

where $f(M)$ = magnitude probability density function; $\beta = b\ln_{(10)}$ (*b*=slope of *b*-line); M = magnitude; and m_0 and m_{max} = the lower and upper magnitude limits, respectively.

As with the distance distribution, the magnitude distribution can be discretized, with the center value of each interval being used in the attenuation expression. Even for simple problems, a large number of computations are typically required in a PSHA. As a result, computers are normally employed; discretization is conducive of computer code solutions, even for functions where the closed mathematical form is known.

Given the occurrence of an earthquake, the probability that its magnitude and distance fall in the respective intervals needs to be determined. If the lengths or areas of each segment or sub-area for a given discretized source are equal, then there is an equal probability that an earthquake will occur on any of the segments or sub-areas. Therefore, the probability that an earthquake will occur at any one of the discretized source to site distance array values is simply 1/(the number of intervals). Given the occurrence of an earthquake, the probability that its magnitude falls in the *i*th magnitude interval is approximated by multiplying the magnitude probability density function evaluated at the center magnitude for that interval (M_1) by the range of the interval (ΔM), or:

$$P(M_i - \frac{\Delta M}{2} \leq M \leq M_i + \frac{\Delta M}{2}) = f(M_i)\Delta M \tag{2.2}$$

The lower limit magnitude (m_o) used in a PSHA was at first thought by the earthquake community to be inconsequential, but it was later discovered that this was not the case. In selecting m_0, a sensitivity study should be conducted to determine how much influence it has on the results. A body wave magnitude (m_b) of 5.0 is a commonly used lower bound value since it usually defines the threshold of damage to engineered structures, systems or components.

The choice of magnitude for the maximum credible earthquake (m_{max}) has a much greater influence on the results than m_o, and varies from zone to zone. Because the earthquake database in the United States only extends back a couple of hundred years, the maximum historic earthquake for a region or source is usually used as a minimum value for m_{max}. Geologic investigations are also used to determine m_{max}. For example, correlations exist between surface fault length and maximum possible earthquake. Also trenching of a fault may determine the amount and rate of slip and provide a clue as to the associated size of past events.

The average occurrence rate *(v)* of a source is defined as the average number of events between m_o and m_{max}, per unit time. The average occurrence rate is related to the normalized recurrence expression (*b*-line) and the size of the source by the following expression:

$$V = [N(m_o) - N(m_{max})] \bullet L \tag{2.3}$$

where $N(m_o)$ and $N(m_{max})$ = average number of earthquakes (per unit time, per unit size) of magnitudes equal to or greater than m_o and m_{max}, respectively; and L = size of the source.

Step 3: Ground Motion Estimation

Once the source(s) are identified and the recurrence relationship(s) with corresponding values of m_o and m_{max} are established, an attenuation expression needs to be selected to estimate the ground motion at a site. Numerous attenuation relationships exist for various source mechanisms and geologic conditions; one appropriate for the given conditions needs to be selected.

Given the occurrence of an earthquake, the probability that the site PGA will exceed an acceleration (acc) of interest needs to be determined for every combination of discretized magnitude and distance for each identified source or stated in mathematical notation: $P(\text{PGA} > acc/EQ{:}\ R_j, \text{M}_i)$. Note that the vertical line and subsequent symbols read "given the occurrence of an earthquake at distance R_j, M_i." Usually, the uncertainty in an attenuation expression (resulting from the scatter in the data from which it is derived) is assumed to be log-normally distributed. This assumption provides a means of quantifying the probability that given the occurrence of an earthquake of magnitude M at distance R the PGA is above (or below) an acceleration level of interest. The inclusion of this uncertainty in the analysis requires a large computation effort, but must be accounted for if a true PSHA is to be conducted. Also note that there is uncertainty resulting from the choice of attenuation relationship used. This uncertainty can be formally treated as a node in the logic tree.

Step 4: Construction of the Hazard Curve

A hazard curve combines all of the above information into one plot. The distance and magnitude distributions and ground motion probability expression for a given source are combined to produce a statement of the probability that given the occurrence of a seismic event with a magnitude of interest anywhere on the kth source, the site PGA will exceed an acceleration of interest or

$$\text{P}_\text{k}(\text{PGA} > \text{acc} \mid \text{EQ}_\text{k}) = \iint \text{P}(\text{PGA}) > \text{acc} \mid \text{EQ}_\text{k}{:}\ \text{M, R})\ \text{f(M) f(R) dMdR} \tag{2.4}$$

Restated in summation notation, which is applicable to the discretized distributions:

$$\text{P}_\text{k}(\text{PGA} > \text{acc} \mid \text{EQ}_\text{k}) = \sum_j \sum_i \text{P}(\text{PGA} > \text{acc} \mid \text{EQ}_\text{k}{:}\ \text{M}_\text{i}, \text{R}_\text{j})\ \text{f}(\text{M}_\text{i},\text{R}_\text{j})\ \text{f}(\text{M}_\text{i})\ \Delta\text{M}\ \text{f}(\text{R}_\text{j}) \tag{2.5}$$

The above expression is the probability that the site PGA will exceed an acceleration of interest given the occurrence of an earthquake, and does not take into account the probability that an earthquake will occur.

A forecasting model needs to be selected to express the occurrence probability of future seismic events (i.e., correlation of events in space and time). Forecasting models are generally based on the theory of stochastic (or random) processes and not on the extrapolation of past data, but may employ data for the evaluation of parameters (e.g., v, the average occurrence rate). A widely used model is the homogenous Poisson model.

The Poisson model of occurrence can be written as:

$$P(nmt) = \frac{(vt)^n e^{-vt}}{n!} \tag{2.6}$$

where $P(n,t)$ = probability of having exactly n events in a future time period t; and v = average occurrence rate. This expression reduced to $P(0,1) = \exp(-v)$ for the probability that no events will occur in a one-year time period. Therefore, the probability that at least one event will occur in the year is: $p = 1 - \exp(-v)$. When this expression is combined with the probability that the site PGA will exceed an acceleration of interest (given the occurrence of an earthquake), the annual probability that the site PGA will exceed an acceleration of interest (acc) resulting from the occurrence of an earthquake at the kth source becomes:

$$\text{P}_\text{k}(\text{PGA} > \text{acc}) = 1 - \text{e}^{(\text{-vp})\text{k}} \tag{2.7}$$

Where p is $P_k(\text{PGA} > acc/EQ_k)$. For values of $vp \ll 1$, the above expression is approximately:

$$\text{P}_\text{k}(\text{PGA} > \text{acc}) = (\text{vp})_\text{k} \tag{2.8}$$

or in expanded summation form:

$$\text{P}_\text{k}(\text{PGA} > \text{acc}) = \text{v}_\text{k} \sum_j \sum_i \text{P}_\text{k}(\text{PGA} > \text{acc} \mid \text{EQ}_\text{k})\ \text{f}(\text{M}_\text{i})\Delta\text{M}\ \text{f}(\text{R}_\text{j})\Delta\text{R} \tag{2.9}$$

The following expression may be used to combine the hazard from each source, resulting in a site hazard curve:

$$P(PGA > acc) = 1 - \prod_k (PGA < acc)_k \quad (2.10)$$

or:

$$P(PGA > acc) = 1 - \prod_k \{1 - (PGA > acc)_k\} = 1 - \prod_k (1 - P_k) \quad (2.11)$$

where P = exceedence probability due to all sources; P_k = exceedence probability due to the kth source; and Π = series product.

The seismic hazard curves generally have the shape shown in Figure 2.3, represent the acceleration of all seismic events which may affect the site. The hazard curve for a particular earthquake below about 10^{-4}/yr probability of exceedence would typically show a rapid drop off. Below about 10^{-4}/yr, it is thought that limits on rock fracture capacities and the linear rather than point nature of a large magnitude seismic fault rupture may control seismic acceleration at a site. The random variable of site ground acceleration, whether it is peak ground acceleration or spectral acceleration at a particular frequency and level of damping, is generally assumed to have the skew-positive characteristic of a log normal, Wiebull or extremal distribution. Usually the log normal distribution is assumed for seismic acceleration random variable hazards. Therefore, the mean value would be larger than the median value. For most hazard curves mean value estimates are used as the basis for design. However, for seismic hazard curves the median value is often used because of the much larger uncertainty that exists in mean versus median value estimations.

Logic Trees and Uncertainty

The logic tree is an explicit method of documenting and quantifying the uncertainty inherent in the hazard model selected. A logic tree allows the consideration of a variety of source definitions, recurrence relationships, m_o and m_{max}, attenuation expressions, forecasting models, etc. Each branch at the right end of the tree represents a unique seismic hazard curve for a specified set of assumptions.

All the branches that extend from nodes are assigned a weight that quantifies the likelihood of that alternative being correct; the sum of all the weights attached to a node is unity. The assignment of the weights is

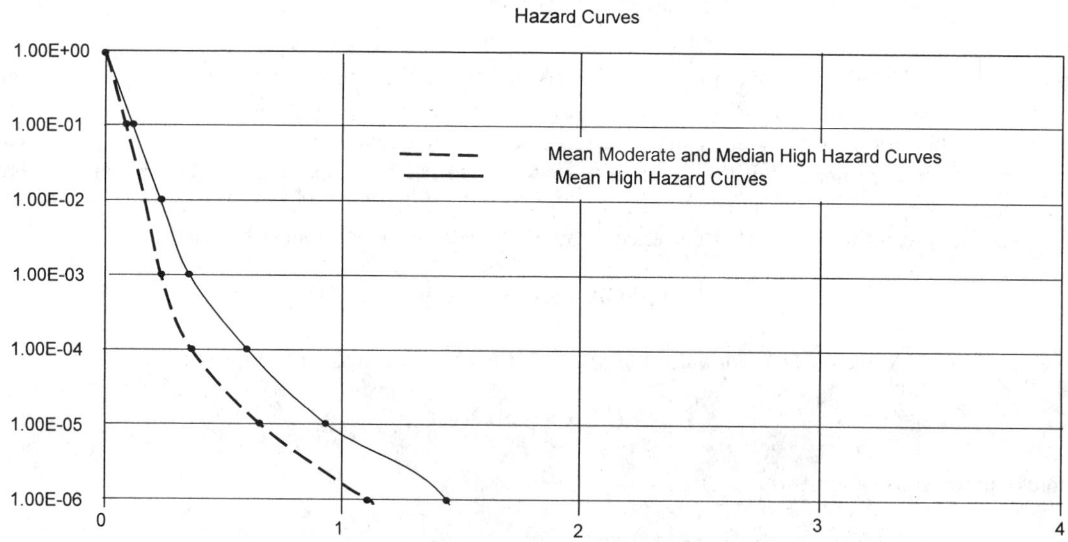

FIGURE 2.3. TYPICAL SEISMIC HAZARD CURVES (PEAK GROUND ACCELERATION) FOR MEDIAN AND MEAN SEISMIC HAZARD CONSIDERING ALL EARTHQUAKES

a formalized method of documenting subjective inputs. The product of all the weights of the branches associated with a hazard model expresses the confidence in the associated hazard curve. From the family of hazard curves associated with the various combinations of assumptions, the mean, median and 84th percentile hazard curves can be calculated.

Assumptions Inherent in PSHA

Any time nature is modeled mathematically, simplifying assumptions have to be made. As a result, the validity of a mathematical model is limited to scenarios where the assumptions are reasonable. The use of any engineering procedures without an understanding of the associated assumptions may lead to erroneous results; PSHA is not an exception. The following discussion points out some of the assumptions involved in PSHA.

When conducting a PSHA, assumptions are inherent in the selection of both the magnitude recurrence relationship and the earthquake occurrence forecasting model. Although the selection of these two models is distinct, there are some overlapping assumptions. The recurrence model is discussed first.

The Gutenberg-Richter B-Line

The Gutenberg-Richter *b*-line is the most commonly used recurrence model. The *b*-line provides information concerning the magnitude distribution, given the occurrence of an event, and the average occurrence rate (v) of earthquakes for a given region. However, the *b*-line does not provide information concerning the time distribution of events. For example, if the average occurrence rate of earthquakes for a given region is computed to be 12 per year, no information is provided as to whether the time distribution of the events is constant (e.g., one per month, year in, year out) or clustered (e.g., five the first month, seven the second month, and zero in the remaining 10 months, or none for five years, and then a swarm). The assumed time distribution is inherent in the selected forecasting model.

A direct comparison cannot be made among *b*-lines proposed by different studies even for the same region, because each is sensitive to both the size of the area used to model the source and the time interval from which the analyzed data was recorded. One inherent assumption in selecting the size of an area used to model a source zone is that it be of uniform earthquake potential. Therefore, as the size of an area model increases, higher activity portions of the zone are "diluted" by lower activity portions, but in computing the seismic hazard of a site, it is assumed earthquakes can occur only in areas identified as sources. As a result, a small area with a characteristically more active *b*-line may present a hazard equal to that of a large area with a characteristically less active *b*-line. The resulting hazard curves of PSHA using different area sizes would have to be compared to determine how sensitive the PGA is to these changes. The uncertainty in the selection of an appropriate sized model can be incorporated as a node on the logic tree, but the assigned confidence in each model is a subjective input.

The time interval from which the analyzed records are taken for use in the construction of the *b*-line should be as large as possible, but in the United States the historic database only extends back a couple of hundred years. When a small region or individual fault is examined, the time interval is generally much shorter, with few (if any) large events having occurred. Various methods have been proposed to make up for the insufficient amount of historical data. In some cases, the slope and *y*-intercept of the *b*-line are determined from small magnitude events and simply extend as a straight line for larger magnitudes. Also, geologic evidence and the slip rate of the fault often are used to complement the database. The assumptions inherent in extending the *b*-line or using geologic data have been questioned by some seismologists. One should be aware that no one method of filling the gap resulting from an insufficient amount of data is universally accepted.

Poisson Forecasting Model

Assumptions also are involved in the selection of a probabilistic earthquake forecasting model. The Poisson model is the most commonly used forecasting model, but others, such as clustering and renewal, also may be employed. Two common assumptions in using the Poisson model are the constant-in-time average occurrence rate (v) of earthquakes and the "memory-less" behavior of occurrences. The use of a constant occurrence rate becomes unreasonable when determining the hazard for small time intervals relative to the observed clustering return period.

2.4.2 Extreme and Rare Wind and Missile Velocities

Exceedence Probability and Return Period for Extreme Straight Winds

The basic probability of extreme winds needed for structural deign is the exceedence probability P_E, which is the probability that a given wind speed will be exceeded within a one-year period [9]. The reciprocal of exceedence probability is called the return period or recurrence interval, namely,

$$T_R \text{ return period} = \frac{1}{P_E} \tag{2.12}$$

where P_E and T_R can be determined from high wind data collected at a wind velocity measuring site over a period of time.

The data for a site are arranged in descending order of wind speed as shown in column 2 of Table 2.3. The values of exceedence probability listed in the last column of the Table is calculated from

$$P_E = \frac{r}{N+1} \tag{2.13}$$

where r = rank number (column 1); and N = total number of years – 10 in this case. Note that $r/(N+1)$ instead of r/N was used to calculate the exceedence probability in order to avoid P_E=1 (certainty) for $r = N$.

The values of V versus P_E in Table 2.3 can be plotted on the selected probability density function probability paper.

The probability density function of wind speed V, designated as $f(v)$, is the probability per unit wind speed. From this definition, the probability of wind speed between any value V and an infinitesimally larger value $V + dV$ is $F(V)\,dV$.

The probability for wind speed to be less than V is:

$$F(V) = \int_O^V f(V)dV \tag{2.14}$$

from which

$$f(V) = \frac{dF(V)}{dV} \tag{2.15}$$

where $f(V)$ = probability density function.

The exceedence probability P_E is related to the probability distribution function by:

$$P_E(V) = 1 - F(V) = 1 - \int_O^V f(V)dV \tag{2.16}$$

TABLE 2.3. CALCULATION OF EXCEEDENCE PROBABILITY FOR ANNUAL FASTEST-MILE 3-SECOND WIND GUST AT WIND SPEED MEASURING SITE (1990–1999)

Wind Speed Ranking *R*	Annual Fastest-Mile Wind *V* (mph)	Year of Occurrence	Exceedence Probability P_E
1	63	1992	0.09
2	61	1998	0.18
3	58	1990	0.27
4	58	1991	0.36
5	57	1993	0.45
6	56	1994	0.55
7	56	1996	0.64
8	56	1997	0.73
9	49	1995	0.82
10	49	1999	0.91

This equation shows that the exceedence probability can be easily calculated from either the probability distribution function or the probability density function.

The wind speed used in the structural design of hazardous facilities is the extreme (maximum or fastest) values for a given period such as 1,000 to 100,000 years or 10^{-3} to 10^{-6} probability of exceedence. For conventional structures, design-basis wind speeds are typically defined at 2×10^{-2} or 10^{-2}/yr level. To model extreme wind speeds, one should, if possible, distinguish straight winds (i.e., thunderstorm or weather-front-related nontornadic winds) from hurricane and tornado winds because they follow different probability laws.

The extreme wind date for straight winds can be adequately represented by the following probability density function:

$$F(V) = \exp\left[-\exp\left(-\frac{a+V}{b}\right)\right] \tag{2.17a}$$

and

$$a = 0.450s - \bar{V}$$
$$b = 0.780s \tag{2.17b}$$

where V = annual fastest wind speeds; $\bar{V}$ = mean of V; and s = standard deviation of V.

It can be proved that for this type of distribution the wind speed V corresponding to a return period T_R (in years) is:

$$V = \bar{V} + 0.78\ (\ln T_R - 0.577)s \tag{2.18}$$

The probability density function given by Eq. (2.17) is referred to in the literature under different names such as Extremal Type-I distribution, Gumbel distribution or Fisher-Tippett distribution. The ASCE standard on wind load [6] uses this distribution to determine the maximum fastest 3-second gust wind speed corresponding to different return periods. For instance, based on the 10-year data in Table 2.3, the mean wind speed $\bar{V}$ and the standard deviation are 56.3 mph and 4.47 mph, respectively. Substituting these values into Eq. (2.18) yields a wind for the selected return period T_R. The wind speed contours for the United States for 2×10^{-2}/yr probability of exceedence are defined in Figure 6.1 of Reference. [6], and were determined essentially from the foregoing approach, using the extreme wind data of 129 stations that had reliable wind records [9] referenced in the [6] commentary.

Exceedence Probability and Return Period for Rare Winds

a. Hurricane Wind Probability

The extremal type-I distribution has been used widely for extreme straight wind analysis not only in nonhurricane regions of the U.S., but also in hurricane regions. However, it should be realized that when analyzing the annual fastest wind data for any site in a hurricane region, the data are a mixture of rare hurricane cyclonic winds and extreme straight winds. They cannot be modeled properly by a single probability density function no matter what probability law is used. It is often not feasible to model the two separately because hurricanes do not strike the same location each year. There may be insufficient data for hurricanes at a given station, and besides, the fastest hurricane wind speed for most years may be zero. Proper modeling of the hurricane wind speed would require a much longer record than most stations have.

For stations that have a relatively long (over-50-year) record of hurricane wind speeds, one can list the highest speed of each hurricane measured at a station in a way similar to that in Table 2.3, except that some years may have more than one value listed, whereas some other years may have none. Such a listing is called a partial duration series, as shown in Table 2.4, which is different from the annual series listed in Table 2.3.

TABLE 2.4. RELATION BETWEEN T_R T'_R FOR A PARTIAL DURATION SERIES

T_R (Annual Series)	T'_R (Partial Duration Series)
1.1	0.417
1.5	0.91
2	1.44
5	4.48
10	9.49
25	24.5
50	49.5
100	99.5
500	499.5

Suppose r is the rank (order) of a hurricane wind speed in a partial duration series containing M hurricanes in a record of N years. The average number of times per year that this hurricane wind speed will be exceeded is:

$$m = \frac{r}{N} \tag{2.19}$$

The reciprocal of m is the average number of years between hurricanes with wind speeds greater than that of the rth ranked hurricane. While P_E in Eq. (2.20) is probability, m in Eq. (2.19) is not probability. The value of m can be greater than 1.0 if, on the average, more than one hurricane per year pass through the station, as is the case in some typhoon regions of the world such as Taiwan.

The exceedence probability based on the partial duration series can be calculated from

$$P_E = \frac{r}{nN} \tag{2.20}$$

where n = average number of hurricanes per year for the record, namely, $n = M/N$.

To use partial duration series data to determine the relationship between V and T_R Eq. (2.18) is rewritten as:

$$V = \bar{V}' + 0.78\,(\ln T'_R - 0.5777)^{s'} \tag{2.21}$$

where the primed quantities are values calculated from the partial duration series. The quantity T'_R is related to T_R as follows:

$$T'_R = -\frac{1}{\ln[1-(1/T_R)]} \tag{2.22}$$

Values of T'_R calculated from Eq. (2.22) are given in Table 2.4. Note that the vales of T'_R approach that of T_R as the return periods get large.

Eqs. (2.21) and (2.22) can be used simultaneously to determine the relationships between V and T_R.

The variation of hurricane wind speeds along the Gulf and Atlantic coastlines of the United States has been determined by using Monte Carlo simulation. The result is very useful, and it has been used in *ASCE 7-02 Standard* [6] to specify the values of wind speeds along the hurricane coastline of the United States.

b. Tornado Wind Probability

Tornado wind speed records are very rare because tornadoes tend to affect very small areas and the wind instrumentation is usually destroyed when in the path of an F2 or larger tornado. There are several kinds of tornado probabilities. The strike probability P_1 is the probability that a tornado will strike a given point or location within a given year. The speed probability P_2 is the probability that a tornado will have a wind speed greater than a certain value *V*. The probability *P* that a tornado with wind speed greater than *V* will strike a given location in a year is the product of P_1 and P_2, namely [9]:

$$P(V) = P_1 P_2 \tag{2.23}$$

The strike probability P_1 is calculated from

$$P_1 = n' \frac{A_d}{A_o} \tag{2.24}$$

where A_d = average damage area of a tornado; A_o = a reference area such as 100 square miles or a 1 to 5 degree longitude-latitude square; and n' = average number of tornadoes per year in the area A_o. A recent refinement of the average damage area of a tornado is to include the potential to strike any portion of a building of a finite size in that area.

Figure 2.4 gives the values of P_1 for different regions of the United States based on tornado data up to the 1970s [10]. The speed probability P_2 is estimated from a large number of damage surveys using the Fujita Scale (Table 2.5) to determine the wind speed of each tornado [11]. Figure 2.5 gives the values of P_2 based on the average condition of tornadoes occurring anywhere in the U.S. From Figures 2.4 and 2.5 and Eq. (2.23), the probability of having a tornado with a wind speed greater than 150 mph at a particular location in the state of Kansas during a given year is:

$$P(V > 150 \text{ mph}) = P_1 \times P_2 = 336 \times 10_{-5} \times 0.1 = 3.4 \times 10_{-4}$$

From Figure 2.4, more than 50% of tornadoes have maximum gust speeds lower than 100 mph, and only 10% of tornadoes have maximum speeds higher than 150 mph. Because the wind speeds used in deriving Figure 2.4 estimated rather than measured values, any wind speed predicted from this graph can easily be in error by 20%. A more recent study of tornado probabilities was completed by the Lawrence Livermore National Laboratory in 2000 [12]. Based on this study the occurrence of tornadoes appears to have increased about 10 percent probably due to the better means available for detecting tornadoes.

The exceedence probability of tornado winds may be compared to that of winds as shown in Figures 2.6, 2.7, 2.8 and 2.9 for the low moderate and high tornado regions defined in the U.S. [13, 14].

TABLE 2.5. RANGES OF FUJITA-RECOMMENDED WIND SPEED *(F)* PATH LENGTH *(P_L)* AND PATH WIDTH *(P_W)*

Scale	*F* (mph)	P_L (miles)	P_W (yds)
Negative	**Less than 40**	**Less than 0.3**	**Less than 6**
0	40–72	0.3–1.0	6–17
1	73–112	1.0–3.1	18–55
2	113–157	3.2–919	56–175
3	158–206	10–31	176–556
4	207–260	32–99	*0.3–0.9
5	261–318	100–315	*1.0–3.1
6	319–380	316–999	*3.2–9.9

* In miles

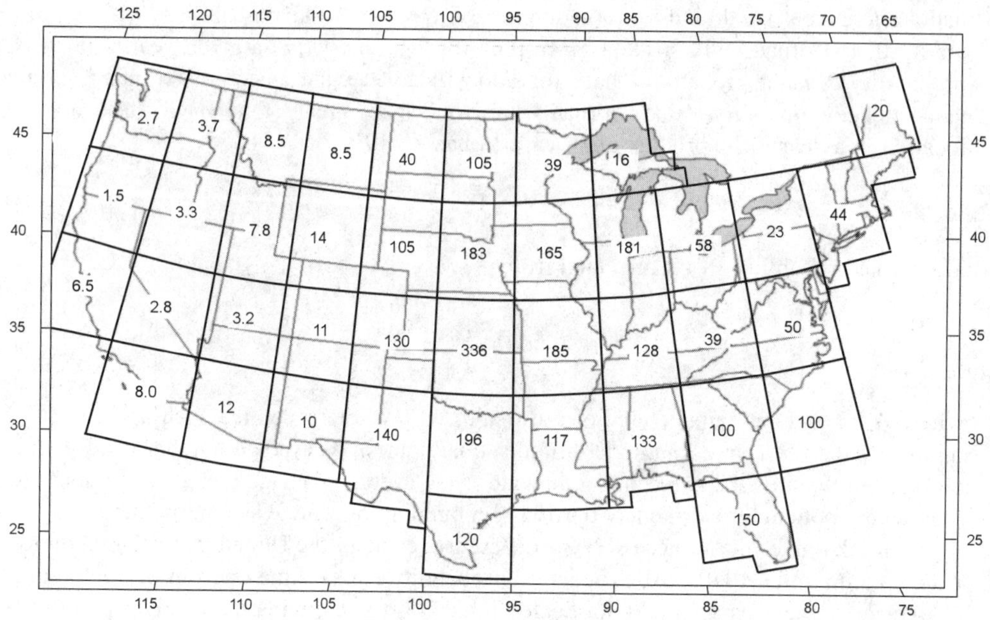

FIGURE 2.4. TORNADO STRIKE PROBABILITY WITHIN FIVE-DEGREE SQUARES IN THE CONTIGUOUS UNITED STATES (UNITS ARE 10–5 PROBABILITY PER YEAR)

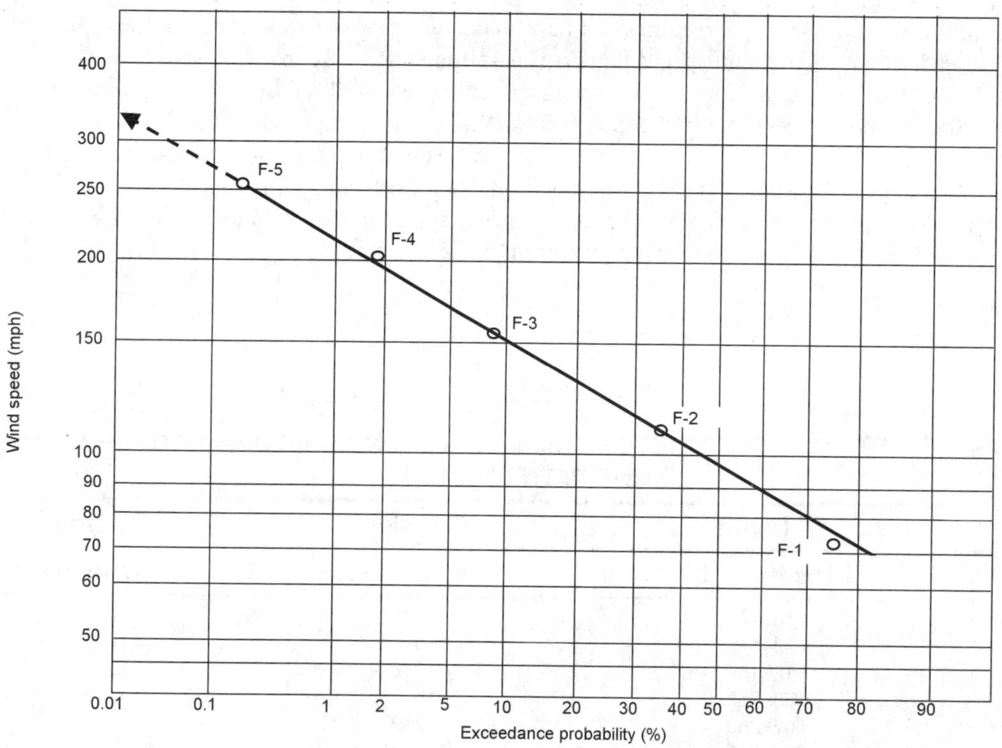

FIGURE 2.5. TORNADO WIND SPEED PROBABILITY P_2 (SAME AS THE FRACTION OF TORNADOES HAVING A WIND SPEED HIGHER THAN V. NOTE: WIND SPEED USED IN THIS GRAPH IS GUST SPEED; FROM MARKEE, BERKERLEY AND SANDERS, 1974) [9]

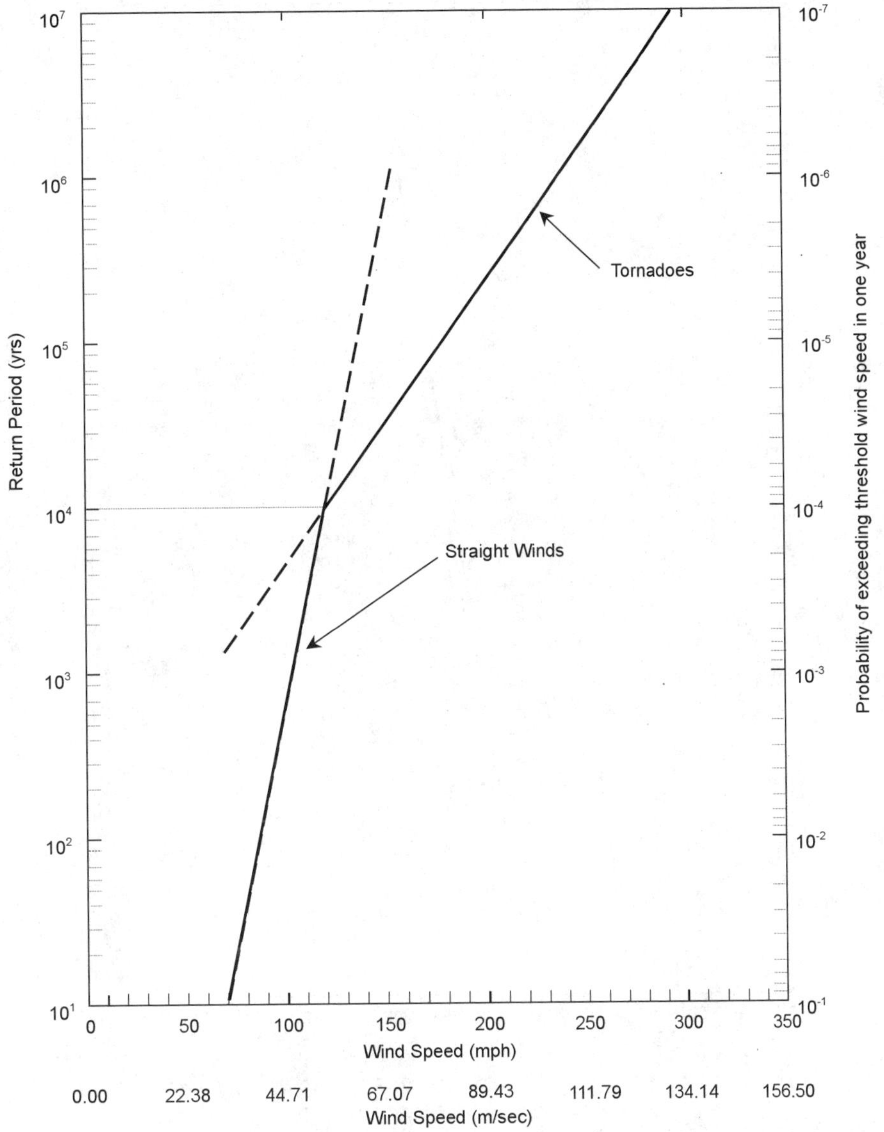

FIGURE 2.6. WIND SPEEDS AT A REGION I OR HIGH TORNADO HAZARD SITE

Wind-Borne Missile Velocities

Wind-borne missile velocities are generally defined as a deterministic function of maximum wind velocities. As a result, probabilities are not used directly in determining wind-borne missile velocities. Procedures applicable to wind-borne missile velocity determination are found in [15].

2.4.3 Extreme and Rare Flood and Precipitation Parameters

Extreme and Rare Floods and Precipitation

Flood

Extreme and rare flood parameters, usually in the form of maximum flood height, are computed in much the same way as extreme and rare wind velocities on a probabilistic basis. Maximum flood stream, river or lake, bay, estuary and sea storm surges are recorded on a yearly basis in much the same way that maximum yearly

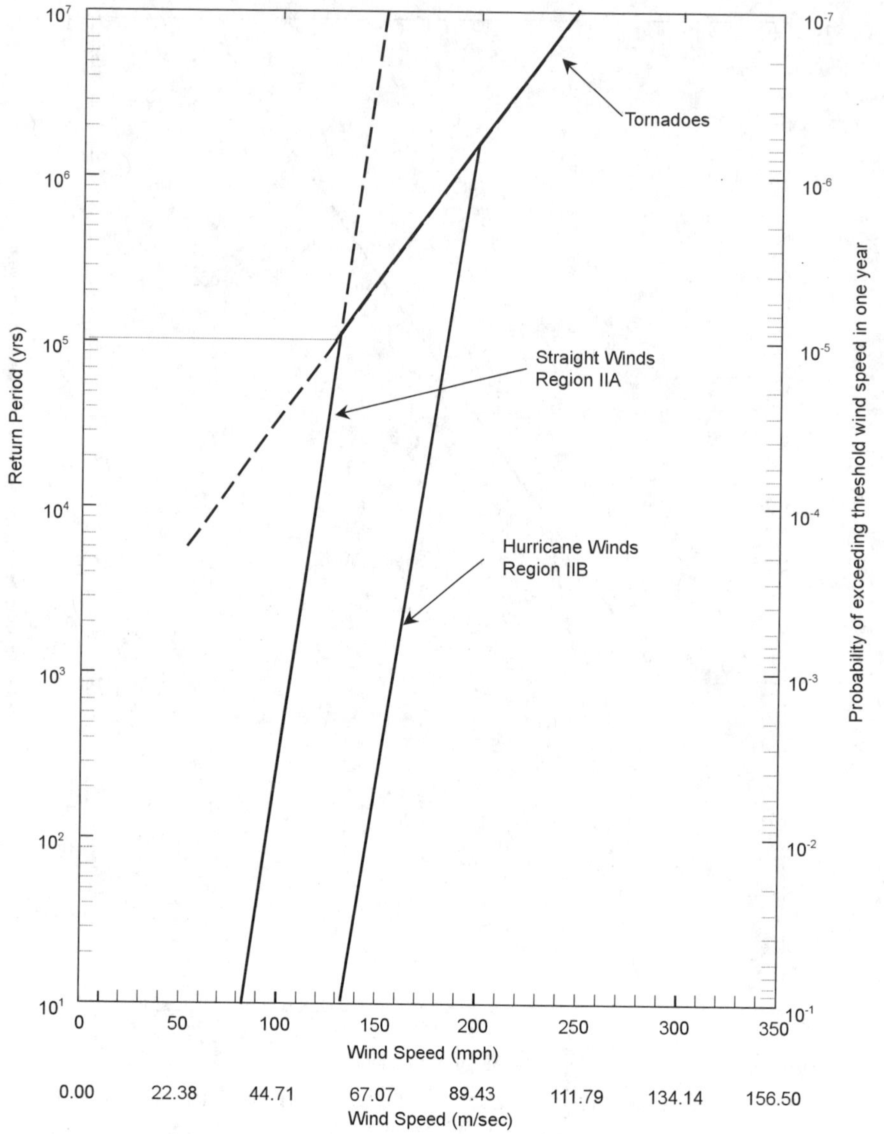

FIGURE 2.7. WIND SPEEDS AT A REGION II OR MODERATE TORNADO HAZARD SITE

3-second gust wind velocities are recorded. Using the same statistical distribution as that for extreme wind velocities, it is possible as a function of probability of exceedence or return periods to predict flood heights for return periods well beyond the observation period. The usual approach to flood design is to locate the hazardous facilities above the probabilistically predicted flood level. For this purpose it is common to use a 10^{-4}/yr to 10^{-5}/yr probability of exceedence.

A procedure, again similar to rare hurricane wind events, is used to predict rare flood and precipitation events using statistical data developed from a partial duration series. Included in this database would be the same types of events that contribute to rare wind plus other phenomena such as upstream dam failures, under water landslides, tsunamis and seiches.

Precipitation

Local or site precipitation (rain, snow or ice) are also determined probabilistically in the same manner. However, design for these phenomena cannot be avoided by site elevation as in the case for flooding.

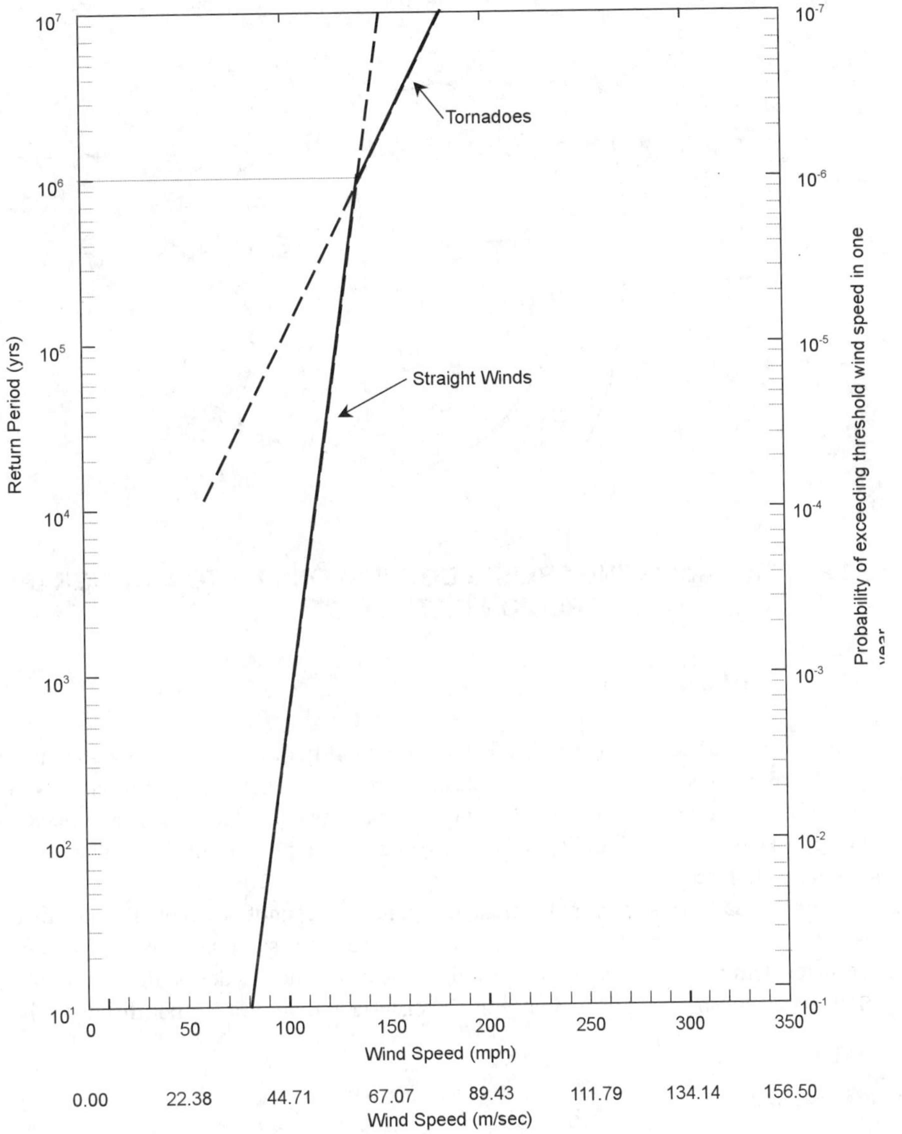

FIGURE 2.8. WIND SPEEDS AT A REGION III OR LOW TORNADO HAZARD SITE

Typically, storm sewers are designed for a 25-year return period precipitation. Therefore, site topography must be evaluated to assure surface runoff for precipitation at a 10,000- to 100,000-year return period without loss of safety function of site facilities. If safety-related structures have flat roofs, precipitation may impose excessive roof loads due to ponding effects and potential plugging of primary roof drains.

2.5 APPLICATION OF PROBABILITY CONCEPTS FOR EVALUATION OF MAN-INDUCED ACCIDENTAL EXTREME LOADS

2.5.1 Introduction

Accidental design-basis extreme loads can and have in some instances been evaluated on a probabilistic basis. As in the case of natural phenomena hazards, these loads can be evaluated probabilistically against the

FIGURE 2.9. TORNADIC WIND SPEEDS CORRESPONDING TO A GEOGRAPHICAL REGION OF THE U.S.

performance goals defined in Section 1.6.1.2.2. When the probability of the event is less than the established performance goal there is no need to define it as a design-basis event. Unlike natural hazards that are typically probabilistically defined as a continuous distribution, the man-induced hazards are usually defined as single discrete events. Hence, the probability of failure is the simple product of the probability of the event times the probability of failure given the event.

Man-induced extreme loads caused by malevolent acts generally cannot be defined probabilistically. They are either considered a design-basis event or not, as evaluated by a safeguards group threat assessment. Design to resist malevolent acts is further complicated by the reluctance to define them in terms that can be addressed publicly as design-basis loads and acceptance criteria and are not further discussed here.

2.5.2 Aircraft Hazard

The probability of an accidental aircraft crash is considered to be less than about 10^{-7} per year as shown in Table 2.6 if the distances from the facility meet all the requirements listed below [16]:

1. The plant-to-airport distance D is between 5 and 10 statute mile (8 to 16 km), and the projected annual number of operations is less than 500 D^2, where D is expressed in miles or the plant-to-airport distance D is greater than 10 statue miles (16 km), and the projected annual number of operations is less than 1,000 D^2, where D is expressed in miles. An operation is usually considered as a takeoff or a landing.
2. The plant is at least 5 statute miles (8 km) from the edge of military training routes, including low-level training routes, except for those associated with a usage greater than 1,000 flights per year, or where activities (such as practice bombing) may create an unusual stress situation.
3. The plant is at least 2 statute miles (3 km) beyond the nearest edge of a federal airway, holding pattern or approach pattern.

If these limitations are not met the aircraft hazard assessment consists of the following steps:

TABLE 2.6. PROBABILITY OF AIRCRAFT CRASH PER SQUARE MILE PER AIRCRAFT MOVEMENT

Distance from End of Runway (miles)	Probability (× 10⁸) Crash per Square Mile per Aircraft Movement		Aircraft Movement
	U.S. Air Carrier	General Aviation	Military Aviation
0–1	16.7	84	7.0
1–2	4.0	15	1.7
2–3	0.96	6.2	0.7
3–4	0.68	3.8	0.36
4–5	0.27	1.2	0.3
5–6	0	NA	NA
6–7	0	NA	NA
7–8	0	NA	NA
8–9	0.14	NA	NA
9–10	0.14	NA	NA

NOTE: Statistics shown are for fatal aircraft crashes, which are assumed to be the most destructive.

1. Define aviation uses: Data describing aviation uses in the airspace near the proposed site—including airports and their approach paths, federal airways, Federal Aviation Administration (FAA) restricted areas and military uses—are developed. Basically there are three types of aircraft movements: (1) general; (2) commercial; and (3) military.
2. Define airways: For situations where federal airways or aviation corridors pass through the vicinity of the site, the probability per year of an aircraft crashing into the plant (P_{FA}) should be estimated. This probability will depend on a number of factors such as the altitude and frequency of the flights, the width of the corridor and the corresponding distribution of past accidents.

   One way of calculating P_{FA} is by using the following expression:

   $$P_{FA} = C \times N \times A/w \tag{2.25}$$

   where: C = in-flight crash rate per mile for aircraft using airway; W = width of airway (plus twice the distance from the airway edge to the site when the site is outside the airway) in miles; N = number of flights per year along the airway; and A = effective area of plant in square miles.

   General aviation movements generally do not use airways and their crash probabilities would have to be added to the airway crash probabilities.

   This gives a combination of probabilities upper bound on aircraft impact probability if care is taken in using meaningful and conservative values for the individual factors. For a commercial aircraft, a value of $C = 4 \times 10^{-10}$ per aircraft mile has been used. For heavily traveled corridors (greater than 100 flights per day), a more detailed analysis may be required to obtain a proper value for this factor, which is addressed in [20].
3. Evaluate civilian and military airports and heliports crash statistics. The probability of an aircraft crashing into the site as a function of its proximity should be evaluated.

   The probability per year of an aircraft crashing into the site based on airport operation cases (P_A) may be calculated by using the following expression.

   $$P_A = \sum_{i=l}^{L} \sum_{j=l}^{M} C_j N_{ij} A_j \tag{2.26}$$

where: M = number of different movements; L = number of flight trajectories affecting the site; C_j = probability per square mile of a crash per aircraft movement, for the jth aircraft; N_{ij} = number (per year) of movements by the jth aircraft along the ith flight path; and A_j = effective plant area, including its shadow and skid characteristics for the jth aircraft.

The manner of interpreting the individual factors in the above equation may vary on a case-by-case basis because of the specific conditions of each case or because of changes in aircraft accident statistics.

Values for C_j currently being used are taken from the data summarized in Table 2.6.

The probability per year of a potentially damaging crash at the site due to operations at the facility under consideration (P_M) is then given for this case by the following expression:

$$\mathrm{PM} = \mathrm{C} \times \mathrm{A} \tag{2.27}$$

where C = total probability of an aircraft crash per unit area per year in the vicinity of the site due to all cases: and A = effective area of one unit of the plant.

In the NRC Standard Review Plan Section 3.5.1.6 "Aircraft Hazards," item 7 [16], *a,* effective plant area is defined as follows: "A shadow area of the plant elevation upon the horizontal plane based on the assumed crash angle…"; this definition leads to a contradiction that may be demonstrated as follows:

Assume that there is a horizontally disposed plane of area A (Figure 2.10). Probability of aircraft impact on it is P_{hor}(A). Consider now the same plane but vertically disposed, e.g., a building wall (Figure 2.11). Probability of impact on it is denoted P_{vert}(A). The question is whether the latter probability is more or less than P_{hor}(A)?

Note that the horizontal plane can be struck by an aircraft from any point of the upper half-sphere. But impact on vertical plane is possible only from its face side, as the opposite side is situated in the dead space. Consequently, the vertical plane can be struck only from a half of the upper half-sphere. Thus, the probability of impact on the vertical plane cannot be more than that on the horizontal plane.

But what follows from the SRP 3.5.1.6 rule? If an aircraft flies at an angle $\alpha < 45°$ to the horizontal than the shadow area $A_{shad} > A$ (Figure 2.11). In accordance with the SRP rule, it means that probability of impact on the vertical plane is more than on a horizontal one.

Moreover, if an aircraft flies horizontally, i.e., $\alpha = 0$, then the effective area becomes infinitely large regardless of building dimensions.

This fault arises because when calculating failure on the open horizontal plane the probability rate is not the same as on the shadow of the vertical wall. This is because failure on the first is possible from any side, but on the latter it is possible only through the "window" bounded by the vertical wall outline.

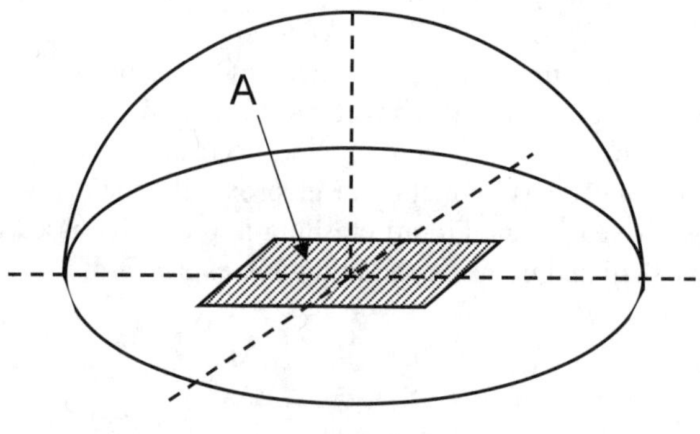

FIGURE 2.10. AIRCRAFT IMPACT AREA

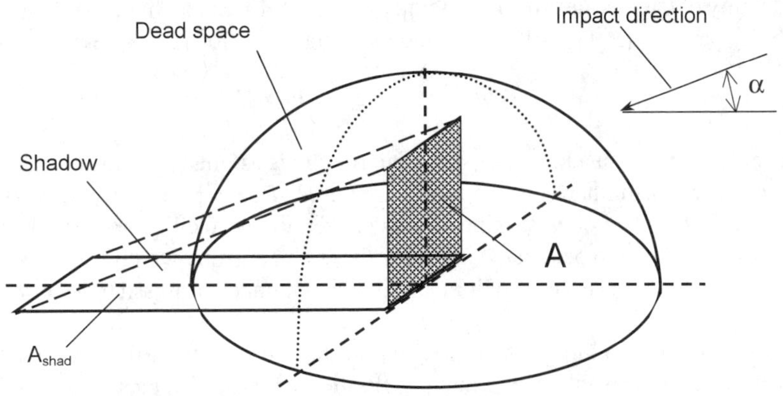

FIGURE 2.11. AIRCRAFT IMPACT SHADOW

In reality, probability of impact on NPP building depends on its dimensions as well as on its shape and probability distribution of approach angles [18, 19]

In [18], formula are given to calculate the probability of impact on surfaces of different shape (horizontal and inclined plane, cylinder with vertical axis, spherical segment). Additionally, the way randomness of other parameters (aircraft velocity, mass and class) can be accounted for and how they influence the loads to be specified for building structures and systems design is demonstrated there.

4. Effect of holding patterns: Holding patterns are race-track-shaped courses at specified altitudes, associated with one or more radio-navigational facilities, where aircraft "circle" while awaiting clearance to execute an approach to a landing at an airport or to continue along an airway. Holding patterns that are sufficiently distant from the plant need not be considered. Otherwise, traffic in the holding pattern should be converted into equivalent aircraft passages taking into account the characteristics, including orientation with respect to the plant, of the holding pattern.

2.5.3 Evaluation of Facility Upset and Accident Loads

Anticipated Transient Operation Upset Loads

These types of loads would be considered in the severe load category with a greater than 10^{-1}/yr probability of occurrence. System start-up and shutdown and safety relief discharge would typically be considered in this category. Specific loads are not defined probabilistically. While transient operation loads are not normally considered extreme loads, they may be caused by and act concurrently with extreme loads. For example, a large earthquake typically causes shutdown of rotating equipment. Such a shutdown may induce a safety relief valve (SRV) to discharge. The SRV loads on the system would then have to be considered concurrent with the earthquake loads. Such concurrent consideration is typically assumed when transient operation occurs within 30 seconds of the initiating event.

High Energy System Rupture

High energy system ruptures are typically of two types: (1) rupture due to overpressurization or cracking of a fluid system; and (2) rotating equipment failure. Rotating equipment failures are typically defined on a probabilistic basis, while fluid system failures are treated deterministically.

The potential consequences of a high energy system rupture are: (1) impulse load; (2) compartment pressurization (in a confined space); and (3) missiles. Of these three phenomena, only the missile phenomenon is generally treated probabilistically.

The total probability of the missiles striking a vulnerable critical area of a high hazard facility is estimated. The total probability per year (P_T) may be estimated by using the following expression [20]:

$$P_T = P_E \times P_{MR} \times P_{SC} \times P_P \times N \tag{2.28}$$

where P_E = probability per year of a high energy system rupture is estimated. If this estimate is below 10^{-7}/yr probability of occurrence no further consideration is required; P_{MR} = probability of missiles reaching the facility; P_{SC} = probability of missiles striking a critical area of the facility; P_P = probability of missiles exceeding the energies required to penetrate vital areas (based on existing wall thickness), or producing secondary missiles that could damage vital equipment; and N = number of missiles generated by the design-basis event.

P_P may be assumed to be equal to one as a first step in the analysis. If P_T thus calculated is greater than 10^{-7} per year, then missile effects that would include missile size, shape, impact area, weight, energy, material properties and trajectory should be evaluated.

Heavy Load Drop

Heavy load drops considered as a design-basis event are usually defined on a deterministic basis. However, probability is sometimes used to define other concurrent loads. (For example, if a crane or hoist system is in operation more than 2% of the time concurrent consideration of lift design loads and earthquake is required.) Postulated heavy load drops are usually associated with cranes or hoists, which have not been designed to be failure-proof as defined in [21].

2.6 USE OF CODE-DEFINED IMPORTANCE FACTORS TO DEVELOP PROBABILISTIC HAZARD CURVES

2.6.1 Basic Development

The ASCE 7-05 and IBC-2003 codes have generally used importance factors to modify the conventional structure design loads applicable to essential or hazardous facility safety-related SSC. These factors generally define a halving of the probability of load exceedence to be used in design.

2.6.2 Earthquake Hazard Curve

The basic importance factor used with the seismic design of structures is 1.25. This applies to all but hazardous structures and to safety related systems and components where an importance factor of 1.5 is specified in *ASCE Std. 7-05* [6]. If the 1.25 factor is selected as the basis for defining the slope of the hazard curve between a 2×10^{-3} and 10^{-3}/yr probability of exceedence, it is possible to define the effective seismic acceleration as a function of probability of exceedence as shown in Table 2.7, assuming the slope of the hazard curve remains the same in doubling of the return period.

2.6.3 Wind Hazard Curve

The relationship between importance factor for wind and probability of exceedence is more clearly shown in the commentary to the *ASCE-05* code. The evaluation of wind velocity data for a large number of recording points in the U.S. indicates a slope of the hazard curve between 2×10^{-2} and 2×10^{-3} as 1.07. In the 1995 edition of the code, this value was taken as 1.10 when hurricane wind data was included, but since has been reduced to 1.07 for all straight and hurricane wind data. The importance factor can be used to define the slope of the hazard curve of 1.1 in going from a 2×10^{-2} to 10^{-2}/yr probability of exceedence. The procedure to obtain wind load as a function of probability level is to use a multiplication factor of 1.1 applied to the 50-year return period map given in *ASCE 7-02 Standard* and multiplying by a 1.1 factor for each doubling of the return period thereafter.

TABLE 2.7. COMPUTING SEISMIC 5.0 HZ, 5% DAMPED RESPONSE SPECTRA ACCELERATIONS BASED ON A 1.25 IMPORTANCE FACTOR AS DEFINED BY THE IBC-2003 STANDARD

Probability of Exceedence	Spectral Acceleration at 5% Damping
2×10^{-3}/yr	0.2 g
10^{-3}/yr	0.25 g
5×10^{-4}/yr	0.312 g
2.5×10^{-4}/yr	0.39 g
1.25×10^{-4}/yr	0.49 g

Note: To obtain mean peak ground accelerations associated with the spectral acceleration, divide by 2.5.

2.6.4 Snow (Precipitation) Hazard Curve

Similar to wind, if the slope of the hazard curve (importance factor) in going from a 2×10^{-2} to 10^{-1} probability of exceedence is also 1.1, then the resultant slope factors for various probabilities of exceedence or return periods is the same for wind using the snow load map in *ASCE Standard 7-02* for a 50-year return period as the base.

2.7 Probabilistic Behavior and Design

In the preceding sections of this chapter the use of probability was limited to definition of loads being defined as a design basis. However, in current accepted design practice, such loads thus defined are considered deterministically as prescribed in existing mechanical-structural design codes and standards. Capacity to resist these loads could also be defined probabilistically and would affect the failure probability of the SSC.

It is anticipated that within the next 100 years or so, as more data and practical probabilistic evaluation procedures become available structural designers and design standards will consider the probabilistic nature of both the load and the capacity to resist such loads explicitly in design. Instead of prescribing an allowable stress as a design limitation, the numerical probability of failure permitted might be specified instead as a code stress limit.

While the theoretical bases for such technology applicable to mechanical component design has been defined for at least the past 25 years, and are currently used to some degree in aerospace design, they have yet to be applied to heavy industry mechanical system and component design.

2.8 REFERENCES

1. ASME, 2003, *Risk Based Methods for Equipment Life Management: An Application Handbook,* CRDT Vol. 47, American Society of Mechanical Engineers, New York, NY.
2. Center for Chemical Process Safety, 2000, *Guidelines for Chemical Process Quantitative Risk Analysis,* 2nd *Ed.,* American Institute of Chemical Engineers,
3. Reactor Safety Study, 1975, "An Assessment of Accidental Risk in U.S. Commercial Nuclear power Plants, Washington 1400," U.S. Nuclear Regulatory Commission,
4. *DOE Std. 1021,* 1996, "Natural Phenomena Hazard Performance Categorization Guidelines for Structures, Systems and Components," U.S. Department of Energy, Washington, D.C.
5. Frunkel, A., et. al., 1997, "1997 Seismic Hazard Maps for the Continental United States," *U.S. Geological Survey Open File Report 97-131-F,*

6. *ASCE Std. 7-02,* 2002, "Minimum Design Loads for Buildings and Other Structures," American Society of Civil Engineers, Reston, VA.
7. Green, R. A. and Hall, W. J., 1994, "An Overview of Selected Seismic Hazard Analysis Methodologies," *Civil Engineering Studies, Structural Research Series No. 592,* University of Illinois,
8. Abdalla, J. A., Mohamedzein, Ye-A. and Wahab, A. B., 2001, "Probabilistic Hazard Assessment of Sudan and Its Vicinity," *Earthquake Spectra,* EERI Vol. 17, No. 3. Englewood, NJ.
9. Liu, H., 1991, *Wind Engineering – A Handbook for Structural Engineers,* Prentice Hall,
10. WASH 1400, 1974, "Technical Basis for Interim Regional Tornado Criteria," *Rep.,* U.S. Atomic Energy Commission, May,
11. Fujita, T. and Pearson, 1973, "Results of FPP Classification of 1971 and 1972 Tornadoes," *Preprints, Proc., Eight Conference on Severe Local Storms,* Denver, CO, 142–145.
12. Boissonnade, A. et. al., 2000, "Development of Probabilistic Tornado Wind Hazard Model for the Continental United States, Vol. 1, Main Report," *UCRL-ID 140922 Vol. 1,* Lawrence Livermore National Laboratory, Livermore, CA.
13. Coats and Murry, R. C., "Natural Phenomena Hazard Modeling Project: Extreme Wind/Tornado Models for Department of Energy Sites," *UCRL 53526 Rev. 1,* Lawrence Livermore National Laboratory, 1995.
14. *ANS 2.3,* 2000, "Draft Standard for Estimating Tornado, Hurricane and Extreme Straight Wind at Nuclear Facility Sites," American Nuclear Society, Palo Alto, CA.
15. "Tornado Missile Simulation and Design Methodology," 1981 *NP-2005, Vols. 1 and 2,* Electric Power Research Institute, La Grange Park, IL.
16. Standard Review Plan 3.5.1.6, 1981, "Aircraft Hazards," *Rev. 2, NUREG-0800,* Nuclear Regulatory Commission, Washington D.C.
17. *DOE Std. 3014-96,* 1996, "Accident Analysis for Aircraft Crash into Hazardous Facilities," U.S. Department of Energy, Washington, D.C.
18. *Paper # 1644,* 2001, "Probabilistic Assessment of NPP Building Structures Strength under Aircraft Impact," *Proc. Transact. 16th Structural Mechanic in Reactor Technology Conf.,* Washington, D.C.
19. *Paper # 1645,* 2001, "Probabilistic Assessment of NPP Building Vibrations Caused by Aircraft Impact," *Proc. Transact. 16th Structural Mechanic in Reactor Technology Conf.,* Washington, D.C.
20. Stevenson, J. D. (ed), 1980, *Structural Analysis and Design of Nuclear Plant Facilities, Manual of Engineering Practice No. 58,* American Society of Civil Engineers, Reston, VA.
21. *ASME NOG-1-1989,* 1989, "Rules for Construction of Overhead and Gantry Cranes (Top. Running Bridges, Multiple Girder, NOG-6100)," American Society of Mechanical Engineers, New York, NY.

CHAPTER

3

LOAD PHENOMENA

3.1 INTRODUCTION

Extreme loads applied to mechanical SSC often involve time-dependent structure fluid interaction. Over the past 5 to 10 years enormous strides have been made in the ability to more accurately define these loads in the design of mechanical components. Primarily due to the strides made in the development of finite-element and difference algorithms and resultant computer programs, these loads particularly associated with system transient time-dependent loads at charges in directions and areas can be determined with an accuracy not previously achieved. To date these analyses have generally been restricted to elastic or linear response systems. However, it is expected that they will rapidly become available for inelastic nonliner system analysis more characteristic of mechanical systems at or near failure.

Unfortunately, these new analytical techniques tend to be relatively expensive both with respect to the cost of the computer software and the cost of inputting and evaluating the output results. In addition, it tends to be difficult to evaluate the validity of the results. This chapter provides approximate first-order load definitions for various types of extreme loads typically applicable to mechanical SSC. These design inputs may be used where more exact solutions are not required, or can be used to independently verify more exact solutions.

3.2 STATIC

Extreme static loads are occasionally included as extreme loads in design. They are typically associated with in-service live loads significantly larger than those defined in applicable standards for normal industrial systems or components or considered as the design basis. Typically included in this category for mechanical equipment are:

- overpressure due to failure of relief valve or barriers to connected high-pressure service
- movable equipment with weight loads much in excess of those assumed in design
- maintenance loads
- wind loads
- snow loads
- rain loads

The loads just identified are also considered primary loads, which must satisfy the laws of equilibrium. Another set of static loads that may have to be considered in the design of mechanical components are loads induced by restraint of free end displacement and thermal gradients. These are usually referred to as secondary loads. Both types of loads can give rise to localized or peak loads or effects, which result in localized high stresses or strains above yield that can be the source of crack propagation and fatigue or ratchet-induced failure. Such peak load effects tend to be repeated rather than dynamic in nature.

It should also be noted that design-basis loads are usually made up of a number of loads in a particular load combination. Design to resist extreme loads will also include normal load contributions from dead, live, pressure, etc., as discussed in Section 3.4.

3.3 DYNAMIC

Dynamic loads applied to the design of mechanical systems and components are of three basic types: cyclic, impulse and impact. In addition, a fourth potential cyclic-type load exists: the vibratory motion category. Although vibratory motion is not usually considered in the original design basis, it may be observed during steady-state or transient operation of sufficient magnitude such that remediation may be required. This vibratory effect may be of such magnitude or frequency to cause premature fatigue or ratchet failure of metal components. In such instances, reconfiguration or modified support of the mechanical system or component may be required to reduce the vibratory effects to within tolerable limits during the operating conditions of the particular plant system or component. As originally defined in ASME B&PVC Code Section III, the operating life of the plant includes any design-basis normal, abnormal, emergency or faulted conditions defined in the design specifications. These operating conditions should not be confused with service levels A, B, C and D that are currently defined in the code for design purposes. It is possible to have different service level design conditions for the same facility operating condition, depending on the required response of a component.

There are many standards or guidelines available to provide criteria for the design of mechanical (metal) systems and components to resist dynamic loads. The primary guidance is found in Appendix N Dynamic Analysis Methods of the *ASME Boiler and Pressure Vessel Code,* Section III [1]. This guideline is defined as a Non-Mandatory Appendix to ASME B&PVC Section III and at the present time, it is primarily concerned with cyclic dynamic loads associated with earthquakes and fluid flow-induced vibration of structures contained within the cross-flow of fluids. For completeness, this chapter has been expanded to consider other types of dynamic loads—not currently addressed explicitly by Appendix N—that the design of a pressure-retaining nuclear component may need to consider.

Another standard that complements ASME Section III, Appendix N guideline is *ASCE Standard 4-98* [2], which is used to define earthquake motions in the building foundation and its structure that house or support the mechanical system or component. From this standard, the specific earthquake motion transmitted through a building foundation and its structure as an earthquake load-design input to a mechanical system or component can be developed. Both the ASME Section III and the ASCE Standard deal primarily with earthquake cyclic-type dynamic loads.

Other guidelines for dynamic analysis of mechanical systems and components are those associated with dynamic-impulse and impact-type loadings. This guidance is developed as *ANS Standard 58.2* [3] for (impulse) and *ANS Standard 58.3,* Appendix B [4] (for impact) in addition to Appendix II of the *ASME B31.1* code [5]. Yet another guideline was developed to address impulse and impact loads:—it is the report of the ASCE Committee on Impact and Impulse Loads [6]. *ASME OM-S/G,* Part 3, contains a standard for determining the effect of vibratory motion from operation that is independent of the cause of vibration [7]. Finally, there is a new *ASME B&PVC Section VIII* Div. 3 code [8] text, which addresses construction of pressure vessels under impulsive and impact loads.

By using the standards and guidelines of references cited above, it is possible to develop a fairly clear understanding of current design or analysis procedures and methods for first-order (approximate methods) mechanical systems and components designed to resist dynamic loads. Such procedures and methods are for engineers who possess a background in structural dynamics without recourse to complex, specialized computer codes. For impulse-type loads, any analytical rigor beyond a single simplified rectangular- or triangular-forcing function generally will require the use of hydrodynamic computer codes that use control volumes to more accurately define the time dependence and magnitude location and direction of multiple forcing functions [9].

3.3.1 Types of Dynamic Loads

Cyclic Loads

Earthquake or Other Building Filter-Cyclic Loads

Cyclic loads are characterized as alternating plus or minus loads or displacements that are applied to mechanical systems or components. The earthquake-associated loads are caused by the inertial response of an SSC to the earthquake-induced motion of its supports or the foundations of the supports. The response of a system or component when using modal analysis is highly dependent on the percent critical damping of the system. In addition, the system or component elastic response depends on the closeness of the time-dependent forcing-function frequency (which defines the input) to the dominant response frequency of the seismically loaded system or component. If the system or component has a natural frequency close to that of the applied forcing function, resonance can occur, resulting in large amplification of an equivalent static load applicable to the mechanical system or component. Such amplification effectively may be as high as 100 times the peak cyclic-force level at very low-damping and near-resonance (system or component natural mode response frequency is very near the cyclic-load frequency).

Fluter or Vortex-Shedding Loads

When a smooth-surface tubular structure is in a fluid-flow stream, vortices in the flow can become established as a result of the tube interference with the fluid flow. When the velocity and density of the fluid are such that vortices become established at or near the same dominant frequency of the tube, a resonant condition develops that can produce cyclic loads perpendicular to the direction of flow. At resonance, the amplitude and number of cycles can cause failure of the tube. This particular loading phenomenon for vessels and steel cylinders is covered in references [1] and [10].

Impact Loads

Dynamic-impact loads are characterized as an external force or moment applied suddenly to a target SSC relative to the response period of the target. These loads typically have a finite amount of kinetic energy associated with a missile or fluid slug of known mass, velocity, area and duration striking a target structure, mechanical system or component because of a sudden change in fluid-flow conditions. For relatively slow-moving missiles, the kinetic energy or momentum of the missile must be absorbed by the target by potential strain energy, which is defined by strain limits of the material and structural configuration of the target (including its supports). In addition to evaluating the impact energy or momentum transfer, it may be necessary to evaluate the potential for local-penetration or perforation effects on the target. Relatively high-velocity, small-contact-area missile impacts typically absorb the missile-impact energy by local crushing of the missile or target as well as strain energy in the target. In such cases, the force and duration of the missile impact are defined by the crushing strength of the missile and/or target.

In the penetration-type analysis, it is common to use local-penetration formulas developed from experiments discussed and referenced in the references [4, 5 and 6]. However, with the relatively recent advent of the nonlinear, dynamic, finite-element computer codes, it has become possible to compare computed strains in the target directly with the inelastic plastic strains necessary to cause rupture to determine design adequacy and the potential for target penetration.

Impulse Loads

Similar to impact loads, impulse loads are external loads suddenly applied to the target; however, typically they are relatively long-lasting compared to the response period of the target. In this category of loads are fluid jets from postulated rupture of pressurized systems and exterior or interior blast waves from an explosion on a target or shield structure. Also included are the effects of the injection of high-energy gases or vapors into a liquid body (i.e., pressure-suppression) containment type of loading is associated with relief-valve discharge and fluid-hammer (i.e., steam and water hammer) phenomena. The fluid-hammer phenomena are generally associated with the sudden termination of fluid flow by a stop valve, but they may also be associated with a slug of liquid being accelerated by the differential fluid pressure introduced into a piping system as a transient-load effect. These differential pressures and associated shock waves generally result in pressure and velocity changes varying with time at each change in area or direction in the system or component, thereby generating time-dependent forcing functions at multiple points in the system or component, as discussed in Chapter 8. In general, the rigorous solution of this problem cannot be made without recourse to control-volume fluid-dynamics computer codes [9]. For first-order design from postulated fluid-impulse load, a single forcing function is typically applied at the discharge nozzle, a postulated break location and orientation, or a postulated liquid-slug location. General guidance from the simplified fluid-impulse design can be found in [3].

Simple use of acceleration criteria alone—without any reference to the duration of acceleration as acceptance criteria—should be avoided; acceleration alone without duration or shape of the impulse does not define the dynamic-input load.

3.3.2 Basic Equations of Motion and Time-Dependent Loading Functions

Cyclic Loads

The basic equations of motion can be found in [11] or other similar introductory texts on structural dynamics. Those equations involve the following parameters:

$$m\ddot{x} + c\dot{x} + kx = f(t) \tag{3.1}$$

where m = mass of the system or component; k = stiffness of the system or component; c = damping coefficient, due to energy absorption in the material and system; $f(t)$ = dynamic forcing function; x = displacement of the system or component for a particular degree of freedom; $\dot{x}$ = velocity of the system or component for a particular degree of freedom, dx/dt; and $\ddot{x}$ = acceleration of the system or component for a particular degree of freedom, d^2x/dt^2

The actual formulation and method of solution of the general differential equation of motion depends considerably on the nature and location of the forcing function $f(t)$. For cyclic loads, the system or component is usually represented as a multi-degree-of-freedom system with multidirectional input in which the force applied to a given mass point is a function of that mass point's inertial properties in response to a base-input motion rather than an external force applied to the system or component. This type of loading lends itself to solution by the modal response spectrum method of analysis.

Associated with the solution of the basic equation of motion, ignoring damping and the forcing function, are the following relationships:

$$\omega = \sqrt{\frac{k}{m}} \tag{3.2}$$

where ω = natural frequency of the system or component (in radians/sec)

and

$$f = \frac{1}{2\pi}\sqrt{\frac{k}{m}} \tag{3.3}$$

where f = natural frequency of the system or component (in cycles/sec or Hz); and π = 3.1416

It has been found for cyclic (earthquake and vibratory) design loads in which the dynamic force is the result of inertial response to a base motion loading, the solution is expedited by variable substitution, such as a modal equation of motion, that becomes:

$$\ddot{A}_i + 2w_i\beta_i A_i + \omega_i^2 A_i = P_i\ddot{x}_s \tag{3.4}$$

where A_i = modal displacement (rotation) of mode i relative to the base; A_i = modal velocity (rotational) of mode i relative to the base; $\ddot{A}_i$ = modal acceleration (rotational) of mode i relative to the base; ω_i = natural frequency in mode i (in radians/sec); β_i = percent critical damping in the ith mode; $\ddot{x}_s$ = absolute translational (angular) acceleration of the base; and P_i = modal participation factor in the ith mode.

$$P_i = \frac{\sum_{r=1}^{n} m_r \Phi'_{ri}}{\sum_{r=1}^{n} m_r \Phi^2_{ri}} \tag{3.5}$$

Finally, the inertial force applied to the rth mass in the ith mode becomes:

$$f_{ri} = m_r a_{ri} = m_r (P_i S_{ai} \Phi_{ri}) \tag{3.6}$$

where a_{ri} = inertia acceleration applied to the r mass in the ith mode; m_r = the rth mass; P_i = modal participation factor in the ith mode; f_{ri} = inertia force applied to the rth mass in the ith mode; S_{ai} = response spectra acceleration in the ith mode (damping is built into this parameter; response spectra are a plot of the maximum response of a series of Single Degree of Freedom to the earthquake time-history input; Φ'_{ri} = modal displacement of the rth mass in the ith mode in the direction of the applied earthquake motion; and Φ_{ri} = all modal displacements of the rth masses in the ith mode.

The Φ_{ri} are Eigen vectors; the $w^2{}_i$ are the Eigen values solution of the modal equation of motion.

Chapter 5, relative to seismic loads, provides a more detailed discussion of the use of modal analysis applied to cyclic-type support motions.

Impact Loads

For relatively hard or nondeformable missile impact, the overall deflection of a target structure or barrier during impact is considered negligible when compared to penetration in which the work done by the missile as it penetrates the structure is equal to the initial kinetic energy of impact. Thus

$$A_t = 0.5\frac{W}{g}v_0^2 \tag{3.7}$$

where A_t = total work done by the missile as it penetrates the structure; g = acceleration of gravity; W = weight of missile; v_o = initial velocity of missile at impact; and x = penetration depth (typically determined by empirical penetration formula as discussed in Chapter 7).

The assumption that the velocity reduces linearly to 0 is used to determine the time of impulse or duration of the impact force.

$$t = 2x/v_o \tag{3.8}$$

and acceleration

$$a = v_o/g \tag{3.9}$$

and impact force

$$F = \frac{W}{g}a = \frac{Wv_o}{gt} \tag{3.10}$$

The value of F and t define a rectangular-impulse loading applied to a deformable target structure impacted by a rigid, nondeformable missile.

For relatively soft or deformable missile impacts, the forcing function can be determined as follows:

$$F = P + uv^2 \tag{3.11}$$

where P = missile crushing-strength at impact interface; u = mass/unit length of the uncrushed missile at impact interface; v = velocity of the uncrushed portion of missile relative to the barrier; and F_t = force applied to the rigid structure.

As it acts on the target structure barrier, the force F_t can be determined by integrating the motion of the missile to determine the point of missile in contact with the barrier and its instantaneous velocity.

The first-order evaluation of impact or impulse dynamic loads acting on a pressure-relieving component is usually taken as a rectangular or triangular forcing function applied perpendicular to the plane of a relief-valve discharge or postulated break. The system or component frequency excited by the forcing function represented as an SDOF system is used to determine an equivalent static load. Both elastic and inelastic response of the SDOF system is available [11] Responses to blast-type loads are discussed in Chapter 7.

Impulse Loads

In first-order analysis, the impulse load is often considered applicable to a single mass point at the discharge point or postulated break location of a pressurized component ejecting fluid; an equivalent static load is determined by the response of the mass point, including its stiffness restraint to the time-dependent loading [5, 8]. This equivalent static external load is then applied to the node of a static model of the pressurized component to obtain resultant internal forces and moments acting on the component.

$$F_e = D_f C_T K \left[\frac{W}{g} V_e + (P - P_0) A \right] \tag{3.12}$$

where C_r = fluid thrust coefficient: 1.26 for saturated steam, and 2.0 for sub-cooled liquid; F_e = equivalent static reaction force at exit or break plane; W = mass flow rate; V_e = exit velocity of fluid; g = gravitational acceleration; D_f = dynamic load factor for suddenly applied loads; P = static pressure at exit; P_0 = ambient pressure; A = exit area; and K = exit orifice coefficient depending on the shape of the nozzle or opening (it can vary between 2.0 and 1.0, and it is usually taken equal to 1.0 for pipe-break impulsive loads).

The term P_0 is often set to zero because it is taken as atmospheric pressure, 14.7 psai (100 kPa) or somewhat higher if it is discharging into a partially enclosed space.

The resultant impulsive fluid jet may also load any structure in its path. The total force of the jet is assumed equal to F_e with a dispersion angle of 10°.

A more detailed discussion of the pressure relief valve-discharge loading can be found in [5] and Chapter 7.0. Explosive blast also generates impulse-type loads. They are discussed in more detail in [12] and Chapter 7.

3.4 CLASSIFICATION OF LOADS

Loads on mechanical SSC are typically classified as normal, severe or extreme.

Normal loads include the following:

- Dead weight, D
- Live load, L
- Design pressure, P_D
- Design temperature, T_D

which occur during normal or steady-state operation.

Severe loads typically include the following:

- ASCE-7 Wind, W (50- to 100-year return load)
- ASCE-7 Earthquake, E_o (500- to 1,000-year return load)
- ASCE-7 Snow (50- to 100-year return period)
- ASCE-7 Rain (50- to 100-year return period)
- Relief valve discharge, R_D (deterministic)

Extreme loads typically include the following:

- Wind, hurricane and tornado load, *W'* (1,000- to 1,000,000-year return period)
- Snow, *S'* (10,000- to 100,000-year return period)
- Rain, P_R (10,000- to 100,000-year return period)
- Flood, F_L (10,000- to 100,000-year return period)
- Earthquake, E_S (2,000- to 100,000-year return period)
- Accidental man-induced evens, M_A (10,000- to 1,000,000-year return period)
- Man-induced malevolent events, M_M (deterministic)
- High-energy system rupture, R_A (deterministic)
- Heavy load drop, C_D (deterministic)

Associated with these individual loads are load combinations of simultaneously occurring loads. For mechanical SSC, these load combinations are typically defined in the SSC design specification.

3.5 LOADS CONSIDERED IN DESIGN OF HAZARDOUS FACILITIES

Safety-related SSC in hazardous nuclear facilities generally consider three categories: normal, severe and extreme, identified here for design purposes.

Petrochemical safety-related SSC designs are typically limited to normal and severe categories of loads, except that on a case-by-case basis, external explosion and hurricane or tornado may be considered a design basis. Hazardous facility safety-related SSC in particular, such as those in a large LNG storage facility, might also consider the extreme load category.

Biomedical facility safety-related SSC are generally limited to the normal and severe load categories.

REFERENCES

1. *ASME Boiler and Pressure Vessel Code,* 2001, "Rules for Construction of Nuclear Facility Components-Dynamic Analysis Methods," Appendix N Div. 1, Section III, American Society of Mechanical Engineers, New York, NY.
2. *ASCE Standard SEI/ASCE 7-02,* 2003, "Minimum Design Loads for Buildings and Other Structures," American Society of Civil Engineers, Reston, VA.
3. *ANS 58.2,* 1988, "Design Basis for Protection of Light Water Nuclear Power Plants Against the Effects of Postulated Pipe Rupture," American Nuclear Society,
4. *ANS 58.3 Appendix B,* 1992, "Physical Protection for Nuclear Safety-Related Systems and Components – Appendix B Plant Design Against Missiles," American Nuclear Society,
5. *ASME B31.1 2001-Appendix II,* 2001, "Power Piping – Non Mandatory Approach Rules for the Design of Safety Valve Installation," American Society of Mechanical Engineers, New York, NY.
6. *Civil Engineering and Nuclear Power,* 1980, "Report of the ASCE Committee on Impactive and Impulsive Loads," American Society of Civil Engineers, Reston, VA.

7. *ASME OM/SG,* 2001, "Code for Operation and Maintenance of Nuclear Power Plants," American Society of Mechanical Engineers, New York, NY.
8. "Rules for Construction of Vessels to Resist Impact and Impulse Loads," 2004, *Draft B&PVC,* American Society of Mechanical Engineers, New York, NY.
9. *ANSYS Computer Program Release 5.3,* 1996, ANSYS, Inc.,
10. *ASME STS-1-2000,* 2001, "Steel Stacks – An American National Standard," American Society of Mechanical Engineers, New York, NY.
11. Biggs, S.M., *Introduction to Structural Dynamics,* 1964, McGraw Hill Book Co., New York, NY.
12. U.S. Army Corps. of Engineers, 1957, "Design of Structures to Resist the Effects of Atomic Weapons," *Manual EM1110-345-415.*

CHAPTER

4

ACCEPTANCE CRITERIA

4.1 INTRODUCTION

Acceptance or behavior criteria for mechanical components are typically given in terms of allowable stresses as a function of both specified minimum yield and ultimate strength (stress). These allowable stresses have been developed as a function of the experience with pressure vessel and piping failures over the past 100 years of service.

It is expected that over the next 100 years, allowable stress criteria will gradually give way to more direct measures of failure probability, as discussed in Chapter 2.

Currently there are four operating service levels and their associated behavior criteria applicable to nuclear safety-related mechanical structures, systems and components (SSC) [1]. For the petrochemical and biomedical industries there are only two such states defined. Recently *ANS Standard 2.26* defined thc characteristics of limit or damage states for mechanical SSC *Structures, Systems and Components,* [2] on which Table 4.1 is based. In Table 4.2 is an approximate relationship between ASME B&PVC Section III service limits and the ANS defined limit states. In Table 4.3 are listed the recommended inelastic energy absorption factors to be used for limit state A in Table 4.2.

TABLE 4.1. DAMAGE OR LIMIT STATES FOR MECHANICAL DISTRIBUTION SYSTEMS AND COMPONENTS

SSC Type	Limit State A	Limit State B	Limit State C	Limit State D
Pressure vessels and piping designed to ASME code	Tanks, pressure vessels and piping systems that are at the threshold of loss of leak-tight and structural integrity. Approximately 10–50% probability of rupture and collapse. Repair may require replacement of vessel or piping.	Tanks, pressure vessels and piping systems that sustain deformation and damage. Respond well in the inelastic region. Approximately a 1% probability of containing fluid release or collapse. In situ repair of vessel may be possible.	Tanks, pressure vessels and piping systems that can sustain damage and permanent deformation. Approximately a 10^{-3} probability of rupture or collapse. Generally can remain in service until minor damage is repaired.	Tanks, pressure vessels and piping systems that remain essentially elastic except for very limited inelastic response due to thermal loads. Limits typically used to ensure operability or functionality of active components. Less than 10^{-4} probability of rupture or collapse.

Note: Anchor bolts designed to code allowables generally will be less damaged than indicated by the Limit State because of the greater conservatism inherent in the standard design procedures (e.g., factor of safety of 3 or 4 against ultimate capacity for expansion anchors under limit state B loads, where ultimate capacity ratio to limit State B for the component itself is typically a factor of 1.5 to 2.0).

TABLE 4.2. RELATIONSHIP BETWEEN THE LIMIT OR DAMAGE STATS AND ASME B&PVC SECTION III

Limit States	ASME B&PVC Section III Service Levels
D	A & B
C	C
B	D
A	D[1]

1. Inelastic energy absorption factor (F_μ) of Table 4.3 is applicable.

The primary codes and standards that define acceptance criteria for mechanical systems and components are Sections I, III, IV and VIII of the *ASME Boiler and Pressure Vessel Codes* and *ASME B31.1* and *B31.3* for piping as well as *STS-1* codes for steel stacks. The *B31.3* and *B&PVC Section VIII* codes are used primarily for the petrochemical and biomedical industries as well as for nonreactor nuclear systems and components, which are regulated by the U.S. Department of Energy [3]. ASME B&PVC Section III is used for

TABLE 4.3. EQUIPMENT AND DISTRIBUTION SYSTEMS INELASTIC ENERGY ABSORPTION FACTOR, F_μ[1]

	Factor, F_μ
	Limit State A[2]
EQUIPMENT:	
Vessel	1.50
Heat exchanger	1.50
Coolers	1.50
Chillers	1.50
Tanks – vertical	1.25
Tanks – horizontal	1.50
Pumps	1.50
Fans	1.50
Valves	1.50
Dampers	1.50
DISTRIBUTION SYSTEMS	
Butt-jointed groove-welded pipe	1.75
Socket-welded pipe	1.50
Threaded pipe	1.25
Instrument tubing	1.50
HVAC duct	1.50
EQUIPMENT SUPPORTS[3]	2.00

1. These inelastic energy absorption factors (F_μ) are applicable to equipment functioning in a passive mode. For active components, the limit state is restricted to limit state *D*, where $F_\mu = 1.0$
2. Except as discussed in 3 (below), the allowable behavior limits for passive components are based on the ASME B&PVC Section III allowables for service level *D*. It should be noted that ASME B&PVC Section III service level *D* stresses range from 1.0 to 2.0 σ_y. In cases where allowable stresses are limited to σ_y the equipment F_μ can be increased by a 1.6 multiplier.
3. These components are normally designed to AISC allowables, which are typically limited to 0.8 to 1.0 σ_y for limit state B; hence, they are allowed higher inelastic energy absorption factors as described in 2 (above), as compared to ASME code allowables, where allowable stresses can be as high as 2.0 σ_y.

the nuclear power industry reactor safety systems and components regulated by the U.S. Nuclear Regulatory Commission.

Other standards are applicable to tanks, air handling components, cranes, etc., but they generally follow similar limits as presented in *ASME B&PVC* and *B31* codes.

It should be understood that the *ASME B&PVC Section III* and *VIII* and *B31* codes only address structural and leak-tight integrity. Operability or functionality of those systems and components, which are required to move or change state (active) when subjected to extreme loads, are not addressed by *ASME B&PVC* and *B31* codes. To a limited degree operability and functionability are addressed by *ASME QME-1* code [4].

This chapter is meant to summarize the contents of sections III and VIII of *ASME B&PVC* and *B31.1* and *B31.3* codes relative to design acceptance criteria applicable to normal, severe and extreme type loads as described in Chapter 3. No attempt should be made to design these codes without a thorough understanding of the code being used as well as its many detailed caveats and limitations.

4.2 ASME B&PVC SECTION III

This code section is written primarily for nuclear power plant SSC associated with nuclear reactor safety and other high-risk nuclear SSC that involve public health and safety. Other high-risk nuclear activities are associated with transport of highly radioactive material or processing and storage of spent nuclear fuel required to meet regulatory requirements of the Nuclear Regulatory Commission [5].

There are several classes of components defined in the ASME B&PVC Section III Standard, including the following:

- Division 1
  - Class 1 – vessels, pumps, valves and piping – subsection NB
  - Class 2 – vessels, pumps, valves, piping and tanks – subsection NC
  - Class 3 – vessels, pumps, valves, piping and tanks – subsection ND
  - Class MC, metal nuclear reactor containment – subsection NE
  - Class 1, 2 and 3 – metal supports – subsection NF
  - Class CS – reactor core supports – subsection NG
  - Class NH – elevated temperature – subsection NH
- Division 2
  - Class CC – nuclear reactor concrete containment – CC[1]
- Division 3
  - Class WB – transportation cask containment – WB
  - Class WC – storage cask containment – WC

It should be understood that the owner or certificate holder design specification assigns the classification and associated acceptance criteria to components in their facilities as a function of their importance to safety.

For classes 1, 2, MC, WB and WC class 1, 2 and 3 component supports, there are two types of analytical procedures permitted: design by analysis and design by rule.

4.2.1 Design by Analysis – Section NB-3200, NC-3200, WB-3200 and WC-3200

Design by analysis allowable behavior criteria are defined by a Hopper diagram as shown in Figures 4.1, 4.2 and 4.3 and Table 4.3 for design and service levels A, B and C for class 1 components. Service levels A correspond to normal design loads associated with normal operating pressure, dead weight and normal operating temperature. The design condition is limited to primary stresses and design pressures, which are somewhat higher than the maximum operating pressure.

[1] There also exists a suspended subsection CB, which is applicable to concrete reactor vessels.

Stress Category	Primary		
	General membrane	Local membrane	Bending
Description (for example, see Table 4.3)	Average primary stress across solid section. Excludes discontinuities and concentrations. Produced only by mechanical loads	Average stress across any solid section. Considers discontinuities but not concentrations. Produced only by mechanical loads	Component of primary stress proportional to distance from centroid of solid section. Excludes discontinuities and concentrations. Produced only mechanical loads.[1]
Symbol[2]	P_m	P_L	P_b

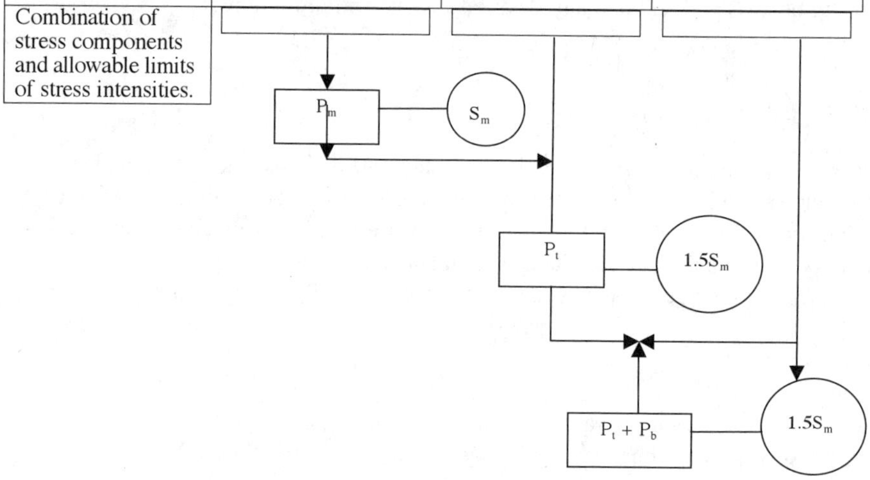

Legend
Use Design Loads

1. Bending component of primary stress for piping shall be stress proportional to the distance from centroid of pipe crossection.
2. The symbols, P_m, P_L, P_b do not represent single quantities, but rather sets of six σ_h, σ_ℓ, σ_t, $\sigma_{h\ell}$, σ_{ht}, $\sigma_{\ell t}$ where l represents stress in the longitudinal, h is the loop and t through the plate or wall thickness directions.

FIGURE 4.1. STRESS CATEGORIES AND LIMITS OF STRESS INTENSITY FOR DESIGN CONDITIONS – HOPPER DIAGRAM

Service level B permits a small increase in allowable stress, but still maintains the component response in the elastic range for primary stresses and ensures elastic shakedown for primary plus secondary stresses with no resulting damage to the component. Service levels C and D provide higher allowables, and behavior of the component is permitted into the inelastic range. Allowables for service level D can be found in Appendix F to *ASME B&PVC Section III.*

It should be noted that in design by analysis the stresses considered are stress intensities (S_m) rather than directional (longitudinal or hoop) or principal stresses. Stress intensities are determined as the largest of the following:

$$S_m = \begin{matrix} |\sigma_1 - \sigma_2| \\ |\sigma_2 - \sigma_3| \\ |\sigma_3 - \sigma_1| \end{matrix} \tag{4.1}$$

Stress Category	Primary: General Membrane	Primary: Local Membrane	Primary: Bending	Secondary: Expansion	Secondary: Membrane plus Bending	Peak
Description (for example see Table 4-1)	Average primary stress across solid section. Excludes effects of discontinuities and concentrations. Produced by pressure and mechanical loads.	Average stress across any solid section. Considers effects of discontinuities but not concentrations. Produced by pressure and mechanical loads, including inertia earthquake effects.	Component of primary stress proportional to distance from centroid of solid section. Excludes effects of discontinuities and concentrations. Produced by pressure and mechanical loads, including inertia earthquake effects. [Note (1)]	Stresses which result from the constraint of free end displacement. Considers effects of discontinuities but not local stress concentration (not applicable to vessels).	Self-equilibrating stress necessary to satisfy continuity of structure. Occurs at structural discontinuities. Can be caused by pressure, mechanical loads or differential thermal expansion. Excludes local stress concentrations.	(1) Increment added to primary or secondary stress by a concentration (notch). (2) Certain thermal stresses which may cause fatigue but not distortion.
Symbol [Note 2]	P_m [Note (3)(7)]	P_L [Note (3)(7)]	P_b [Note (3)]	P_e	Q	F
Combination of stress components and allowable limits of stress intensities.						

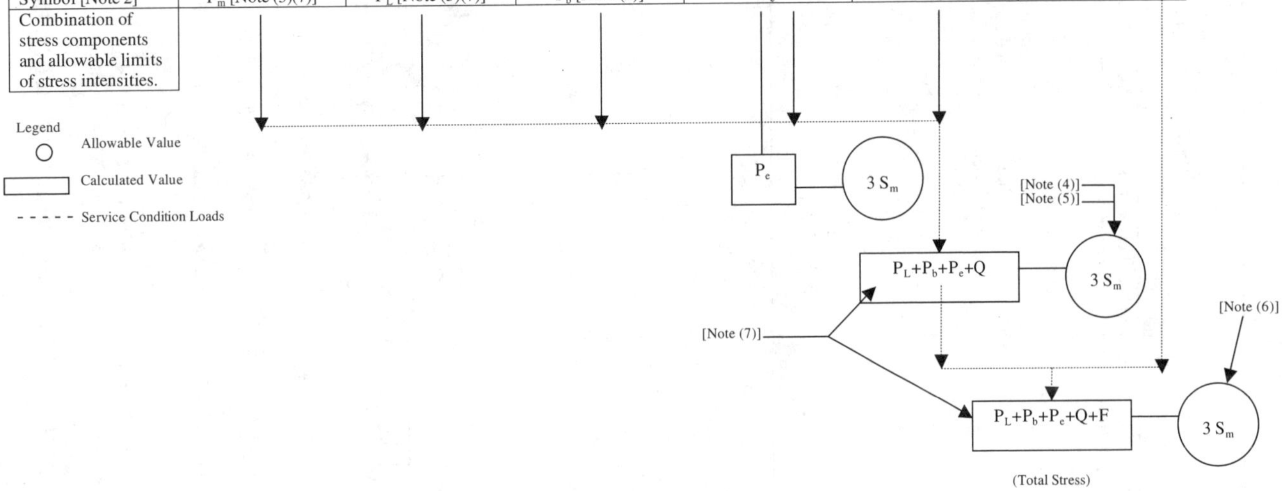

NOTES:

(1) Bending component of primary stress due to mechanical loads for piping shall be the stress proportional to the distance from the centroid of pipe cross section. For piping the calculation of P_b stresses are not required for reversing dynamic loads (including inertia earthquake effects). See NB- 3223(b)(2).

(2) The symbols P_m, P_L, P_b, P_c, Q and F do not represent single quantities, but sets of six quantities representing the six stress components σ_h, σ_ℓ, σ_t, $\sigma_{h\ell}$, σ_{ht}, and $\sigma_{\ell t}$.

(3) For level B service limits for primary stress intensities generated by Level B service loadings, see NB-3223(a)(1).

(4) When the secondary stress is due to a temperature transient at the point at which the stresses are being analyzed or for restraint of free and deflection, the value of S_m shall be taken as the average of the tabulated S_m values for the highest and the lowest temperatures of the metal during the transient. When part or all of the secondary stress is due to mechanical load, the value of S_m shall not exceed the value for the highest temperature during the transient. When part or all of the secondary stress is due to mechanical load, the value of S_m shall not exceed the value for the highest temperature during the transient.

(5) Special rules for exceeding 3 S_m are provided in NB-3223.5.

(6) S_a is obtained from the fatigue curves (Figs. I-9). The allowable stress intensity for the full range of fluctuation is $2S_c$.

(7) The stresses in category Q are those parts of the total stress that are produced by thermal gradients, structural discontinuities, etc., and they do not include primary stresses that may also exist at the same point. However, it should be noted that a detailed stress analysis frequently gives the combination of primary and secondary stresses directly and, when appropriate, the calculated value represents the total of P_m+P_b+Q and not Q alone. Similarly, if the stress in category F is produced by a stress intensity P_m and has a notch with a stress concentration factor K, then $P_m \leq S_m$, P_b Q, Q=0, F=P_m (K-1), and the peak stress intensity equals P_m+P_m(K-1)=KP_m. However, P_L is the total membrane stress that results from mechanical loads, including discontinuity effects, rather than a stress increment. Therefore, the P_L value always includes the P_m contribution.

FIGURE 4.2. STRESS CATEGORIES AND LIMITS OF STRESS INTENSITY FOR LEVEL A AND LEVEL B SERVICE LIMITS – HOPPER DIAGRAM

Stress Category	Primary [Notes (1) and (2)]			Secondary [Notes (1) and (2)] Membrane Plus Bending	Peak [Notes (1) and (2)]
	General Membrane	Local Membrane	Bending		
Description (for examples see Table NB-3217-1)	Average primary stress across solid section. Excludes discontinuities and concentrations. Produced only by mechanical loads.	Average stress across any solid section. Considers discontinuities but not concentrations. Produced only by mechanical loads.	Component of primary stress proportional to distance from centroid of solid section. Excludes discontinuities and concentrations produced only by mechanical loads.	Self-equilibrating stress necessary to satisfy continuity of structure. Occurs at structural discontinuities. Can be caused by mechanical load or by differential thermal expansion. Excludes local stress concentrations.	(1) Increment added to primary or secondary stress by a concentration (notch). (2) Certain thermal stresses which may cause fatigue but not distortion of vessel shape.
Symbol [Note (1)]	P_m NB-3213.6 and NB-3213.8 [Note (3)]	P_L NB-3213.10	P_b NB-3213.7 and NB-3213.8	Q NB-3213.9	F NB-3213.11

Combination of stress components and allowable limits of stress intensities.

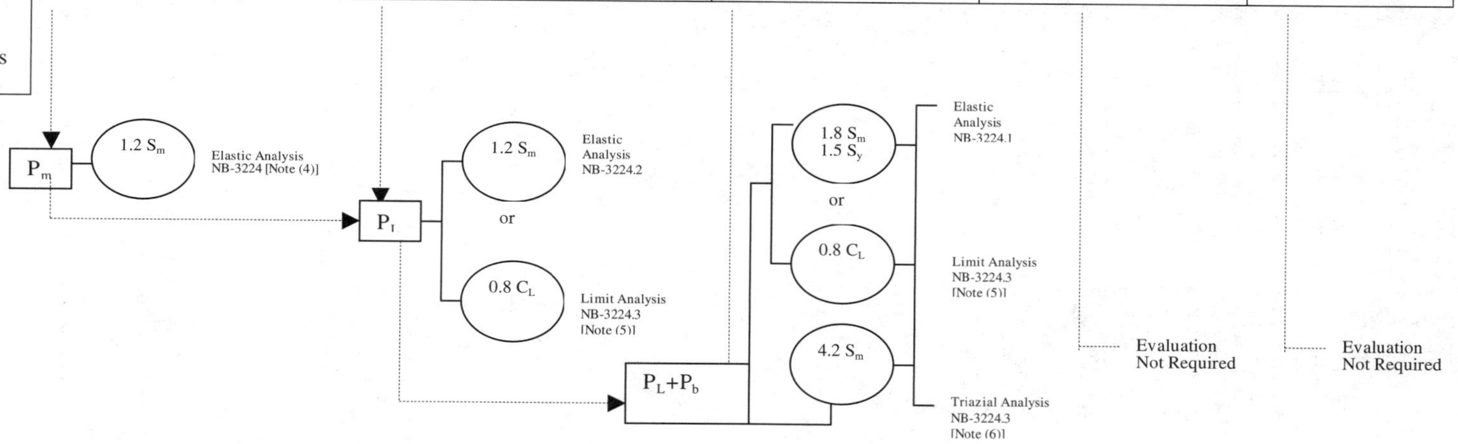

General Note: For configurations where compressive stresses occur, the stress limits shall be revised to take into account critical buckling stresses [NB-321 1(c)].

Notes:

(1) The symbols P_m, P_L, P_b, Q and F do not represent single quantities, but rather sets of six quantities representing the six stress components s_h, s_ℓ, s_t, $s_{h\ell}$, $s_{h\ell}$, and $s_{\ell t}$.
(2) For piping, alternative requirements are provided in NB-3224.7.
(3) The limits shown are for stresses resulting from pressure in combination with other mechanical loads. For ferritic materials, the P_m elastic analysis limits for pressure loadings alone shall be equal to the greater of 1.1 S_m or 0.9 S_m.
(4) Use the greater of the values specified.
(5) C_L, the collapse load calculated on the basis of the lower bound theorem of limit analysis and yield strength values specified in Section II, Part D, Subpart 2, Table Y-1.
(6) The triaxial stresses represent the algebraic sum of the three primary principal stress (s_1 + s_2, + s_3) for the combination of stress components.

FIGURE 4.3. STRESS CATEGORIES AND LIMITS OF STRESS INTENSITY FOR LEVEL C SERVICE LIMITS – HOPPER DIAGRAM

TABLE 4.4. CLASSIFICATION OF STRESS INTENSITY IN CLASS 1 VESSELS FOR SOME TYPICAL CASES

Vessel Part	Location	Origin of Stress	Type of Stress	Classification
Cylindrical or spherical shell	Shell plate remote from discontinuities	Internal Pressure	General membrane Gradient through plate thickness	P_m Q
	Junction with head or flange	Axial thermal gradient	Membrane Bending	Q Q
		Internal pressure	Membrane Bending	P_L Q[1]
Any shell or head	Any section across entire vessel	External load or moment, or internal pressure	General membrane averaged across full section	P_m
		External load or moment	Bending across full section	P_m
	Near nozzle or other opening	External load or moment, or internal pressure	Local membrane Bending Peak (fillet or corner)	P_L Q F
	Any location	Temperature difference between shell and head	Membrane Bending	Q Q
Dished head or conical head	Crown	Internal pressure	Membrane Bending	P_m P_b
	Knuckle or junction to shell	Internal pressure	Membrane Bending	P_L[2] Q
Flat head	Center region	Internal pressure	Membrane Bending	P_m P_b
	Junction to shell	Internal pressure	Membrane Bending	P_L Q[1]
Perforated head or shell	Typical alignment in a uniform pattern	Pressure	Membrane (averaged through cross section) Bending (averaged through width of ligament, but gradient through plate) Peak	P_m P_b F
	Isolated or atypical ligament	Pressure	Membrane Bending Peak	Q F F
Nozzle (NB-3227.5)	Within the limits of reinforcement defined by NB-3334	Pressure and external moments, including those attributable to restrained free end displacements of attached piping	General membrane Bending (other than gross structural discontinuity stresses) averaged through nozzle thickness	P_m P_m
Nozzle (NB-3227.5)	Outside the limits of reinforcement defined by NB-3334	Pressure and external axial, shear and torsional loads other than those attributable to restrained free end displacements of attached piping	General membrane stresses	P_m
		Pressure and external loads and moments other than those attributable to restrained free end displacements of attached piping	Membrane Bending	P_L P_b
		Pressure and all external loads and moments	Membrane Bending Peak	P_L Q F

(continues)

TABLE 4.4. (Continued)

Vessel Part	Location	Origin of Stress	Type of Stress	Classification
	Nozzle Wall	Gross structural discontinuities	Local membrane	P_L
			Bending	Q
			Peak	F
		Differential expansion	Membrane	Q
			Bending	Q
			Peak	F
Cladding	Any	Differential expansion	Membrane	F
			Bending	F
Any	Any	Radial temperature distribution[3]	Equivalent linear stress[4]	Q
				F
			Nonlinear portion of stress distribution	F
Any	Any	Any	Stress concentration (notch effect)	F

General Note: Q and F classification of stresses refers to other than Design Condition
NOTES:
(1) If the bending moment at the edge is required to maintain the bending stress in the middle to acceptable limits, the edge bending is classified as Pb. Otherwise, it is classified as Q.
(2) Consideration shall also be given to the possibility of wrinkling and excessive deformation and excessive deformation in vessels with a large diameter-thickness ratio.
(3) Consider possibility of thermal stress ratchet.
(4) Equivalent linear stress if defined as the linear stress distribution which has the same net bending moment as the actual stress distribution.

where σ_1 and σ_2 = principal stresses in or parallel to the midplane of the shell, wall or plate of the component; and σ_3 = principal stress perpendicular to the midplane of the shell, wall or plate of the component. Where σ_1, σ_2 and σ_3 are compressive in nature they are taken as a negative value.

As can be seen by the Hopper diagrams shown in Figures 4.2 and 4.3, the Section III code distinguishes between primary stresses, which must satisfy equilibrium, and secondary stresses, which are deformation or displacement limited, and peak stresses, which are associated with local stresses at points of discontinuity or stress concentration. Secondary and peak stresses are generally only of importance to design in repeated loading fatigue or ratcheting analysis.

4.2.2 Design by Rule NX and WX-3100 and NX and WX-3300

Design by rule is generally applicable to the design of simple or well-defined structural geometries such as cylinders and spheres or flat plates with pressure; the major component of load and are contained in sub-articles of code NX and WX-3100 and 3300. These rules were generally developed before the widespread use of finite element or finite difference analytical computer algorithms and procedures were available to analyze and design pressure retaining components. Given the current power and availability of such computer programs, it is anticipated that design by analysis (NX-3200) will normally be used as the basis of pressure retaining component design supplemented by the specific component design rules (i.e., NX-3300, NX-3400, NX-3500, NX-3600, etc.).

In general, design by rules has been based on directional stress components (longitudinal and loop), particularly as they relate to compressive stresses. It is highly unlikely that class 1 pressure retaining components will be designed by rule because they do not adequately address stresses developed by factors other than dead weight and pressure for a limited number of simple geometries.

Design by rules allowable stresses for class 2 and 3 components are given in Tables 4.5 and 4.6. The σ values in these tables are defined as follows:

σ_m = general membrane stress (psi). This stress is equal to the average stress across the solid section under consideration. It excludes discontinuities and concentrations and is produced only by pressure and other mechanical loads.

TABLE 4.5. CLASSIFICATION OF STRESS INTENSITY IN CLASS 1 PIPING, TYPICAL CASES

Piping Component	Locations	Origin of Stress	Classification	Discontinuities Considered: Gross	Discontinuities Considered: Local
Pipe or tube, elbows and reducers. Intersections and branch connections, except in crotch regions	Any, except crotch regions of intersections	Internal pressure	P_m	No	No
			P_L and Q	Yes	No
			F	Yes	Yes
		Sustained mechanical loads, including weight Nonreversing dynamic loads	P_b	No	No
			P_L and Q	Yes	No
			F	Yes	Yes
		Expansion	P_e	Yes	No
			F	Yes	Yes
		Axial thermal gradient	Q	Yes	No
			F	Yes	Yes
		Reversing dynamic loads	1		
Intersections, including tees and branch connections	In crotch region	Internal pressure, sustained mechanical loads, expansion and nonreversing dynamic loads	P_L & Q^2	Yes	No
			F	Yes	Yes
		Axial thermal gradient	Q	Yes	No
			F	Yes	Yes
		Reversing dynamic loads	1		
Bolts and flanges	Any	Internal pressure, gasket compression and bolt load	P_m	No	No
			Q	Yes	No
			F	Yes	Yes
		Thermal gradient	Q	Yes	No
			F	Yes	Yes
		Expansion	P_e	Yes	No
			F	Yes	Yes
Any	Any	Nonlinear radial thermal gradient	F	Yes	Yes
		Linear radial thermal gradient	F	Yes	No
		Anchor point motions, including those resulting from earthquake	Q	Yes	No

1. The stress intensity resulting from this loading has special requirements that must be satisfied.
2. Analysis is not required when reinforced in accordance with applicable NB-3600 requirements.

TABLE 4.6. DIRECTIONAL COMPONENT STRESS LIMITS FOR CLASS 2 AND 3 VESSELS DESIGN AND SERVICE LOADING[1]

Service Limit	Stress Limits[2]
Design and Level A	σ_m ″ 1.0 S
	(σ_m or σ_L) + σ_b ″ 1.5 S
Level B	σ_m ″ 1.10 S
	(σ_m or σ_L) + σ_b ″ 1.65 S
Level C	σ_m ″ 1.5 S
	(σ_m or σ_L) + σ_b ″ 1.8 S
Level D	σ_m ″ 2.0 S
	(σ_m or σ_L) + σ_b ″ 2.4 S

1. These limits do not take into account either local or general buckling, which might occur in thin-wall vessels.

σ_L = local membrane stress (psi). This stress is the same as σ_m, except that it includes the effect of discontinuities.

σ_b = bending stress (psi). This stress is equal to the linear varying portion of the stress across the solid section under consideration. It excludes discontinuities and concentrations and is produced only by pressure and other mechanical loads.

S = allowable stress value [6] given in Table 1A, 1B, and 3 of ASME B&PVC, Section II, Part D, Subpart 1. The allowable stress shall correspond to the highest metal temperature in the section under consideration during the condition under consideration.

4.2.3 Piping Design – Article NX-3600

Class 1 – Subsection NB-3600

Piping design requirements of *ASME B&PVC Section III, Class 1* is contained in article NB-3600. These requirements are more rigorous with respect to how thermal stresses and fatigue stresses are considered then for class 2 or class 3 piping in articles NC-3600 and ND-3600, respectively. It should also be noted that both *ASME B31.1* and *B31.3* piping when designed by analysis (as opposed to the use of tables and charts as discussed in Chapter 5) is very similar to *ASME B&PVC Section III, Class 2, NC-3600* rules.

The basic design equations, which must be met for Class 1 piping are as follows:

Minimum wall thickness:

$$t_m = \frac{PD_o}{2(S_m + Py)} + A \tag{4.2}$$

$$t_m = \frac{Pd + 2A(S_m + Py)}{2(S_m + Py - P)} \tag{4.3}$$

where A = an additional thickness to provide for material removed in threading, corrosion or erosion allowance, and material required for structural strength of the pipe during erection, as appropriate [in. (mm)]; D_o = outside diameter of the pipe [in. (mm)]. For design calculations, the specified outside diameter of pipe disregarding outside tolerances shall be used to obtain the value of t_m; d = inside diameter, [in. (mm)]; P = internal design pressure [psi (kPa)]; S_m = maximum allowable stress intensity for the material at the design temperature taken from *ASME B&PVC Section II, Part D, Subpart 1,* Tables 2A and 2B [psi (kPa)] [6]; t_m = the minimum required wall thickness [in. (mm)] [Eq. (4.3) is valid only if $d = D_o\text{-}2t_m$. If pipe is ordered by its nominal wall thickness, the manufacturer's tolerance on wall thickness must be taken into account.]; and $y = 0.4$

To satisfy design condition (primary stress intensities):

$$B_1 \frac{PD_o}{2t_n} + B_2 \frac{M_A}{Z} \le 1.5S_h \tag{4.4}$$

where B_1, B_2 = primary stress indices for the specific product under investigation; P = internal design pressure [psi (kPa)]; D_o = outside diameter of pipe [in. (mm)]; t_n = nominal wall thickness [in. (mm)]; M_A = resultant moment loading on cross section due to weight and other sustained loads [in.-lb (kN·m)]; and Z = elastic section modulus of pipe [in.^{3} (mm^3)].

To satisfy primary plus secondary stress intensity range:

$$S_n = C_1 \frac{P_o D_o}{2t} + C_2 \frac{D_o}{2I} M_i + C_3 E_{ab} x \left| \alpha_a T_a - \alpha_b T_b \right| \le 3S_m \tag{4.5}$$

All load sets, for which level A service limits are to be evaluated, shall satisfy the fatigue requirements of NB-3653.1 through NB-3653.6 and the thermal stress ratchet requirement of NB-3653.7.

If for one or more pairs of load sets Eq. (4.5) is not met, the piping product may still be satisfactory, provided that the conditions of NB-3653.6 are met, or that the requirements of NB-3200 are satisfied.

The nomenclature used in Eq. (4.5) is defined as follows: C_1, C_2, C_3 = secondary stress indices for the specific component under investigation (NB-3680); $d_a(d_b)$ = inside diameter on side a(b) of a gross structural discontinuity or material discontinuity [in. (mm)]; E_{ab} = average modulus of elasticity of the two sides of a gross structural discontinuity or material discontinuity at room temperature [psi (kPa)] (Section II, Part D, Subpart 2, Tables TM) [6]; M_i = resultant range of moment that occurs when the system goes from one service load set to another [in.-lb. (kN·m)]. Service loads and combinations thereof shall be provided in the design specification. In the combination of moments from load sets, all directional moment components in the same direction shall be combined before determining the resultant moment (i.e., resultant moments from different load sets shall not be used in calculating the moment range M_i). Weight effects need not be considered in determining the loading range since they are noncyclic in character. If the method *o* analysis is such that only magnitudes without relative algebraic signs are obtained, the most conservative combination shall be assumed. If a combination includes reversing dynamic loads, M_i shall be either: (1) the resultant range of moment due to the combination of all loads considering one-half the range of the reversing dynamic loads; or (2) the resultant range of moment due to the full range of the reversing dynamic loads alone, whichever is greater; S_n = stress intensity range of primary plus secondary stresses; P_o = range of service pressure [psi (kPa)]; $T_a(T_b)$ = range of average temperature on side a(b) of gross structural discontinuity or material discontinuity [°F (°C).]. For generally cylindrical shapes, the averaging of T (NB-3653.2) shall be over a distance of $\sqrt{d_a t_a}$ for T_a and over a distance of $\sqrt{d_b t_b}$ for T_b; $t_a t_b$ = average wall thickness through the length $\sqrt{d_a t_a}$ ($\sqrt{d_b t_b}$) [in. (mm)]. A trial and error solution for t_a and t_b may be necessary; and $\alpha_a(\alpha_b)$ = coefficient of thermal expansion on side a(b) of a gross structural discontinuity or material discontinuity, at room temperature (Section II, Part D, Subpart 2, Tables TE).

Satisfy peak stress intensity range

$$S_p = K_1 C_1 \frac{P_o D_o}{2t} + K_2 C_2 \frac{D_o}{2I} M_i$$

$$+ \frac{1}{2(1-\nu)} K_3 E\alpha |\Delta T_1| + K_3 C_3 E_{ab} \tag{4.6}$$

$$x |\alpha_a T_a - \alpha_b T_b| + \frac{1}{1-\nu} E\alpha |\Delta T_2|$$

Note: This simplified analysis is intended to provide a value of S_p that conservatively estimates the sum of P_L+P_b+P_e+Q+F as required in Figure NB-3222-1.

The nomenclature used in Eq. (4.6) is defined as follows: $E\alpha$ = modulus of elasticity (E) times the mean coefficient of thermal expansion (α) both at room temperature [psi/°F (kPa/°C)]; K_1, K_2, K_3 = local stress indices for the specific component under investigation (NB-3680); $|\Delta T_2|$ = absolute value of the range for that portion of the nonlinear thermal gradient through the wall thickness not included in ΔT_1 as shown below [°F (°C)]; and $|\Delta T_1|$ = absolute value of the range of the temperature difference between the temperature of the outside surface T_o and the temperature of the inside surface T_i of the piping product assuming moment generating equivalent linear temperature distribution [°F (°C)].

For a quantitative definition of $|\Delta T_1|$ and $|\Delta T_2|$ see NB-3653.2(b) below. All other terms are as defined for Eq. 4.5.

Quantitative definitions of $|\Delta T_1|$ and $|\Delta T_2|$.

The following nomenclature is used: T_i = value of $T(y)$ at inside surface [°F (°C)]; = T(-t/2); T_o = value of $T(y)$ at outside surface, [°F (°C)]; $T_o = T(t/2)$; $T_j(y)$, $T_k(y)$ = temperature, as a function of radial position, for load set *j* and load set *k*, respectively [°F (°C)]; $T(y)$ = temperature distribution range from condition *j* to condition

k, [ºF (ºC)]; $T(y) = T_k(y) - T_j(y)$; t = thickness of the wall of the pipe or element [in. (mm)]; and y = radial position in the wall, measured positive outward from the midthickness position (-t/2 " y " t/2) [in. (mm)].

Other alternate equations as defined in the code may be used if the base equations just defined cannot be met. In addition, there is a requirement to evaluate thermal stress ratcheting as defined by the code.

Class 2 and 3 – Subsections NC-3600 and ND-3600

The design equations for *ASME B&PVC* class 2 and 3 piping are as follows: Minimum wall thickness: Same as for class 1 except allowable stress S is substituted for allowable stress intensity.

Longitudinal Stress: It should be noted that for piping the manner in which torsional moments in pipe are converted to an equivalent bending stresses, the stresses actually being evaluated are principal stresses.

The effects of pressure, weight and other sustained mechanical loads must meet the requirements of Equation (4.6):

$$S_{SL} = B_1 \frac{PD_o}{2t_n} + B_2 \frac{M_A}{Z} \leq 1.5S_h \tag{4.7}$$

B_1, B_2 = primary stress indices for the specific product under investigation; P = *internal* design pressure [psi (kPa)]; D_o = outside diameter of pipe [in. (mm)]; t_n = nominal wall thickness [in. (mm)]; M_A = resultant moment loading on cross section due to weight and other sustained loads [in.-lb (kN·m)] (NC-3653.3); Z = section modulus of pipe [in.^{3} (mm^3)]; and S_h = basic material allowable stress at design temperature [psi (kPa)].

The effects of pressure, weight and other loads, including nonreversing dynamic loads, for which level B service limits are designated, shall meet the requirements of Eq. 4.8

$$S_{OL} = B_1 \frac{P_{\max} D_o}{2t_n} + B_2\left(\frac{M_A + M_B}{Z}\right) \leq 1.8S_h \tag{4.8}$$

but not greater than 1.5S_y. Terms are the same as in NC-3652, except: P_{max} = peak pressure [psi (kPa)]; M_B = resultant moment loading on cross section due to nonreversing dynamic loads; S_y = material yield strength at temperature consistent with the loading under consideration [psi (kPa)]; and S_h = material allowable stress at temperature consistent with the loading under consideration [psi (kPa)].

The effects of thermal expansion shall meet Eq. (4.9);

$$S_E = \frac{iM_C}{Z} \leq S_A \tag{4.9}$$

Terms are the same as previously defined, except: M_C = range of resultant moments due to thermal expansion [in.-lb.]; also include moment effects of anchor displacements due to nonreversing dynamic loads if anchor displacement effects were omitted from Equation (4.10) (NC-3653.1); S_A = allowable stress range fro expansion stresses (NC-3611.2) [psi (kPa)]; and i = stress intensification factor (NC-3673.2).

The effects of any single nonrepeated anchor movement shall meet the requirements of Eq. (4.10):

$$\frac{iM_D}{Z} \leq 3.0S_c \tag{4.10}$$

Terms are same as in NC-3653.2, except:

M_D = resultant moment due to any single nonrepeated anchor movement (e.g., predicted building settlement) [in.-lb].

The effects of pressure, weight, other sustained loads and thermal expansion shall meet the requirements of Eq. (4.11):

$$S_{TE} = \frac{PD_o}{4t_n} + 0.75i\left(\frac{M_A}{Z}\right) + i\left(\frac{M_C}{Z} \leq\right)(S_h + S_A) \tag{4.11}$$

0.75i shall not be less than 1.0.

The effects of reversing dynamic loads must meet the requirements of Eq. (4.12):

$$S_R = \frac{iM_R}{Z} \leq 2.0S_A \tag{4.12}$$

Terms are the same as in NC-3653.2, except: M_R = range of resultant moments due to inertia and anchor motion effects of reversing dynamic loads.

Thermal stresses are not considered in service levels C and D except as they apply to piping supports. The allowable stress limits for class 2 and 3 piping in service level C is 2.25*S* and 3.0*S* for service level D.

4.2.4 Support Design

Introduction

Basically there are three types of mechanical component supports: plate and shell, linear and standard supports. Standard supports, are normally procured as category items. Design criteria for standard supports generally apply to piping and are found in reference [8]. Plate shell supports design procedures and allowables stress are similar to those used for ASME pressure retaining vessels. The linear type supports typically consist of hot-rolled structural web and flange shapes and follow the design procedure for building structural steel defined by American Institute of Steel Construction specifications [7].

The third type are standard supports contained in MSS-SP-58 [8]. The specific reference for *ASME B&PVC* Section III supports is subsection NF [9].

Acceptance Criteria

A summary of the allowable stress for design service Levels A, B, C and D test loads are contained in Tables 4.7 to 4.10 for component supports other than piping.

TABLE 4.7. CLASSIFICATION OF STRESSES IN CLASS 2 AND 3 VESSELS FOR SOME TYPICAL CASES[1]

Vessel Part	Location	Origin of Stress	Type of Stress	Classification
Cylindrical or spherical shell	Shell plate remote from discontinuitie	Internal pressure	General membrane	σ_m
			Gradient through plate thicknes	Q
	Junction with head or flange	Axial thermal gradient	Membrane	Q
			Bending	Q
		Internal pressure	Membrane	σ_L
			Bending	Q^2
Any shell or head	Any section across entire vessel	External load or moment, or internal pressure	General membrane averaged across full section; stress component perpendicular to cross section	σ_m
		External load or moment	Bending across full section; stress component perpendicular to cross section	σ_m
	Near nozzle or other opening	External load or moment, or internal pressure	Local membrane	σ_L
			Bending	Q
			Peak (fillet or corner)	F
	Any location	Temperature difference between shell and head	Membrane	Q
			Bending	Q

(continues)

TABLE 4.7. (Continued)

Vessel Part	Location	Origin of Stress	Type of Stress	Classification
Dished head or conical head	Crown	Internal pressure	Membrane Bending	σ_m σ_b
	Knuckle or junction to shell	Internal pressure	Membrane Bending	σ_L[3] Q
Flat head	Center region	Internal pressure	Membrane Bending	σ_m σ_b
	Junction to shell	Internal pressure	Membrane Bending	σ_L Q[2]
Perforated head or shell	Typical alignment in a uniform pattern	Pressure	Membrane (averaged through cross section)	σ_m
			Bending (averaged through width of ligament, but gradient through plate) Peak	σ_b F
	Isolated or atypical ligament	Pressure	Membrane Bending Peak	Q F F
Nozzle	Within the limits of reinforcement	Pressure and external moments, including those attributable to restrained free end displacements of attached piping	General membrane Bending (other than gross structural discontinuity stresses) averaged through nozzle thickness	σ_m σ_m
Nozzle	Outside the limits of reinforcement	Pressure and external axial, shear and torsional loads other than those attributable to restrained free end displacements of attached piping	General membrane stresses	σ_m
		Pressure and external loads and moments other than those attributable to restrained free end displacements of attached piping	Membrane Bending	σ_L σ_b
		Pressure and all external loads and moments	Membrane Bending Peak	σ_L Q F
	Nozzle wall	Gross structural discontinuities	Local membrane Bending Peak	σ_L Q F
		Differential expansion	Membrane Bending Peak	Q Q F
Cladding	Any	Differential expansion	Membrane Bending	F F
Any	Any	Radial temperature distribution[3]	Equivalent linear stress[5] Nonlinear portion of stress distribution	Q F
Any	Any	Any	Stress concentration (notch effect)	F

1. Q and F classification of stresses refers to that other than design condition.
2. If the bending moment at the edge is required to maintain the bending stress in the middle to acceptable limits, the edge bending is classified as σ_b. Otherwise, it is classified as Q.
3. Consideration shall also be given to the possibility of wrinkling and excessive deformation and excessive deformation in vessels with a large diameter-thickness ratio.
4. Consider possibility of thermal stress ratchet.
5. Equivalent linear stress if defined as the linear stress distribution, which has the same net bending moment as the actual stress distribution.

TABLE 4.8. ELASTIC ANALYSIS STRESS CATEGORIES AND STRESS LIMIT FACTORS FOR CLASS 1 PLATE-AND SHELL-TYPE SUPPORTS DESIGNED BY ANALYSIS – COMPONENT SUPPORTS

	Stress Limit Factors for Loading Categories[1]					
Stress Category	**Design**	**Service Level A**	**Service Level B[2]**	**Service Level C[3]**	**Service Level D**	**Test Loadings**
Primary stresses[4, 5]	$K_m = 1.0$ $K_v = 1.0$ $K_{bk} = 1.0$	$K_m = 1.0$ $K_v = 1.0$ $K_{bk} = 1.0$	$K_m = 1.33$ $K_v = 1.33$ $K_{bk} = 1.33$	$K_m = 1.5$ $K_v = 1.5$ $K_{bk} = 1.5$	Use Appendix F	$K_m = 1.33$ $K_v = 133$ $K_{bk} = 1.33$
	but stress " ½ of critical buckling stress	but stress " ½ of critical buckling stress	but stress " ½ of critical buckling stress	but stress " ½ of critical buckling stress		but stress " ½ of critical buckling stress
Primary plus secondary stresses[5]	Evaluation is required for critical buckling for all loading categories.					
Peak stress	Evaluation not required.					

Note:
K_m = stress limit factor applicable to the design allowable membrane stress intensity or membrane plus bending stress intensity (see NF-3221.1 and NF-3221.2).
K_v = stress limit factor applicable to the design allowable shear stress (see NF-3223.2).
K_{bk} = stress limit factor applicable to the design allowable membrane stress intensity or membrane plus bending stress intensity (compression only).

1. Control of deformation is not ensured by these stress limit factors. When required by design specification, deformation control must be considered separately.
2. K_m, K_v, K_{bk} = 1.0 for design snubbers.
3. Stress shall not exceed 0.7 S_u.
4. For service levels A, B, C and D, stresses induced on the supports by restraint of free-end displacement and anchor motions of piping shall be considered as primary stresses.
5. Thermal stresses within the support as defined by NF-3121.11 need not be evaluated. For service levels A and B, primary plus secondary stresses shall be limited to a range of 2 S_y or S_u at temperature, whichever is less.

TABLE 4.9. ELASTIC ANALYSIS STRESS CATEGORIES AND STRESS LIMIT FACTORS FOR CLASSES 2, 3 AND MC PLATE-AND-SHELL-TYPE SUPPORTS DESIGNED BY ANALYSIS – COMPONENT SUPPORTS

	Stress Limit Factors for Loading Categories[1]					
Stress Category	**Design**	**Service Level A**	**Service Level B[2]**	**Service Level C[3]**	**Service Level D**	**Test Loadings**
Primary stresses[4, 5]	$K_m = 1.0$ $K_v = 1.0$ $K_{bk} = 1.0$	$K_m = 1.0$ $K_v = 1.0$ $K_{bk} = 1.0$	$K_m = 1.33$ $K_v = 1.33$[6] $K_{bk} = 1.33$	$K_m = 1.5$ $K_v = 1.5$[6] $K_{bk} = 1.5$	Use Appendix F	$K_m = 1.33$ $K_v = 133$[6] $K_{bk} = 1.33$
	but stress ″ ½ of critical buckling stress	but stress ″ ½ of critical buckling stress	but stress ″ ½ of critical buckling stress	but stress ″ ½ of critical buckling stress		but stress ″ ½ of critical buckling stress

Note:

K_m = stress limit factor applicable to the design allowable membrane stress or membrane plus bending stress (see NF-3251.1 and NF-3261).

K_v = stress limit factor applicable to the design allowable shear stress (see NF-3252.2 and NF-3261).

K_{bk} = stress limit factor applicable to the design allowable membrane stress or membrane plus bending stress intensity (compression only).

1. Control of deformation is not ensured by these stress limit factors. When required by design specification, deformation control must be considered separately.
2. K_m, and K_{bk} = 1.0 for design snubbers.
3. Stress shall not exceed 0.7 S_u.
4. For service levels A, B, C and D, stresses induced on the supports by restraint of free-end displacement and anchor motions of piping shall be considered as primary stresses.
5. Thermal stresses within the support as defined by NF-3121.11 need not be evaluated.
6. Shear stress shall not exceed 0.42 S_u.

TABLE 4.10. ELASTIC ANALYSIS STRESS CATEGORIES AND STRESS LIMIT FACTORS FOR CLASSES 1, 2, 3 AND MC LINEAR TYPE DESIGNED BY ANALYSIS – COMPONENT SUPPORTS

	Stress Limit Factors for Loading Categories[1]					
Stress Category	**Design**	**Service Level A**	**Service Level B**[2]	**Service Level C**[3]	**Service Level D**	**Test Loadings**
Primary stresses[4, 5]	$K_S = 1.0$ $K_v = 1.0$ $K_{bk} = 1.0$	$K_S = 1.0$ $K_v = 1.0$ $K_{bk} = 1.0$	$K_S = 1.33$ $K_v = 1.33$[6] $K_{bk} = 1.33$	$K_S = 1.5$ $K_v = 1.5$[6] $K_{bk} = 1.5$	Use Appendix F $4K_{bk} = 1.33$	$K_S = 1.33$ $K_v = 133$[6]
			but stress ″ ⅔ of critical buckling stress	but stress ″ ⅔ of critical buckling stress		but stress ″ ⅔ of critical buckling stress
Primary plus secondary stresses[5]	Evaluation is required for critical buckling for all loading categories. The requirements of this sub-article shall be met for this evaluation.					
Peak stress	Evaluation not required.					

Note:
K_S = stress limit factor applicable to the design allowable tensile and bending stresses (see NF-3321.1 and NF-3322.2, MF-3350 and NF-3360).
K_v = stress limit factor applicable to the design allowable shear stress (see NF-3221.1, NF-3322.1, NF-3350 and NF-3360).
K_{bk} = stress limit factor applicable to the design allowable compressive axial and bending stresses to determine buckling limit.

1. Control of deformation is not ensured by these stress limit factors. When required by design specification, deformation control must be considered separately.
2. K_s, K_v, K_{bk} = 1.0 for design snubbers.
3. Stress shall not exceed 0.7 S_u.
4. For service levels A, B, C and D, stresses induced on the supports by restraint of free-end displacement and anchor motions of piping shall be considered as primary stresses.
5. Thermal stresses within the support as defined by NF-3121.11 need not be evaluated. For service levels A and B, primary plus secondary stresses shall be limited to a range of $2S_y$ or S_u at temperature, whichever is less.
6. Shear stress shall not exceed 0.42 S_u.

TABLE 4.11. SUGGESTED DEAD WEIGHT PIPING SUPPORT SPACING

Nominal Pipe Size (in DN)	Suggested Maximum Span [in ft (m)]	
	Water Service	Steam, Gas or Air Service
1 (25)	7 (2.1)	9 (2.7)
2 (50)	10 (3.0)	13 (3.9)
3 (80)	12 (3.6)	15 (4.5)
4 (100)	14 (4.2)	17 (5.1)
6 (150)	17 (5.1)	21 (6.4)
8 (200)	19 (5.7)	24 (7.3)
12 (300)	23 (7.0)	30 (9.1)
16 (400)	27 (8.2)	35 (10.6)
20 (500)	30 (9.1)	39 (11.8)
24 (600)	32 (9.7)	42 (12.8)

1. Suggested maximum spacing between piping supports for horizontal straight runs of standard and heavier piping with a maximum operating temperature of 750°F (398°C).
2. Does not apply where span calculations are made or where there are concentrated loads between supports such as flanges, valves, specialties, etc.
3. The spacing is based on a maximum combined bending and shear stress of 1.5 ksi (10 300 kPa) and insulated piping filled with water, or the equivalent weight of steel pipe for steam, gas or air service, and the pitch of the line is such that a sag of 0.125 in. (3.0 mm) between supports is permissible.

For piping systems the suggested dead and live load support spacing between supports is given in Table 4.11. Lateral support span spacings for piping are typically two to three times those for vertical loads. Design acceptance for piping supports as defined in ASME B&PVC Section III, NF is given in Tables 4.12 to 4.14. It is common practice to use the spacing in Table 4.10 with a spacing tolerance of up to +50%. Such a limiting spacing tolerance would limit dead weight bend stress to approximately 2,500 psi. However, if the limiting tolerance is used it may be necessary to provide line slope such that the line can be properly drained when not in use.

4.3 ASME B&PVC SECTION VIII

The rules of *ASME B&PVC* Section VIII are generally applied to pressure vessels and heat exchangers used in the petrochemical and biomedical industries. They are also used in the construction of safety-related pressure vessel and heat exchanger components at nuclear facilities other than those associated with nuclear safety of large reactors [12].

Section VIII of the *ASME Boiler and Pressure Vessel Code* is currently divided into three Divisions: Division 1 covers the rules for construction where fatigue is not a design concern. Typically, Division 1 is also used where design pressures are below 1,500 psi and shell thickness does not exceed 2.0 inches.

Division 2 is typically used in vessels that exceed 2.0 inches in thickness and where fatigue is a design consideration. Fatigue tends to be a design issue where there is thermal stress cycling. Fatigue may also be a concern where in operation attached rotating equipment or flow induced vibration occurs. However, these latter fatigue concerns are usually addressed during startup testing where such vibrations are reduced as necessary by modifications to the installed system to bring vibration induced stresses to acceptable levels [13].

Division 3 is generally used for pressure vessels with design pressures in excess of 10,000 psi.

It should also be noted that the allowable directional (longitudinal and hoop) stresses for Division 1 are a function of the lesser of specified minimum two-thirds of yield stress or ultimate stress divided by 3.5 at temperature.

For division 2 the allowable stresses are expressed in terms of stress intensities (Tresca criteria) and are the lesser of the specified minimum two-thirds yield or one-third ultimate stress at temperature.

For division 3 the limiting stress is two-thirds of specified minimum yield at temperatures with no limit based on ultimate stress. However, a knock-down factor is used on specified minimum yield stress when it exceeds 0.7 times ultimate stress.

TABLE 4.12. ELASTIC ANALYSIS STRESS CATEGORIES AND STRESS LIMIT FACTORS FOR CLASS 1 PLATE-AND SHELL-TYPE SUPPORTS DESIGNED BY ANALYSIS – PIPING SUPPORTS

	Stress Limit Factors for Loading Categories[1]					
Stress Category	**Design**	**Service Level A**	**Service Level B**[2]	**Service Level C**[3]	**Service Level D**	**Test Loadings**
Primary Stresses[4, 5]	$K_m = 1.0$ $K_v = 1.0$ $K_{bk} = 1.0$	$K_m = 1.0$ $K_v = 1.0$ $K_{bk} = 1.0$	$K_m = 1.33$ $K_v = 1.33$ $K_{bk} = 1.33$	$K_m = 1.5$ $K_v = 1.5$ $K_{bk} = 1.5$	Use Appendix F	$K_m = 1.33$ $K_v = 133$ $K_{bk} = 1.33$
	but stress ″ ½ of critical buckling stress	but stress ″ ½ of critical buckling stress	but stress ″ ½ of critical buckling stress	but stress ″ ½ of critical buckling stress		but stress ″ ½ of critical buckling stress
Primary plus secondary stresses[5]	Evaluation is required for critical buckling for all loading categories. The requirements of this sub-article shall be met for this evaluation.					
Peak stress	Evaluation not required.					

Note:
K_m = stress limit factor applicable to the design allowable membrane stress intensity or membrane plus bending stress intensity (see NF-3221.1 and NF-3221.2).
K_v = stress limit factor applicable to the design allowable shear stress (see NF-3223.2).
K_{bk} = stress limit factor applicable to the design allowable membrane stress intensity or membrane plus bending stress intensity (compression only).

1. Control of deformation is not ensured by these stress limit factors. When required by design specification, deformation control must be considered separately.
2. K_m, K_v, K_{bk} = 1.0 for design snubbers.
3. Stress shall not exceed 0.7 S_u.
4. For service levels A, B, C and D, stresses induced on the supports by restraint of free-end displacement and anchor motions of piping shall be considered as primary stresses.
5. Thermal stresses within the support as defined by NF-3121.11 need not be evaluated.

TABLE 4.13. ELASTIC ANALYSIS STRESS CATEGORIES AND STRESS LIMIT FACTORS FOR CLASS 2 AND 3 PLATE-AND-SHELL-TYPE SUPPORTS DESIGNED BY ANALYSIS – PIPING SUPPORTS

	Stress Limit Factors for Loading Categories[1]					
Stress Category	**Design**	**Service Level A**	**Service Level B[2]**	**Service Level C[3]**	**Service Level D**	**Test Loadings**
Primary stresses[4, 5]	$K_m = 1.0$ $K_v = 1.0$ $K_{bk} = 1.0$	$K_m = 1.0$ $K_v = 1.0$ $K_{bk} = 1.0$	$K_m = 1.33$ $K_v = 1.33^6$ $K_{bk} = 1.33$	$K_m = 1.5$ $K_v = 1.5^6$ $K_{bk} = 1.5$	Use Appendix F	$K_m = 1.33$ $K_v = 133^6$ $K_{bk} = 1.33$
	but stress ″ ½ of critical buckling stress	but stress ″ ½ of critical buckling stress	but stress ″ ½ of critical buckling stress	but stress ″ ½ of critical buckling stress		but stress ″ ½ of critical buckling stress

Note:

K_m = stress limit factor applicable to the design allowable membrane stress intensity or membrane plus bending stress intensity (see NF-3251.1 and NF-3261).

K_v = stress limit factor applicable to the design allowable shear stress (see NF-3252.2 and NF-3261).

K_{bk} = stress limit factor applicable to the design allowable membrane stress intensity or membrane plus bending stress intensity (compression only).

1. Control of deformation is not ensured by these stress limit factors. When required by design specification, deformation control must be considered separately.
2. K_m, K_v, K_{bk} = 1.0 for design snubbers.
3. Stress shall not exceed 0.7 S_u.
4. For service levels A, B, C and D, stresses induced on the supports by restraint of free-end displacement and anchor motions of piping shall be considered as primary stresses.
5. Thermal stresses within the support as defined by NF-3121.11 need not be evaluated.
6. Shear stress shall not exceed 0.42 S_u.

TABLE 4.14. ELASTIC ANALYSIS STRESS CATEGORIES AND STRESS LIMIT FACTORS FOR CLASS 1, 2 AND 3 LINEAR-TYPE SUPPORTS DESIGNED BY ANALYSIS – PIPING SUPPORTS

	Stress Limit Factors for Loading Categories[1]					
Stress Category	Design	Service Level A	Service Level B[2]	Service Level C[3]	Service Level D	Test Loadings
Primary stresses[4, 5]	$K_S = 1.0$ $K_v = 1.0$ $K_{bk} = 1.0$	$K_m = 1.0$ $K_v = 1.0$ $K_{bk} = 1.0$	$K_S = 1.33$ $K_v = 1.33$[6] $K_{bk} = 1.33$	$K_S = 1.5$ $K_v = 1.5$[6] $K_{bk} = 1.5$	Use Appendix F	$K_S = 1.33$ $K_v = 133$[6] $K_{bk} = 1.33$
			but stress ″ ⅔ of critical buckling stress	but stress ″ ⅔ of critical buckling stress		but stress ″ ⅔ of critical buckling stress
Primary plus secondary stresses[5]	Evaluation is required for critical buckling for all loading categories. The requirements of this sub-article shall be met for this evaluation.					
Peak stress	Evaluation not required.					

Note:
K_S = stress limit factor applicable to the design allowable tensile and bending stress (see NF-3623 and NF-3653).
K_v = stress limit factor applicable to the design allowable shear stress (see NF-3623 and NF-3653).
K_{bk} = stress limit factor applicable to the design allowable compressive axial and bending stresses to determine buckling limit.

1. Control of deformation is not ensured by these stress limit factors. When required by design specification, deformation control must be considered separately.
2. K_s, K_v, K_{bk} = 1.0 for design snubbers.
3. Stress shall not exceed 0.7 S_u.
4. For service levels A, B, C and D, stresses induced on the supports by restraint of free-end displacement and anchor motions of piping shall be considered as primary stresses.
5. Thermal stresses within the support as defined by NF-3121.11 need not be evaluated.
6. Shear stress shall not exceed 0.42 S_u.

4.3.1 ASME B&PVC Section VIII Division 1 Rules for Construction of Pressure Vessels

Division 1 design rules are primarily based on pressure being the dominant load. Stresses are computed and evaluated and their allowables are direction component stresses, such as hoop or longitudinal stresses, not principal stresses or stress intensities.

Design of Shells for Internal Pressure

The thickness of shells under internal pressure shall be not less than that computed by the following formulas. In addition, provision shall be made for any of the other loadings listed in UG-22, when such loadings are defined in a design specification or as otherwise indicated by the owner of the component.

Cylindrical shells: The minimum thickness or maximum allowable working pressure of cylindrical shells shall be the greater thickness or lesser pressure as given by Eqs. (4.13) or (4.14).

Hoop stress: When the thickness does not exceed one-half the inside radius, or P does not exceed $0.385SE$, the following formulas apply:

$$t = \frac{PR}{SE - 0.6P} \text{ or } P = \frac{SEt}{R + 0.6t} \tag{4.13}$$

Longitudinal stress: When the thickness does not exceed one-half of the inside radius, or P does not exceed $1.25SE$, the following formulas shall apply:

$$t = \frac{PR}{2SE - 0.4P} \text{ or } P = \frac{2SEt}{R + 0.4t} \tag{4.14}$$

Spherical shells: When the thickness of the shell of a wholly spherical vessel does not exceed $0.356R$, or P does not exceed $0.665SE$, the following formulas shall apply:

$$t = \frac{PR}{2SE - 0.2P} \text{ or } P = \frac{2SEt}{R + 0.2t} \tag{4.15}$$

where: t = minimum required thickness of shell (in.); P = internal design pressure (psi); R = inside radius of the shell course under consideration (in.); S = maximum allowable stress value (psi); and E = joint efficiency for, or the efficiency of, appropriate joint in cylindrical or spherical shells, or the efficiency of ligaments between openings, whichever is less. For welded vessels, use the efficiency specified in UW-12; for ligaments between openings, use the efficiency calculated by the rules given in UG-53.

When necessary, vessels are provided with stiffners, other additional means of support to prevent overstress, or large distortions under the external loadings listed in UG-22 other than pressure and dead load.

Division 1 of *ASME B&PVC* Section VIII also contains detailed rules on how to design cylindrical, spherical and flat plate pressure vessels for external pressure as well as procedures for stiffening such vessels. However, these rules apply only to compressive instability or buckling associated with direction component compressive stress in the shell due to external pressure and gravity. They do not apply to compressive stresses in the shell due to overturning from seismic or wind loads, which require analytical evaluation of potential instability associated with these types of loads.

4.3.2 ASME B&PVC Section VIII Division 2 Alternate Rules for Construction of Pressure Vessels

Division 2 of Section VIII of the *ASME B&PVC Code* addresses pressure vessels of irregular shape or those subjected to unsymmetrical loads or significant loads other than pressure, as well as where buckling due to shell thickness or fatigue and ratcheting (as described in Chapter 9) is a concern. Analysis is typically by finite-element computer programs, although the rules for thickness and reinforcement opening must also be met. Acceptance criteria for this loading are stress intensities as defined by the code. The methods of analysis and allowable stresses, except for service levels C and D of *ASME B&PVC* Section III are the same as for Section III Article NX or WX 3200. Service levels C and D are not defined for Section VIII components.

4.4 ASME B31.3 PROCESS PIPING

4.4.1 Introduction

The construction requirements for process piping are generally applicable to safety-related petrochemical and biomedical piping. They are also applicable to nuclear safety-related piping for nonpower plant nuclear facilities regulated by the U.S. Department of Energy [3] and for safety-related piping associated with nuclear waste processed and stored at nuclear power stations regulated by the U.S. Nuclear Regulatory Commission [12].

4.4.2 Design Criteria

Straight Pipe Under Internal Pressure

Straight pipe under pressure. For $t < D/6$, the internal pressure design thickness for straight pipe shall be no less than that calculated in accordance with Eqs. (4.16) or (4.17):

$$t = \frac{PD}{2(SE + PY)} \tag{4.16}$$

$$t = \frac{P(d + 2c)}{[SE - P(1 - Y)]} \tag{4.17}$$

where the following nomenclature is used in the equations for pressure design of straight pipe: t_m = minimum required thickness, including mechanical, corrosion and erosion allowances; t = pressure design thickness, as calculated in accordance with paragraph 304.1.2 for internal pressure or as determined in accordance with paragraph 304.1.3 for external pressure; c = the sum of the mechanical allowances (thread or groove depth) plus corrosion and erosion allowances. For threaded components, the nominal thread depth (dimension h of ASME B1.20.1 or equivalent) shall apply. For machined surfaces or grooves where the tolerance is not specified, the tolerance shall be assumed to be 0.5 mm (0.02 in.) in addition to the specified depth of the cut; T = pipe, all thickness (measured or minimum per purchase specification); d = inside diameter of pipe. For pressure design calculation, the inside diameter of the pipe is the maximum value allowable under the purchase specification; P = internal design gauge pressure; D = outside diameter of pipe as listed in tables of standards or specifications or as measured; E = quality factor as defined by code; S = allowable stress value for material as defined by code; and Y = coefficient from Table 4.15 valid for $t < Y$ may be interpolated for intermediate temperatures.

TABLE 4.15. VALUES OF COEFFICIENT *Y* FOR *T* < *D*/6

	Temperature [°C (°F)]					
Materials	**″ 482 (900 & lower)**	**510 (950)**	**538 (1,000)**	**566 (1,050)**	**593 (1,100)**	**≥ 621 (1,150 and up)**
Ferretic steels	0.4	0.5	0.7	0.7	0.7	0.7
Austenitic steels	0.4	0.4	0.4	0.4	0.5	0.7
Other ductile metals	0.4	0.4	0.4	0.4	0.4	0.4
Cast iron	0.0	—	—	—	—	—

For $t > D/6$ or for $P/SE > 0.385$, calculation of pressure design thickness for straight pipe requires special consideration of factors such as theory of failure, effects of fatigue and thermal stress.

Straight pipe under external pressure: To determine wall thickness and stiffening requirements for straight pipe under external pressure, the procedure outlined in the BPV Code, Section VIII, Division 1, UG-28 through UB-30 are followed, using as the design length L the running center line length between any two sections stiffened in accordance with UG-29 of Section VIII, Division 1.

4.5 ASME B31.1 POWER PIPING

The design acceptance criteria for *ASME B31.1* piping is essentially the same as for *ASME B31.3* piping, except that the allowable stress for normal stresses is the specified minimum ultimate stress at temperatures divided by four rather than by three. The two-thirds specified minimum yield at temperature also applies. When considering occasional (severe) loads the allowable stress is reduced from a factor of 1.33 to 1.2 times normal allowable.

A comparison of the construction requirements for piping between *ASME B&PVC* Section III and *ASME B31.1* and *B31.3* can be found in Table 4.16.

TABLE 4.16. CONSTRUCTION REQUIREMENT ATTRIBUTES FOR ASME SECTION III CLASSES 1, 2, 3 ASME B31.1 AND B31.3 PIPING

	Technical Requirements			
Attribute	**SC III – Class 1**	**SC III – Class 2**	**SC III – Class 3**	**B31.1**
Loads/load combinations	There are six load conditions considered • Design • Service levels: A, B, C and D • Test Load combinations are defined in the design specification	There are six load conditions considered • Design • Service levels: A, B, C and D • Test Load combinations are defined in the design specification	There are six load conditions considered • Design • Service levels: A, B, C and D • Test Load combinations are defined in the design specification	There are three loading conditions considered: • Design • Service levels: A, B, C and D • Test
Fatigue	Explicit fatigue evaluation process addressed with factors and equations See: NB 3222.4 NB 3228.4 NB 3228.5	No formal fatigue evaluation. Implicit in the stress intensification factor (SIFs). Stress range reduction factor *"f"* for high cycles. Table NC-3611.2(e)-1	No formal fatigue evaluation. Implicit in the SIFs. Stress range reduction factor *"f"* for high cycles. Table ND-3611.2(e)-1	Implicit in the SIF, no formal fatigue evaluation. Stress range reduction factor *"f"* for high cycles (>7,000). Table 102.3.2©
Stress indices or intensifications	Yes—in piping, *B, C, K* indices	Yes—in piping, *B* indices, *i* (SIF)	Yes—in piping, *B* indices, *i* (SIF)	SIF in piping
Flexibility factors	Piping elbows, bends and branch connections have flex factors different from 1.0	Piping elbows, bends and miter bends have flex factors different from 1.0	Piping elbows, bends and miter bends have flex factors different from 1.0	Piping elbows, bends and miter bends have flex factors different from 1.0

(continues)

TABLE 4.16. (Continued)

	Technical Requirements			
Attribute	**SC III – Class 1**	**SC III – Class 2**	**SC III – Class 3**	**B31.1**
Product forms allowed	Standard pressure components per Table NB-3132-1. Nonstandard pressure-containing components can be used if they meet para NB-3649, -3681(d).	Standard pressure components per Table NC-3132-1. Nonstandard pressure-containing components can be used if they meet para NC-3649.	Standard pressure components per Table ND-3132-1. Nonstandard pressure-containing components can be used if they meet para ND-3649.	Standard pressure components per Table 126.1 Nonstandard pressure-containing components can be used if they meet para 104.7.
Material selection	Pressure retaining material shall conform to a Section II spec. Material must be certified per NB-2130 (CMTR) to NCA-3862. Material Organization has Quality System Program (NB-2600, NCA-3800).	Pressure retaining material shall conform to a Section II spec. Material must be certified per NC-2130 (CMTR) to NCA-3862. Material Organization has Quality System Program (NC-2600, NCA-3800).	Pressure retaining material shall conform to a Section II spec. Material must be certified per ND-2130 (CMTR) to NCA-3862. Material Organization has Quality System Program (ND-2600, NCA-3800).	Materials in Appendix allowed at temps in range of tabulated allowables. Usually specified by ASTM A or B specs.
Stress Allowables	Multiples of S_m and S_y Section II, Part D, Subpart 1, Tables 2A and 2B. Stress equations separated into primary, secondary and peak components. Stressed are defined by Tresca Stress Intensities in NB 3200 Design by Analysis. Stresses are defined as directional component stresses in Design By Rule NB 3100 and NB 3300.	Multiples of S_h, S_m, and S_y Section II, Part D, Subpart 1, Tables 1A, 1B and 3. May be modified by casting quality factor (NC-3613.4). Stress equations separated into primary and secondary components. Stress in Design by Analysis same as Class 1. Design by Rule is the same as Class 1. Piping stresses NC 3600 are principal stresses.	Multiples of S_h, S_m, and S_y Section II, Part D, Subpart 1, Tables 1A, 1B and 3. May be modified by weld joint efficiency (ND-3613.4) factor and casting quality factor (ND-3613.5) Stress equations separated into primary and secondary components. Design by Rule only is the same as Class 1. Piping stresses ND 3600 are principal stresses.	Multiples of S_h Allowables from ASME Section II, Part D, App 1 (may be modified by weld joint efficiency factor and casting quality factor). Stress equations evaluate primary and secondary loadings separately. Stresses defined as principal stresses.
Basis for stress allowable **Per Part D App. 1, 1-100 App. 2, 2-110** Austenitic SS is also limited to 0.9 S_y (at temperature)	Minimum of: (a) ⅓ of specified min tensile strength at room temperature (b) ⅓ of tensile strength at temperature (c) ⅔ of specified min yield strength at room temperature (d) ⅔ of specified yield strength at temperature	Minimum of: (a) specified min tensile strength at room temperature divided by 3.5 (b) tensile strength at temperature divided by 3.5 (c) ⅔ of specified min yield strength at room temperature (d) ⅔ of specified yield strength at temperature	Minimum of: (a) specified min tensile strength at room temperature divided by 3.5 (b) tensile strength at temperature divided by 3.5 (a) ⅔ of specified min yield strength at room temperature (b) ⅔ of specified yield strength at temperature	Minimum of: (a) ⅓ of specified min tensile strength at room temperature (b) ⅓ of tensile specified strength at temperature (c) ⅔ of specified min yield strength at room temperature (d) ⅔ of yield strength at specified temperature

(continues)

TABLE 4.16. (Continued)

	Technical Requirements			
Attribute	**SC III – Class 1**	**SC III – Class 2**	**SC III – Class 3**	**B31.1**
Service Levels	Yes, A,B,C,D (NCA-2142.4) K × all stress Vessels, pumps and valves A: K = 1.0 B: K = 1.2 C: K = 1.8 D: K = 2.4 Piping C: K = 2.25 D: K = 3.0	Yes, A,B,C,D K × all stress Vessels, pumps and valves A: K = 1.0 B: K = 1.2 C: K = 1.8 D: K = 2.4 Piping C: K = 2.25 D: K = 3.0	Yes, A,B,C,D K × all stress Vessels, pumps and valves A: K = 1.0 B: K = 1.2 C: K = 1.8 D: K = 2.4 Piping C: K = 2.25 D: K = 3.0	Higher allowables for "occasional loads." K=1.2 Vintage code cases addressed allowables for design, upset, emergency and faulted conditions.
Pipe Supports	Section III Subsection NF	Section III Subsection NF	Section III Subsection NF	Para 120, MSS SP-58
PSI/ISI	PSI – NDE required. Exams performed by certified personnel. Personnel qualification records retained by employer. Qualification procedure in accordance with recommended guidelines of SNT-TC-1A. ISI required on 10-year schedule for 100% of dissimilar metal piping welds and 25% of similar metal piping welds. Services of an authorized inspection agency required.	PSI – NDE required. Exams performed by certified personnel. Personnel qualification records retained by employer. Qualification procedure in accordance with recommended guidelines of SNT-TC-1A. ISI required on 10-year schedule for 7½% of piping welds (but not less than 28 welds per material group). Services of an authorized inspection agency required.	PSI – NDE required. Exams performed by certified personnel. Personnel qualification records retained by employer. Qualification procedure in accordance with recommended guidelines of SNT-TC-1A. ISI – VT-2 required for all pressure-retaining components Services of an authorized inspection agency required.	PSI – NDE (Table 136.4) for temps exceeding 350°F with pressures exceeding 1,025 psig. VT for others. Exams performed in accordance with ASME Section V. NDE personnel certified to a program established by employer (136.1). ISI – No Code required ISI after installation. Initial service leak test at option of owner (137.7).
Pressure testing	Hydrostatic test – 1.25 × design pressure (NB-6220)	Hydrostatic test – 1.25 × design pressure (NC-6220)	Hydrostatic test – 1.25 × design pressure (ND-6220)	Hydrostatic test – 1.5 × design pressure (hoop stress not to exceed 90% of yield strength at temp) Initial service (leak) test
Dimensional Requirements	Table NB-3132-1 (ASME/ANSI "B," MSS SP)	Table NC-3132-1 (ASME/ANSI "B," MSS SP)	Table ND-3132-1 (ASME/ANSI "B," MSS SP)	ANSI "B" Standards, MSS SP (Table 126.1)

(continues)

TABLE 4.16. (Continued)

Administrative Requirements				
Attribute	**SC III – Class 1**	**SC III – Class 2**	**SC III – Class 3**	**B31.1**
Welding requirements	Welders and welding procedures qualified to ASME Section V.	Welders and welding procedures qualified to ASME Section V.	Welders and welding procedures qualified to ASME Section V.	Welders and welding procedures must be qualified (essentially meets QW-322 of ASME Section V).
Weld inspection	Welds are inspected to Section IX.	Welds are inspected to Section IX.	Welds are inspected to Section IX.	
Fabrication, erection and assembly requirements	Forming, fitting, aligning of welded joints (NB-4000). Treatments, tests, repairs and exams required on materials by various code articles certified with a CMTR (NCA-3860).	Forming, fitting, aligning of welded joints (NC-4000). Treatments, tests, repairs and exams required on materials by various code articles certified with a CMTR (NCA-3860).	Forming, fitting, aligning of welded joints (ND-4000). Treatments, tests, repairs and exams required on materials by various code articles certified with a CMTR (NCA-3860).	Basically welding-related requirements along with component bending and forming (127-135).
Quality control and quality assurance	Owner's Certificate of Authorization, Provision of Design Specifications, Design Reports (NCA-3200); N, NPT, NS & NA Certificate Holders (NCA-3500 – 3700); qualified Material Organizations (NCA-3800); QA Program (NCA-4000); agreement with an Authorized Inspection Agency (NCA-5000); Code Symbol stamping and Data Reports (NCA-8000)	See first column	See first column	See welding requirements. QA program, if any, is an owner-defined process that is not required to meet any formal code or standard.
Purchasing requirements Note: 3520 & 61 are for N Cert holder only; similar requirements are listed for NPT, etc.	Requires materials manufacturers, material suppliers and suppliers of contracted services to be qualified. NCA-3520, – 3561 are typical references.	Requires materials manufacturers, material suppliers and suppliers of contracted services to be qualified. NCA-3520, – 3561 are typical references.	Requires materials manufacturers, material suppliers and suppliers of contracted services to be qualified. NCA-3520, – 3561 are typical references.	No code requirements
Design specification requirements	Required for components and supports, P.E. stamped.	Required for components and supports, P.E. stamped.	Required for components and supports, P.E. stamped.	Not formally Required owner determines what this needs to be.
Stress report requirements	Design report required for class 1 components and supports, P.E. stamped, Owner certified to D-spec, (Recon of design dwg) changes	Required for components designed to service loadings > design loadings. NCA-3551	Required for components designed to service loadings > design loadings. NCA-3551	No formal Code requirement. Individual State jurisdictional requirements may apply

(continues)

TABLE 4.16. (Continued)

Administrative Requirements				
Attribute	**SC III – Class 1**	**SC III – Class 2**	**SC III – Class 3**	**B31.1**
Overpressure protection report	Required by NCA-3270 and NB-7200.	Required by NCA-3270 and NC-7200.	Required by NCA-3270 and ND-7200.	No report required. By B31.1 para 122.1.7 (D) follow requirements for safety and relief valves from ASME Section I (PG-67 thru PG-73)
Code stamping and data reports	In accordance with NCA-8000 and Section III, Appendix V.	In accordance with NCA-8000 and Section III, Appendix V.	In accordance with NCA-8000 and Section III, Appendix V.	Per 100.1.2, for BEP, data reports, inspection and stamping as required by ASME Section I.
Repair/ replacement	R/R in accordance with Section XI, IWA-4000, or code cases N-660,-662.	R/R in accordance with Section XI, IWA-4000, or code cases N-660,-662.	R/R in accordance with Section XI, IWA-4000, or code cases N-660,-662.	Retesting required after repair per 137.8
Code legal status	State law recognizes code. Federal law approves per 10CFR50.55a for commercial nuclear power.	State law recognizes code. Federal law approves per 10CFR50.55a for commercial nuclear power. Not required for DOE-regulated facilities that use Section VIII instead of DOE Order 420.	State law recognizes code. Federal law approves per 10CFR50.55a for commercial nuclear power. Not required for DOE regulated facilities that use Section VIII instead of DOE Order 420.	State law recognizes code. Federal law 10CFR830.120 and DOE order 420 apply

Note:
ASME B31.3 is generally the same as B31.1, except the occasional stress-allowable coefficient is increased from 1.2 to 1.33 and weld joint efficiencies may be different from *B31.1*.

REFERENCES

1. *ASME Boiler and Pressure Vessel Code, Section III, Subsection NCA,* 2001 "General Requirement for Division 1 and 2 Rules for Construction of Nuclear Facility Components, Subparagraph NCA-2142.4, Design Service and Test Limit," American Society of Mechanical Engineers, New York, NY.
2. *ANSI/ANS 2.26, Draft 0, Rev. 12,* 2003 "Categorization of Nuclear Facility Systems and Components for Service Design," American Nuclear Society,
3. *DOG G 420.1-1,* 2000, "Nonreactor Nuclear Safety Design Criteria and Explosive Safety Criteria Guide for Use with DOE O 420.1 Facility Safety," U.S. Department of Energy, Washington, D.C.
4. *ASME QME-1-2001,* 2001 "Qualification of Active Mechanical Equipment Used in Nuclear Power Plants," American Society of Mechanical Engineers, New York, NY.
5. *Regulatory Guide 1.26,* 1976 "Quality Group Classification and Standards for Water, Steam and Radioactive Waste Containing Components of Nuclear Power Plants," Rev. 3, U.S. Nuclear Regulatory Commission,
6. *ASME Boiler and Pressure Vessel Code, Section II,* 2001 "Materials," American Society of Mechanical Engineers, New York, NY.

7. AISC, 1989 "Specification for Structural Steel Design Buildings – Allowable Stress Design and Plastic Design," American Institute of Steel Construction,
8. *MSS-SP-58,* "Pipe Hanger and Supports, Material Design and Manufacture," Manufacturers Standardization Society of the Valve and Fitting Industry,
9. *ASME B&PVC Section III, Subsection NF,* 2001 "Supports," American Society of Mechanical Engineers, New York, NY.
10. *ASME Boiler and Pressure Vessel Code, Section III, Division 3, Subsections WA and WB,* 2001, "Containment Systems for Transportation and Storage Packaging of Spent Nuclear Fuel and High Level Radioactive Waste," American Society of Mechanical Engineers, New York, NY.
11. *ASME B&PVC Section VIII,* 2001 "Rules for Construction of Pressure Vessels," American Society of Mechanical Engineers, New York, NY.
12. *Regulatory Guide 1.143,* 2001 "Design Guidance for Radioactive Waste Management Systems, Structures and Components Installed in Light-Water-Cooled Nuclear Power Plants," Rev. 2, U.S. Nuclear Regulatory Commission,
13. *ASME OM-S/G-2000,* 2000, "Standards and Guides for Operation and Maintenance of Nuclear Power Plants – Part 3 Requirements for Preoperational and Initial Start-Up Vibration Testing of Nuclear Power Plant Piping Systems," American Society of Mechanical Engineers, New York, NY.

CHAPTER

5

EARTHQUAKE DESIGN AND ANALYSIS

5.1 INTRODUCTION

5.1.1 Purpose of Seismic Design

The primary purpose of the seismic design of high-hazard facilities is to construct them so that there is no excessive exposure as a result of radiation or toxic substance release to the public and collocated workers. A major earthquake in relatively close proximity to the site may cause major damage to conventionally designed industrial facilities. For this purpose, a more stringent seismic design is usually required than would be required for conventional facilities to minimize potential for loss of structural and leak-tight integrity. For conventional facilities, in some cases ensuring the operability and life safety of its occupants is usually the primary concern, while for high-hazard facilities the confinement of radiological or toxic substances is the primary concern.

5.1.2 Seismic Design Procedures

Seismic design of the various structures, systems and components (SSC) of a high-hazard facility include: (1) determination of the seismic design motion for the site; (2) confirmation of foundation stability of the support media during and following the earthquake; and (3) seismic design of underground and aboveground structures, distribution systems and equipment. It involves many technical fields, such as seismology, civil engineering and mechanical engineering. This book focuses on the mechanical engineering aspects of earthquake resistant design.

It is necessary to ensure that seismic-induced failure of SSC in high-hazard facilities does not cause a major accident resulting in release of significant radiological or toxic substances to cause serious injury or death to the public or collocated workers outside the facility. For this purpose, designs should be undertaken based on the following procedures: (1) determination of the earthquake motion, which may affect the site with a defined probability of occurrence, and thus taken into consideration as a design basis; (2) calculation of the seismic motion input to the safety-related SSC; (3) calculation of resultant seismic load effects (i.e., force, stress, strain, deformation, etc.); (4) combination of seismic with other concurrent load effects; and (5) comparison of total load effects with acceptance criteria to determine design adequacy.

It should be understood that more than one design-basis earthquake level might be defined at a given site as a function of the SSC's importance to safety and associated performance categories [1] or seismic design categories [2, 3].

5.2 EARTHQUAKE PHYSICAL PHENOMENA

5.2.1 Faults

Geologists have found that earthquakes tend to reoccur along faults, which reflect zones of weakness in the earth's crust. A fault is a fracture in the earth's crust along which two blocks of the crust have slipped with respect to each other. Faults are divided into three main groups, depending on how they move: Normal faults occur in response to pulling or tension; the overlying block moves down the dip of the fault plane. Thrust (reverse) faults occur in response to squeezing or compression; the overlying block moves up the dip of the fault plane. Strike-slip (lateral) faults occur in response to either type of stress; the blocks move horizontally past one another. Most faulting along spreading zones in the earth's crust is normal, along subduction zones is thrust and along laterally moving faults is strike-slip. Even if a fault zone has recently experienced an earthquake there is no guarantee that all the stress has been relieved. Another earthquake could still occur. Further, relieving stress along one part of the fault may increase stress in another part. Typical fault motions are shown in Figure 5.1.

5.2.2 Focus and Epicenter Ground Motion

The focal depth of an earthquake is the depth from the earth's surface to the region where an earthquake's energy originates (the focus). Earthquakes with focal depths from the surface to about 70 kilometers (43.5 miles) are classified as shallow. Earthquakes with focal depths from 70 to 300 kilometers (43.5 to 186 miles) are classified as intermediate. The focuses of deep earthquakes are concentrated in the earth's crust and upper

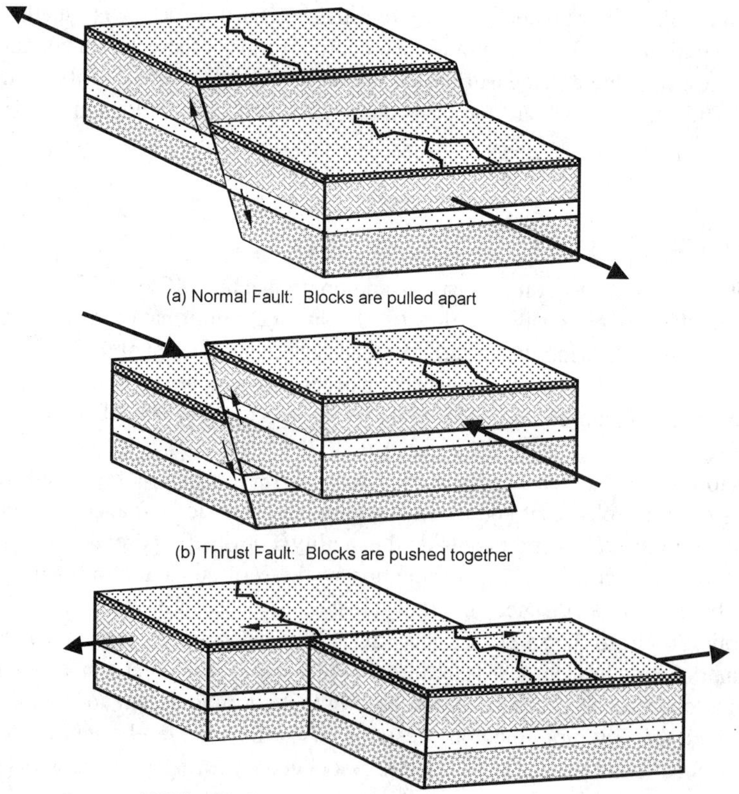

(a) Normal Fault: Blocks are pulled apart

(b) Thrust Fault: Blocks are pushed together

(c) Strike-Slip Fault: Blocks slide past each other

FIGURE 5.1. TYPICAL EARTHQUAKE FAULT MOTIONS.

mantle. The depth to the center of the earth's core is about 6,370 kilometers (3,960 miles), so even the deepest earthquakes originate in relatively shallow parts of the earth's interior.

The epicenter of an earthquake is the point on the earth's surface directly above the focus. The location of an earthquake is commonly described by the geographic position of its epicenter and by its focal depth.

5.2.3 Attenuation of Motion with Distance from the Epicenter

It is interesting to compare the attenuation curves shown in Figure 5.2, which are based largely on the data from shallow-focus earthquakes and records obtained in the western part of the United States, to the attenuation curve proposed for the eastern United States [4]. In the eastern United States, earthquakes have larger focal depths and the attenuation rate for ground motions is significantly less than in the relatively soft basement rock in the (v_S = 1,000 m/sec), where v_S is the shear-wave velocity in the western part of the country. The relative positions of the curves in Figure 5.2 are consistent, with the deeper focus events producing lower accelerations at close distances and leading to higher accelerations at large distance due to the reduced attenuation paths representative of the hard (v_S = 2,000 m/sec) basement rock of the eastern U.S.

5.2.4 Intensity

The oldest useful measure of the "strength" or damage potential of an earthquake is its intensity. Intensity is the measure of damage to works of man, to the ground surface and of human reaction to the shaking. Because earthquake intensity assessments do not depend on instruments, but on the actual observation of effects, intensities can be assigned even to historical earthquakes. In this way, the historical record becomes of utmost importance in modern estimates of seismological risk.

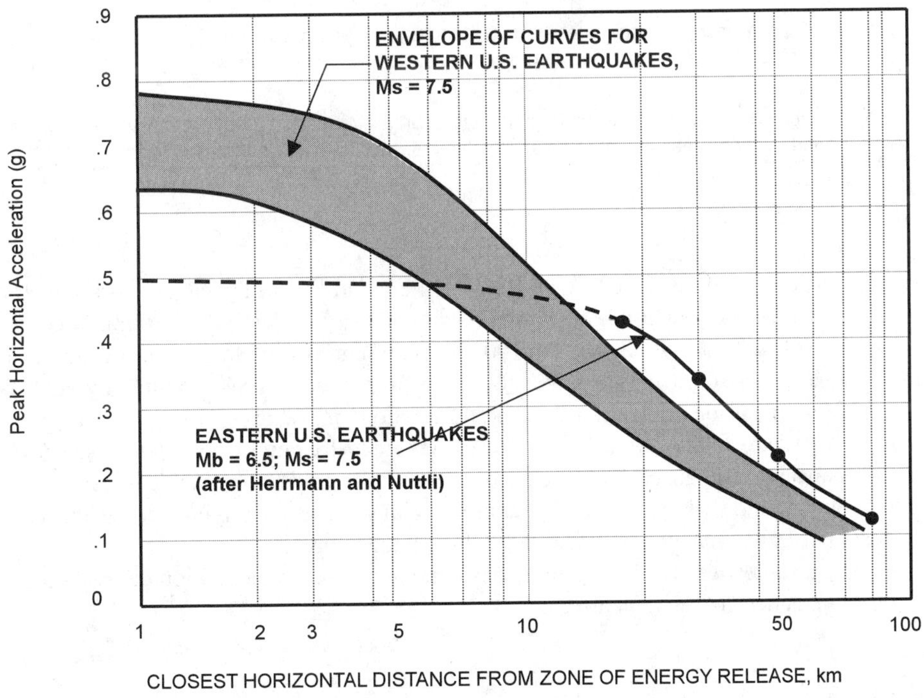

FIGURE 5.2. COMPARISON OF ATTENUATION CURVES FOR WESTERN AND EASTERN U. S. EARTHQUAKES

TABLE 5.1. MODIFIED MERCALLI INTENSITY SCALE (MMI) OF 1931

Sale No.	Description
I	Not felt except by a very few under especially favorable circumstances.
II	Felt only by a few persons at rest, especially on upper floors of buildings. Delicately suspended objects may swing.
III	Felt quite noticeably indoors, especially on upper floors of buildings, but many people do not recognize it as an earthquake. Standing motor cars may rock slightly. Vibration like passing of truck. Duration estimated.
IV	During the day felt indoors by many, outdoors by few. At night some awakened. Dishes, windows, doors disturbed; walls make cracking sound. Sensation like heavy truck striking building. Standing motor cars rocked noticeably.
V	Felt by nearly everyone, many awakened. Some dishes, windows, etc., broken; a few instances of cracked plaster; unstable objects overturned. Disturbances of trees, poles, and other tall objects sometimes noticed. Pendulum clocks may stop.
VI	Felt by all, many frightened and run outdoors. Some heavy furniture moved; a few instances of fallen plaster or damaged chimneys. Damage slight.
VII	Everybody runs outdoors. Damage negligible in buildings of good design and construction; slight to moderate in well-built ordinary structures; considerable in poorly built or badly designed structures; some chimneys broken. Noticed by persons driving motor cars.
VIII	Damage slight in specially designed structures; considerable in ordinary substantial buildings, with partial collapse; great in poorly built structures. Panel walls thrown out of frames structures. Fall of chimneys, factory stacks, columns, monuments, walls. Heavy furniture overturned. Sand and mud ejected in small amounts. Changes in well water. Persons driving motor cars disturbed.
IX	Damage considerable in specially designed structures; well-designed frame structures thrown out of plumb; great in substantial buildings, with partial collapse. Buildings shifted off foundations. Ground cracked conspicuously. Underground pipes broken.
X	Some well-built wooden structures destroyed; most masonry and frame structures destroyed with foundations; ground badly cracked. Rails bent. Landslides considerable from river banks and steep slopes. Shifted sand and mud. Water splashed (slopped) over banks.
XI	Few, if any (masonry), structures remain standing. Bridges destroyed. Broad fissures in ground. Underground pipelines completely out of service. Earth slumps and land slips in soft ground. Rails bent greatly.
XII	Damage total. Practically all works of construction are damaged greatly or destroyed. Waves seen on ground surface. Lines of sight and level are distorted. Objects are thrown into the air.

The first intensity scale was developed by de Rossi of Italy and Forel of Switzerland in the 1880s. This scale, with values from I to X, was used for reports of the intensity of the 1906 San Francisco earthquake, for example. A more refined scale was devised in 1902 by the Italian volcanologist and seismologist Mercalli with a twelve-degree range from I to XII. A version is given in Table 5.1, as modified by H.O. Wood and Frank Neumann to fit conditions in California. The descriptions in Table 5.1 allow the damage to places affected by an earthquake to be rated numerically. These spot intensity ratings can often be separated by lines that form an isoseismal intensity map. Such intensity maps provide crude but valuable information on (1) the distribution of strong ground shaking; (2) the effect of surficial soil and underlying geological strata; and (3) the extent of the source, which may be a line rather than a point source.

Intensity scales are subjective and depend to a considerable degree on population density upon construction conditions and practices of the area affected.

5.2.5 Magnitude

If earthquake sizes are to be compared, a measure is needed that does not depend, as does intensity, on the density of population and type of construction. A strictly quantitative scale that can be applied to earthquakes in both inhabited and uninhabited regions was generated in 1931 by Wadati in Japan and developed by Charles Richter in 1935 in California.

Richter defined the magnitude of a local earthquake as the logarithm to base ten of the maximum seismic wave amplitude in microns (10^{-4} cm) recorded on a Wood-Anderson seismograph located at a distance of 100 km from the earthquake epicenter. This means that every time the magnitude goes up by one unit, the amplitude of the earthquake waves increases 10 times. Since the fundamental period of the Wood-Anderson seismograph is I 0.8 sec, it selectively amplifies those seismic waves with a period ranging from 0.1 to 2 seconds. Because the natural period of many building structures is within this range, the local Richter magnitude is of special value to engineers.

It follows from the definition of the magnitude that it has no theoretical upper or lower limits. However, the size of an earthquake is limited at the upper end by the strength of the rocks of the earth's crust. Since 1935, only a few earthquakes have been recorded on seismographs that have had a magnitude greater than 8.0. At the other extreme, highly sensitive seismographs can record earthquakes with a magnitude lower than minus 2.

Generally speaking, shallow earthquakes have to attain Richter magnitudes of more than 5.5 before significant damage to industrial facilities occurs near the epicenter or intensity MMVI at the site.

At its inception, the idea behind the Richter local magnitude scale (M_L) was a modest one. It was defined for Southern California, shallow earthquakes and epicentral distances less than about 600 km. Today, the method has been extended to apply to a number of types of seismographs throughout the world. Consequently, a variety of magnitude scales, based on different formulas for epicentral distance and ways of choosing an appropriate wave amplitude, have emerged:

Surface-wave magnitude (M_s): Surface waves with a period around 20 sec are often dominant on the seismograph records of distant earthquakes (epicentral distances of more than 2,000 km). To quantify these earthquakes, Gutenberg defined a magnitude scale (M_s), which is based on measuring the amplitude of surface waves with a period of 20 sec.

Body-wave magnitude (m_b): Deep-focus earthquakes have only small or insignificant surface waves. Hence, it has become routine in seismology to measure the amplitude of the *P* wave (particles moving in the direction of wave propogations), which is not affected by the focal depth of the sources, and thereby determine a *P*-wave magnitude (m_b). This magnitude type has also been found useful in continental regions

The moment-magnitude scale (M_w) is the only magnitude scale that does not suffer from a saturation problem for great earthquakes. The reason is that it is directly based on the forces that work at the fault rupture to produce the earthquake and not on the recorded amplitude of specific types of seismic waves. Hence, as might be expected, when moment magnitudes were assigned to the San Francisco earthquake of 1906 and the Chilean earthquake of 1960, the magnitude of the San Francisco earthquake was about 7.9, whereas the magnitude of the Chilean earthquake rose to 9.5. M_s and M_w for some great earthquakes are compared in Table 5.2.

In light of the above discussion, application of different scales have been suggested for measuring shallow earthquakes of various magnitudes [5]:

M_D or M_L	for magnitudes less than 3
M_L or m_b	for magnitudes between 3 and 7
M_s	for magnitudes between 5 and 7.5
M_w	for magnitudes larger than 7.5

5.2.6 Seismic Hazard Curve

Seismic hazard curves for a particular design basis earthquake are typically shown as the probability of exceedence versus peak or spectral acceleration at a given frequency or peak ground acceleration (PGA) as shown in Figure 5.3. It may be the median (50th percentile) value for the distribution of earthquake parameters at a particular probability of exceedence of an assumed log-normal distribution (probability density function) of earthquake acceleration. Other parameters of interest may be the mean, 75, 90, etc., percentile levels. For typical earthquake motion the mean value of the log-normal distribution is approximately the 84th percentile.

TABLE 5.2. MAGNITUDES OF SOME OF THE GREAT EARTHQUAKES

Date	Region	M_s	M_w
July 9, 1905	Mongolia	8.25	8.4
January 31, 1906	Ecuador	8.6	8.8
April 18, 1906	San Francisco	8.25	7.9
January 3, 1911	Turkestan	8.4	7.7
December 16, 1920	Kansu, China	8.5	7.8
September 1, 1923	Kanto, Japan	8.2	7.9
March 2, 1933	Sanrika	8.5	8.4
May 24, 1940	Peru	8.0	8.2
April 6, 1943	Chile	7.9	8.2
August 15, 1950	Assam	8.6	8.6
November 4, 1952	Kamchatka	8	9.0
March 9, 1957	Aleutian Islands	8	9.1
November 6, 1958	Kurile Islands	8.7	8.3
May 22, 1960	Chile	8.3	9.5
March 28, 1964	Alaska	8.4	9.2
October 17, 1966	Peru	7.5	8.1
August 11, 1969	Kurile Islands	7.8	8.2
October 3, 1974	Peru	7.6	8.1
July 27, 1976	China	8.0	7.5
August 16, 1976	Mindanao	8.2	8.1
March 3, 1985	Chile	7.8	7.5
September 19, 1985	Mexico	8.1	8.0
December 26, 2004	Indian Ocean	8.2	9.0

It should be noted that for PGAs tend to increase at a much lower rate once a particular acceleration level is reached. This is the result of underlying rock reaching its fracture capacity and large magnitude earthquakes with fault rupture lines extending 10s to 100s of miles or dimensions that equal or exceed iso-intensity dimensions.

The seismic hazard curves are typically assumed close to linear when plotted on a log-log scale. Thus over at least any tenfold difference in exceedence, frequencies with such hazard curves may be approximated by:

$$K_{(a)} = K_I a^{-K_H} \tag{5.1}$$

where $K_{(a)}$ = annual frequency of exceedence of ground motion level "a"; K_I = appropriate constant to fit the curve; and K_H = a slope parameter defined by:

$$K_H = \frac{1}{\log(A_R)} \tag{5.2}$$

in which A_R = ratio of ground motions corresponding to a tenfold reduction in exceedence frequency; A_R typically ranges between 1.5 and 2.0. It should be noted that the curves shown in Figure 5.3 are for a particular earthquake. For all earthquakes the curves at very low probability are much flater.

5.3 SEISMIC LOAD DEFINITION

Seismic load determination for mechanical SSC can be divided into four methods or procedures. The first and simplest method is termed the static method, applicable to PC-1 and PC-2 SSC as discussed in Chapter 1 ("Functional and Performance Classification of Safety-Related SSC"), and is generally defined in the International Building Code [6] or *SEI/ASCE/7-02* [7]. The next simplest of methods is generally applicable to performance categories 3, 4 and 5 SSC is the equivalent static method. The third is the load developed

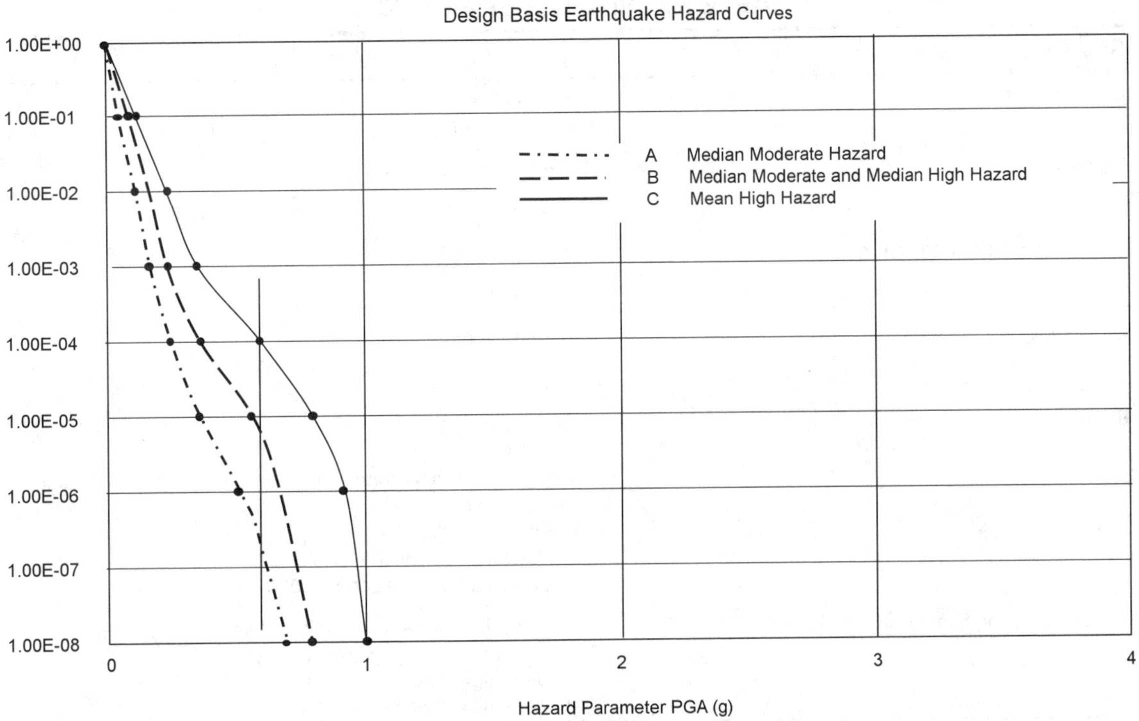

FIGURE 5.3. TYPICAL DESIGN BASIS SEISMIC HAZARD CURVES (PEAK GROUND ACCELERATION) : A IS FOR MEDIAN LOW HAZARD: B IS FOR MEDIAN MODERATE HAZARD AND MEAN LOW HAZARD: AND C IS FOR MEAN HIGH HAZARD SITES

from application of the response spectrum modal analysis method. The fourth, and usually considered the most accurate, hence, the least conservative, the loads determined from the time-history modal analysis or the time-step integration method.

The static and equivalent static analysis methods permit loads to be determined without recourse to computer analysis applied to SSC, except statically. The other two methods require computer analysis and consideration of the dynamic nature of the loading and the structure, distribution system or component and its anchorage or support structural response.

A subset of the response spectrum or time-history methods is where a secondary system component or distribution system finite element model is combined together with the building or supporting structure primary model. This has the effect of decreasing significantly the seismic response of the secondary system and increasing slightly the response of the primary system. This occurs for even relatively small secondary to primary system mass ratios [21]. For example, a mass ratio of one percent would be equal to an additional three percent damping applied to the secondary system.

5.3.1 Static

Static loads form the basis of seismic loads on SSC of performance categories 1 and 2. The seismic loads developed for these categories of SSC are given in Section 9 of reference [7] and Sections 1.6.2.1 and 1.6.2.2 of reference [6]. The Standard in [7] is a voluntary consensus standard developed by the American Society of Civil Engineers. The [6] Standard is a product of the International Code Council, which represents most building code officials in the U.S. Thus it is not officially a consensus standard and is not recognized as such by the American National Standards Institute. For mechanical SSC, the latest edition of the *IBC Code* (2002)

TABLE 5.3. SITE CLASSIFICATION

Site Class	$\bar{v}_s$
A: Hard Rock	>5,000 ft/s (>1,500 m/s)
B: Rock	2,500 to 5,000 ft/s (760 to 1,500 m/s)
C: Very dense soil and soft rock	1,200 to 2,500 ft/s (370 to 760 m/s)
D: Stiff soil	600 to 1200 ft/s (180 to 370 m/s)
E: Soil	<600 ft/s (<180 m/s) Any profile with more than 10 ft of soil having the following characteristics: Plasticity index PI>20 Moisture content w > 40% Undrained shear strength <500 psf
F: Soils requiring site-specific evaluation	1. Soils vulnerable to potential failure or collapse 2. Peats and/or highly organic clays 3. Very high plasticity clays 4. Very thick soft/medium clays

[6] references the procedure of *ASCE-7* [7]. It should also be noted that for other natural phenomena loads such as wind and snow loads, [6] also makes specific reference to the detailed analysis and design requirements of [7].

The following seismic forces and moment design requirements given below are based on the requirements given in [7], section 9.6.1.3.

For other geographical areas 0.2-second period for 5.0% damping free-field ground spectra are developed assuming a site class B foundation condition. Site class B is defined by shear-wave velocity as shown in Table 5.3 from Table 9.4.1.2 of [7].

Seismic Forces

Seismic forces, F_p, applicable to mechanical systems and components, determined in accordance with the [7], are in accordance with:

$$F_p = \frac{0.4a_p S_{DS} W_p}{\left(\frac{R_p}{I_p}\right)}\left(1+2\frac{z}{h}\right) \tag{5.3}$$

and F_p is not required to be taken as greater than:

$$F_p = 1.6 S_{DS} I_p W_p \tag{5.4}$$

And F_p shall not be taken as less than:

$$F_p = 0.3 S_{DS} I_p W_p \tag{5.5}$$

TABLE 5.4. MECHANICAL AND ELECTRICAL COMPONENTS COEFFICIEN

Mechanical and Electrical Component or Element	a_p	R_p
1. General mechanical		
a. Boilers and furnaces	1.0	2.5
b. Pressure vessels on skirts and free-standing	2.5	2.5
c. Stacks	2.5	2.5
d. Cantilevered chimneys	2.5	2.5
e. Other	1.0	2.5
2. Manufacturing and process machinery		
a. General	1.0	2.5
b. Conveyors (nonpersonnel)	2.5	2.5
3. Piping systems		
a. High deformability elements and attachments	1.0	3.5
b. Limited deformability elements and attachments	1.0	2.5
c. Low deformability elements or attachments	1.0	1.5
5. HVAC system equipment		
a. Vibration isolated	2.5	2.5
b. Nonvibration isolated	1.0	2.5
c. Mounted in-line with ductwork	1.0	2.5
d. Other	1.0	2.5
5. Elevator components	1.0	2.5
6. Escalator components	1.0	2.5
7. Trussed towers (freestanding or guyed)	2.5	2.5
8. General electrical		
a. Distributed systems (bus ducts, conduit, cable tray)	2.5	5.0
b. Equipment	1.0	2.5
9. Lighting fixtures	1.0	1.5

where a_p = component amplification factor that varies from 1.00 to 2.50 (select appropriate value from Table 5.4 (Table 9.6.3.2 of [7]); F_p = seismic design force applied at the component's center of gravity or distributed relative to component's mass distribution; I_p = component importance factor that is either 1.00 for PC-1or 1.50 for PC-2 mechanical systems or components; h = average roof height of a structure relative to the base elevation; R_p = component response modification factor that varies from 1.5 to 5.0 (select appropriate value from Table 5.4 (Table 9.6.3.2 of [7]); S_{DS} = design spectral response acceleration at short period (0.2 seconds or less), as determined in Section 9.4.1.2.5 of [7]; W_p = component operating weight; z = height in structure at point of highest attachment of component or system. For items at or below the base, z shall be taken as 0. For items at or above the roof, z is not required to be taken as greater than the roof height.

where:

$$S_{DS} = 2/3\ S_{MS} \tag{5.6}$$

and

$$S_{MS} = F_a S_S \tag{5.7}$$

where F_a is defined in Table 5.5, from Table 9.4.1.2.4(a) of [7]; and S_S = the mapped maximum considered earthquake spectral response acceleration at 0.2 second period as determined from Figure 9.1.1.1(a) of [7].

The force, F_p, shall be applied independently in the horizontal two principal directions and resultant forces and moments or stresses, with plus or minus sensing added to other loads resultants (dead weight, live, pressure load stress) associated with the system or component. The force F_p shall be applied at the center of gravity of the component or mass distribution system between supports, as shown in Figure 5.4(b).

TABLE 5.5. VALUES OF SITE COEFFICIENT F_a AS A FUNCTION OF SITE CLASS AND MAPPED SPECTRAL RESPONSE ACCELERATION AT SHORT PERIODS (S_S)[a]

Site Class	Mapped Spectral Response Acceleration at Short Periods				
	$S_S \leq 0.25$	$S_S = 0.50$	$S_S = 0.75$	$S_S = 1.00$	$S_S \geq 1.25$
A	0.8	0.8	0.8	0.8	0.8
B	1.0	1.0	1.0	1.0	1.0
C	1.2	1.2	1.1	1.0	1.0
D	1.6	1.4	1.2	1.1	1.0
E	2.5	1.7	1.2	0.9	b
F	b	b	b	b	b

a. Use straight line interpolation for intermediate values of mapped spectral acceleration at short period, S_S.
b. Site-specific geotechnical investigation and dynamic site response analyses shall be performed to determine appropriate values.

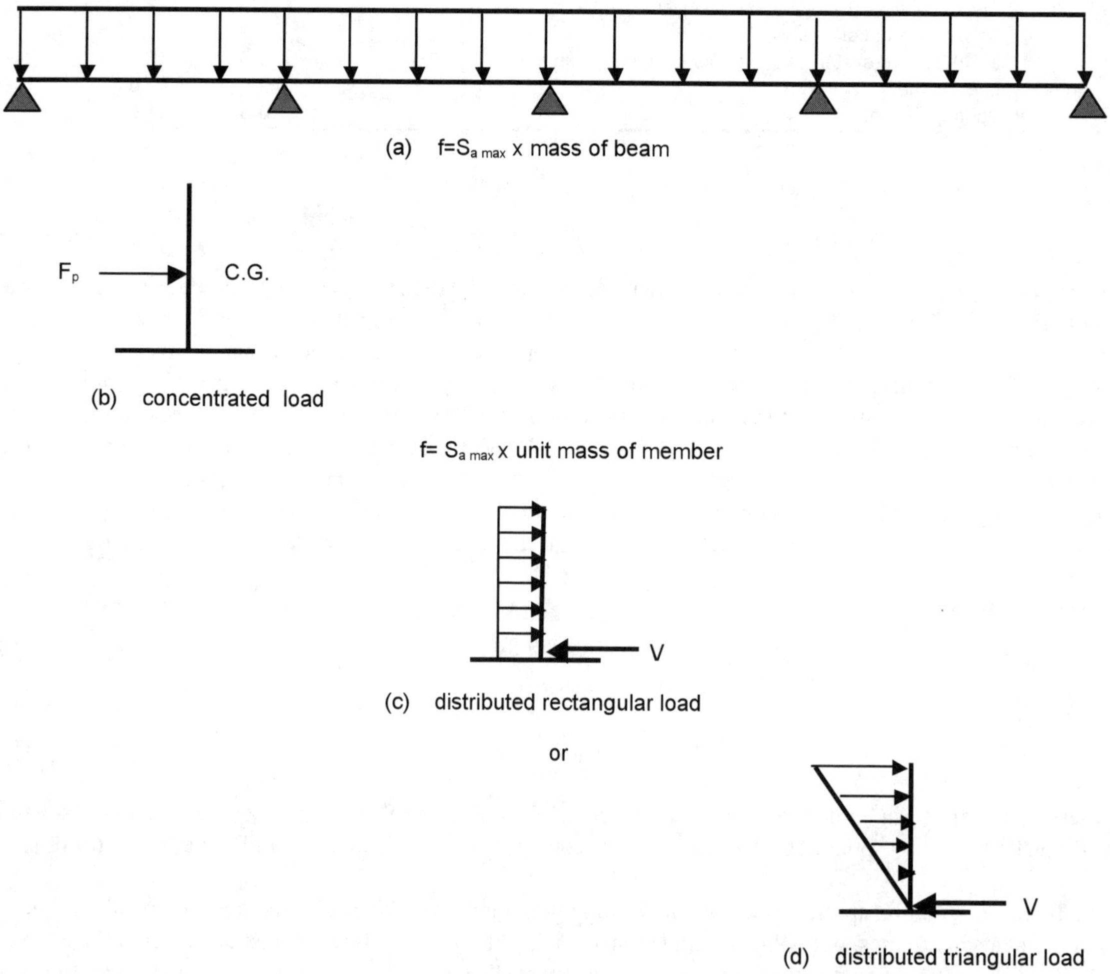

FIGURE 5.4. A TYPICAL LOADING DIAGRAM FOR AN EQUIVALENT STATIC LOAD

Seismic Relative Static Displacements

Seismic relative displacements, D_p, which give rise to secondary stressor and are applicable when the system or component is attached at two or more different elevations h_x and h_y on the same structure A, shall be determined in accordance with the following formulas.

$$D_p = \delta_{xA} - \delta_{YA} \tag{5.8}$$

D_p is not required to be taken greater than:

$$D_p = (h_x - h_y)\frac{\Delta_{aA}}{h_{sx}} \tag{5.9}$$

For two connection points on separate structures, A and B, or separate structural systems, one at level h_x and the other at level h_y, D_p shall be determined as:

$$D_p = |\delta_{xA}| + |\delta_{yB}| \tag{5.10}$$

D_p is not required to be taken as greater than:

$$D_p = \frac{h_x \Delta_{aA}}{h_{sx}} + \frac{h_y \Delta_{aB}}{h_{sx}} \tag{5.11}$$

where D_p = relative seismic displacement that the component must be designed to accommodate; h_{sx} = story height used in the definition of the allowable drift, δ_a, in Table 5.6 (Table 9.5.2.8 of [7]); δ_{xA} = deflection at structure level h_x of structure A, determined by an elastic analysis as defined in Section 9.5.5.7.1 of [7]; δ_{yA} = deflection at structure level h_y of structure A, determined by an elastic analysis as defined in Sections 9.5.5.2.1 of [7]; δ_{yB} = deflection at structure level h_y of structure B, determined by an elastic analysis as defined in Sections 9.5.5.2.1 of [7]; h_x = height of upper support attachment at level x as measured from the base; h_y = height of lower support attachment at level y as measured from the base; Δ_{aA} = allowable story drift for structure A as defined in Table 5.6 (Table 9.5.2.8 of [7]); and Δ_{aB} = allowable story drift for structure B as defined in Table 5.6 (Table 9.5.2.8 of [7].

The displacement criteria addressed in this section apply only to differential support points represented on stick models of the supporting structure. However, the same principles hold on more complex building models that have multiple support points at the same elevation.

The effects of seismic relative displacements shall be considered in combination with relative displacements caused by other loads, such as thermal loads.

It should be understood that stresses caused by relative static displacements induced by earthquake inertia effects are typically termed seismic anchor motion stresses. They are similar to temperature-induced stresses and may be treated as secondary in nature in the affected component or distribution system. Secondary stresses in mechanical components generally have different acceptance criteria from primary stresses induced by dead weight, live loads, pressure and seismic inertia loads in applicable mechanical design standards as discussed in Chapter 4.

Component and Systems Supports and Anchorage

Systems and components in conformity with reference [7] are anchored in accordance with the following:

a. The force in the connected part shall be determined based on the prescribed forces for the component specified in Section 9.6.1.3 of reference [7]. Where component anchorage is provided by shallow expansion anchors, or shallow (low deformability) cast-in-place anchors, a value of R_p=1.50 shall be used as specified in Section 9.6.1.3 of [7] to determine the forces on the connected part.

TABLE 5.6. ALLOWABLE STORY DRIFT, Δ_a[a]

Structure	Seismic Use Group I	II	III
Structures, other than masonry shear wall or masonry wall fame structures, four stories or less with interior walls, partitions, ceilings and exterior wall systems that have been designed to accommodate the story drifts.	$0.025h_{sx}$[b]	$0.020h_{sx}$	$0.015h_{sx}$
Masonry cantilever shear wall structuresc[c]	$0.010h_{sx}$	$0.010h_{sx}$	$0.010h_{sx}$
Other masonry shear wall structures	$0.007h_{sx}$	$0.007h_{sx}$	$0.007h_{sx}$
Masonry wall frame structures	$0.013h_{sx}$	$0.013h_{sx}$	$0.010h_{sx}$
All other structures	$0.020h_{sx}$	$0.015h_{sx}$	$0.010h_{sx}$

a. h_{sx} is the story height below level x.
b. There shall be no drift limit for single-story structures with interior walls, partitions, ceilings and exterior wall systems that have been designed to accommodate the story drifts.
c. Structures in which the basic structural system consists of masonry shear walls designed as vertical elements cantilevered from their base or foundation support, which are so constructed that moment transfer between shear walls (coupling) is negligible.

b. Anchors or supports embedded in concrete or masonry shall be proportioned to carry the lesser of the following:
   1. The design strength of the connected part.
   2. 1.3 times the force in the connected part as given by F_p times R_p.
   3. The maximum force that can be transferred to the connected part by the component structural system.
c. Determination of forces in anchors or supports shall include the expected conditions of installation, including eccentricities, prying effects and torsion.
d. Determination of force distribution of groups of anchors or supports at one location shall include the stiffness of the connected system and its ability to redistribute loads to other anchors or supports in the group beyond yield.
e. Powder-driven fasteners shall not be used for tension load applications in seismic design category D, E or F, as determined in Section 9.4.2.1 of [7], unless approved for such loading by testing.
f. The design strength of anchors in concrete shall be determined in accordance with the provisions of Section 9.6.1.6.6 of [7] or as otherwise specified by ACI Standards.
g. For additional requirements for anchors to steel, see Chapter 22 of reference [6].
h. For additional requirements for anchors or supports in masonry, see Chapter 21 of reference [6].

5.3.2 Equivalent Static

The equivalent static load is usually developed by taking the peak spectral response (usually acceleration) and applying it to the mass distribution of the mechanical SSC. This acceleration times the mass results in a force applied to each distributed or concentrated mass point. Application of these forces to each mass point in the model will result in equivalent static seismic stresses in the members connecting the mass points. It should be noted that the resultant seismic stresses are given a (+) and (–) sensing when combined with other stresses such as dead load, live load or pressure stresses, whichever results in maximum stress and compared with maximum code allowable stresses.

There are two basic structural systems typically associated with mathematical models of SSC associated with mechanical equipment. They are:

- continuous beam models
- cantilever models

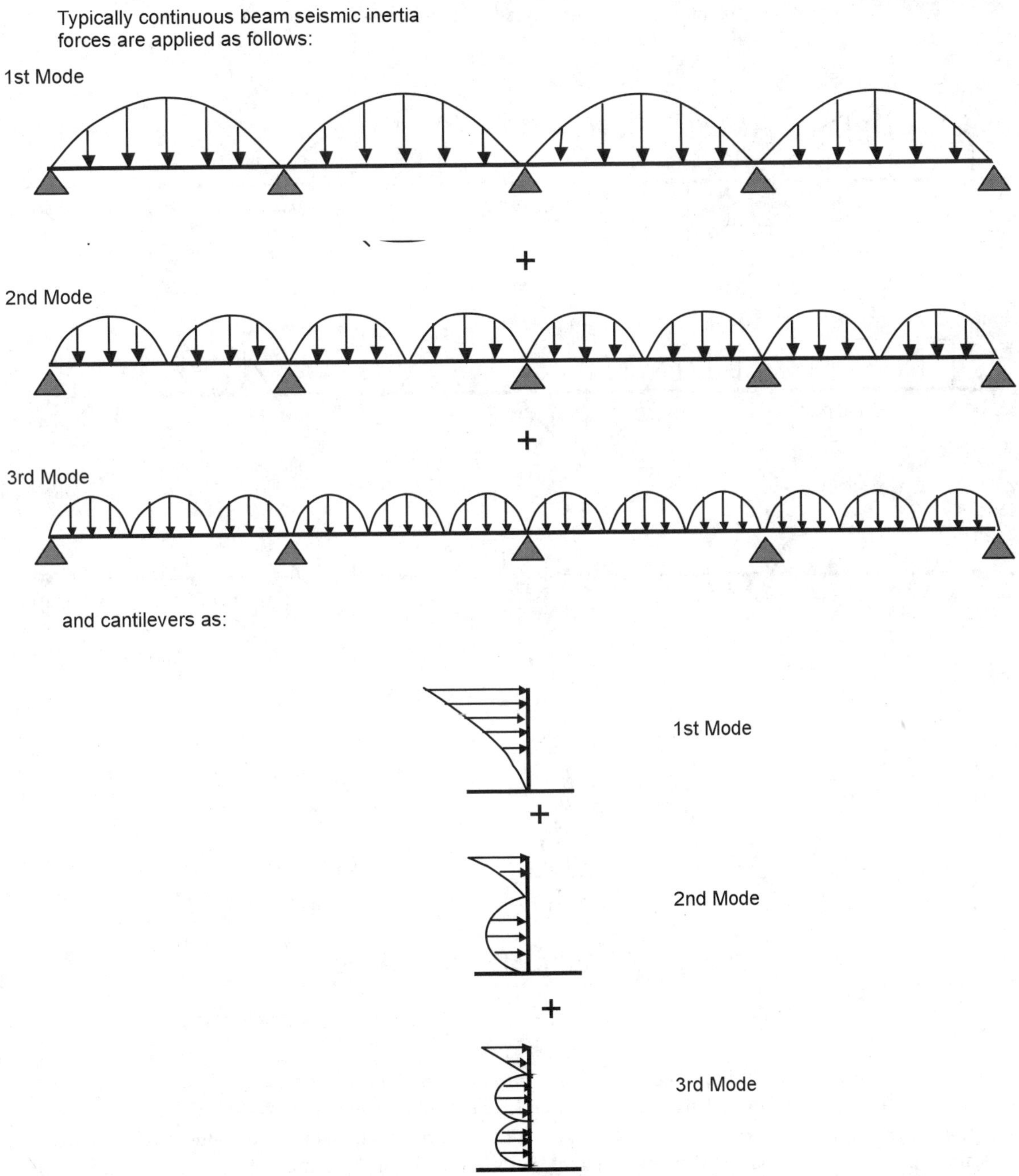

FIGURE 5.5. MODAL FORCE COMBINATIONS USING SQUARE ROOT SUM OF SQUARE

It should be understood that the combination of seismic inertia forces based on a square root sum of squares mode combination of applied external modal forces as shown in Figure 5.5 distorts the mode shapes as shown in Figure 5.6. As a result, a load factor of 1.5 is often applied to accelerations to compensate for this distortion. Lower load factors may be used if the geometry in question justifies such reductions [8, 9].

The cantilever mode is characteristic of vertical components supported only at the base.

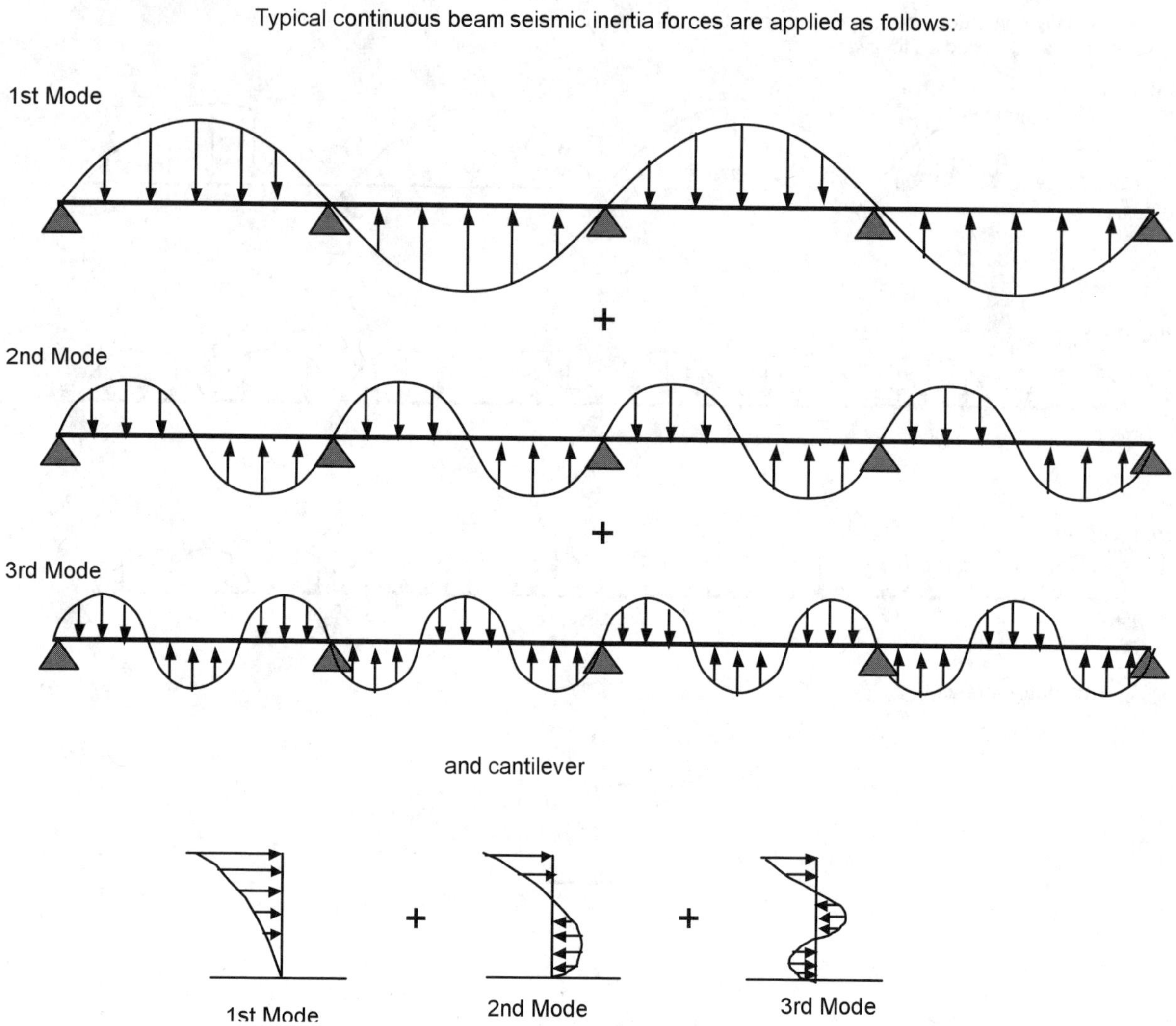

FIGURE 5.6. TYPICAL MODE SHAPES

The continuous beam model in its more general form is for distribution systems (pipe, duct and tubing) continuous over several supports. This type of seismic load definition and analysis is often applied to small bore piping systems typically equal to or less than 6.0 inches (150 mm) in diameter and ductwork 12 inches (300 mm) in diameter or smaller. A more stringent definition of small bore pipe of 2.5 inches (6.25 mm) nominal diameter is also often used. The 6.0 inch (150 mm) nominal diameter is used because piping above this diameter can affect the design adequacy of the building structural steel that supports the pipe.

Cold tubing and small bore pipe less than 3 inches (75 mm) in nominal diameter (design temperature less than 150°F [65°C]) are usually designed by use of spacing tables and charts, which employ an equivalent static loading procedure to define necessary lateral support spacing limitations as discussed in Section 5.7.

In recognition that the dominant first response mode for a cantilever is parabolic in shape, as shown in Figure 5.6, the applied equivalent static load can be approximated as triangular in shape rather than as a rectangle or a concentrated load at the center of gravity, as a better approximation of a cantilever seismic base moment loading. In addition, a small concentrated load may also applied to the Figure 5.6 triangular load to better simulate the dominant first mode parabolic loading.

5.3.3 Dynamic-Response Spectrum Modal Analysis

Dynamic response spectrum modal analysis loading assumes the system being evaluated is responding elastically, otherwise natural frequencies cannot be defined. It provides a more exact definition of the applied load than the static or equivalent static procedure, has either plus or minus directional sensing of seismic stresses, and comprises the combined mode shapes to define load.

In order for the mode shape seismic loading not to be distorted, it would be necessary to apply the seismic forces in each mode to compute seismic stresses for each mode and then combine these stresses on a square root sum of squares basis to get resultant seismic stresses with a plus or minus sensing.

$$a_{ij} = \Gamma_i S_{ai} \phi_{ij} \tag{5.12}$$

where Γ_j = modal participation factor in j mode as described in Chapter 3; a_{ij} = the acceleration of the i mass in the j mode of response; S_{aj} = spectral acceleration of the j mode frequency; ϕ_{ij} = the mode shape of the i mass in the j mode; and ϕ'_{ij} = the mode shape of the i mass in the j mode in the direction of the earthquake.

It should be noted that both or either Γ_j or ϕ_{ij} can be negative, which determines the direction or sensing of the modal force for each mode.

The seismic inertia force f_{ri} applied to the i mass in the j mode is determined:

$$\mathrm{f_{ij}} = \mathrm{m_i}\, a_{\mathrm{ij}} \tag{5.13}$$

Note that $\mathrm{f_{ri}}$ may be positive or negative. To obtain the correct seismic inertia stress or internal forces or moment resultant, these modal inertia forces should be applied to each mass mode in each mode to the analytical modal of the structure. The direction of externally applied forces is determined by the sign of the acceleration, which in turn is determined by the sign of the modal participation factors Γ modes shape ϕ as shown in Eq. (5.14). These resultant modal stresses or internal forces and moments are based on the assumed independence of modes, and are then combined to get the total seismic inertia stresses, internal forces or moments. The combination of independent mode shapes is usually on the basis of square root sum of squares as shown in Eq. 5.14:

$$S_n = \left(\sum_{i=1}^{n} (s^2{}_{ri}) \right)^{1/2} \tag{5.14}$$

where S_{ri} = stress or internal force or moment from the applied modal external force f_{rj}; and S_n = total stress or internal force or moment resulting from the combined modal stresses, internal forces or moments, or alternatively and conservatively the absolute sum of S_n. The value of S_n is assumed either positive or negative when combined with other stress or internal forces or moments to obtain maximum values for a particular load combination.

This formulation assumes an earthquake applied in only one direction. When earthquakes are applied in three orthogonal directions to a multi-mass system, the formulation is somewhat more complicated for the system shown in Figure 5.7, since results from the different directions of earthquakes must now be combined.

The three directional seismic input to a two-mass system can be expressed[1]:

$$\Gamma_{jk} = \sum_{jk=1}^{3} \sum_{i=1}^{2} M_i \phi_{ijk} \Big/ \sum_{k=1}^{3} \sum_{i=1}^{2} M_i \phi_{ij}^2 \tag{5.15}$$

[1] In general, each mass has six degree of freedom, three associated with displacement and three with rotation. In most analysis the rotation degrees of freedom are neglected.

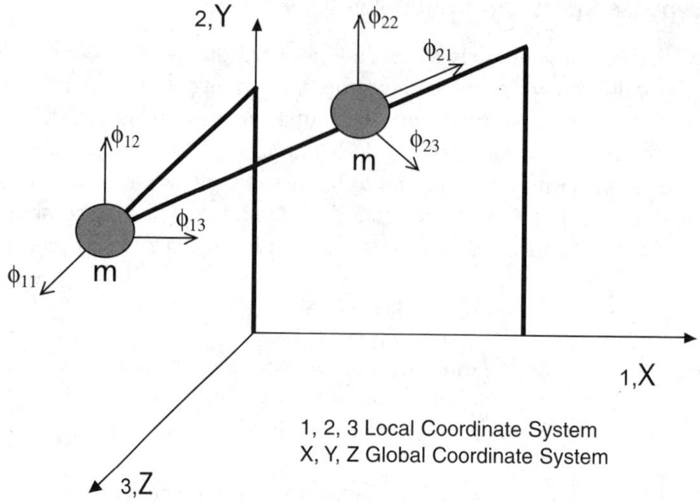

FIGURE 5.7. TWO-DEGREE-OF-FREEDOM COUPLED SYSTEM SUBJECTED TO THREE DIRECTIONS OF EARTHQUAKE

It should be noted that there are as many modes j as there are degrees of freedom (DOF).

There are two masses with three translational DOF in three orthogonal directions for a total of six degrees of freedom. Therefore, there are six j modes, six modal participation factors and each modal vector contains six modal relative displacements elements.

where i = mass 1 to 2; j = mode 1 to 6; and k = direction of applied earthquakes = 1 to 3

There are six mode shapes with a total of 36 modal vector elements.

Given modal frequencies and eigen vector mode shapes as defined in Table 5.7[2].

It should be noted that both or either Γ_{ij} or ϕ_{rj} can be negative, which determines the direction of modal forces for each mode. Γ_j and ϕ_{ij} change their sign simultaneously because the participation factor Γ_i varies in inverse proportion to the modal vector $\{\phi_i\}$.

TABLE 5.7. ASSUMED EIGEN VECTOR QUANTITIES

		Eigen Vectors					
		m_1			m_2		
Mode	**Eigen Values Hz**	**1**	**2**	**3**	**1**	**2**	**3**
1	2	1.0	0.1	−0.1	0.0	0.2	0.1
2	4	0.3	0.6	−0.3	0.2	0.4	−0.2
3	7	−0.2	0.1	0.7	−0.3	0.1	0.0
4	10	0.3	−0.2	−0.1	0.8	−0.3	−0.2
5	15	−0.1	0.4	0.1	0.0	0.6	−0.2
6	35	0.2	−0.3	0.2	−0.1	0.3	0.7

Note: These are made-up values for illustrative purposes and do not represent a physical system.

[2] Table 5.7 does not represent a particular structural system.

For the 1 mode, participation factor with an earthquake in the 1 direction:

$$\Gamma_{11} = \frac{m_1\phi_{111} + m_2\phi_{211}}{m_1\phi^2_{111} + m_1\phi^2_{112} + m_1\phi^2_{113} + m_2\phi^2_{211} + m_2\phi^2_{212} + m_2\phi^2_{213}} \quad (5.16a)$$

$$\Gamma_{11} = \frac{1x1.0 + 1x0.0}{1x1.0^2 + 1x0.1^2 + 1x - 0.1^2 + 1x0.0^2 + 1x0.2^2\, x1x0.1^2} \quad (5.16b)$$

$$\Gamma_{11} = \frac{1}{1{+}.01{+}.01 + 0 + 0.04 + 0.01} \quad (5.16c)$$

$$\Gamma_{11} = \frac{1}{1.07} = +0.93 \quad (5.16d)$$

For the 1 mode with an earthquake in the 2 direction:

$$\Gamma_{11} = \frac{m_1\phi_{112} + m_2\phi_{212}}{m_1\phi^2_{111} + m_1\phi^2_{112} + m_1\phi^2_{113} + m_2\phi^2_{211} + m_2\phi^2_{212} + m_2\phi^2_{213}} \quad (5.17a)$$

$$\Gamma_{12} = \frac{1x0.1 + 1x0.2}{1x1.0^2 + 1x0.1^2 + 1x - 0.1^2 + 1x0.0 + 1x0.2^2 + 1x0.1^2} \quad (5.17b)$$

$$\Gamma_{12} = \frac{0.3}{1.07} = +0.28 \quad (5.17c)$$

For the 1 mode with an earthquake in the 3 direction:

$$\Gamma_{13} = \frac{m_1\phi_{113} + m_2\phi_{113}}{1x1.0^2 + 1x0.1^2 + 1x - 0.1^2 + 1x0.0 + 1x0.2^2 + 1x0.1^2} \quad (5.18a)$$

$$\Gamma_{13} = \frac{1x - 0.1 + 1x0.1}{1x1.0^2 + 1x0.1^2 + 1x - 0.1^2 + 1x0.0 + 1x0.2^2 + 1x0.1^2} \quad (5.18b)$$

$$\Gamma_{13} = \frac{-0.1 + 0.1}{1.07} = \frac{0}{1.07} = 0 \quad (5.18c)$$

In like manner, modal participation factors for modes 2 to 6 and directions 1 to 3 can be determined. Once the modal participation factors for the *j* mode for an earthquake in the *k* direction are determined, it is possible to determine the modal forces applied to each mass in each mode in each direction.

The spectral accelerations, S_{ajk}, are defined for each *j* modal frequency or period and direction as shown in Figure 5.7 and applied to the modal participation factor and modal displacements by Eq. (5.19). Given the spectral accelerations from Figure 5.8 for 5% damping and 2.0 Hz, frequency (0.5 second period) is 0.6g.

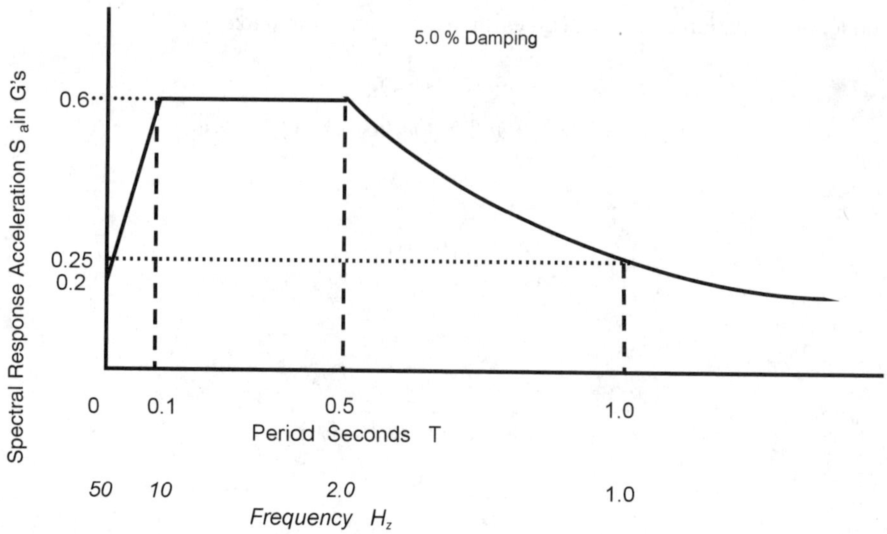

FIGURE 5.8. TYPICAL ACCELERATION GROUND RESPONSE SPECTRUM EXPRESSED AS A FUNCTION OF PERIOD AND FREQUENCY

Therefore, the modal acceleration applied to the i mass in the j mode for an earthquake in the k direction:

$$a_{ijk} = \Gamma_{jk}\, S_{jk}\, \phi_{ijk} \tag{5.19a}$$

$$a_{111} = +0.93 \times 0.6\text{g} \times 1.0 \tag{5.19b}$$

$$a_{111} = 0.55\text{g} \tag{5.19c}$$

Note: The modal participation factor, as well as the modal vector element ϕ_{ijk}, either may be negative or positive, which determines the direction of the nodal force.

The nodal force applied to a 1 kg mass in the 1 mode due to an earthquake in the (1) direction:

$$\text{F}_{111} = 1.0 \times 0.55\text{daN} \tag{5.21}$$

To obtain the resultant stress or internal force acting on the n mass of the system in the 1 mode due to an earthquake in the 1 direction:

$$s_{11} = f\,(F_{111}, F_{211} \ldots F_n 11 \tag{5.21}$$

and

$$S_i = \sum_{j=1}^{6} (S_{j1}^2)^{1/2} \tag{5.22}$$

S_l = stress or internal force resulting from the application of all nodal force F on the system due to an earthquake in the 1 direction.

This is referenced as the square root sum of squares (SRSS) combination of modal stress or internal forces or moments.

S_l is the stress or internal force in the system due to an earthquake in the 1 direction, and

$$S = \sum_{k=1}^{3} (S_k)^{1/2} \tag{5.23}$$

where S = resultant seismic stress or internal forces and moments in the system due to earthquake with either a "+" or "–" sensing.

As an alternative to the use of SRSS procedures for computing total earthquake stresses, the following set of coefficients may be applied to the 1, 2 and 3 directions of earthquake (100%, 40%, 40%), (40%, 100%, 40%), (40%, 40%, 100%) may be applied to the stress resultants from the 3 directions of earthquake and the highest one considering both a "+" or "–" sensing combined with other applicable levels.

5.3.4 Dynamic Time History

The dynamic time history procedure uses either time step integration by solving the equations of motion or time step modal analysis.

Equations of Motion

The response of a multi-degree-of-freedom (MDOF) linear structural system is described by the differential equation of motion expressed in matrix form [10].

$$[M]\{\ddot{x}\} + [C]\{\dot{x}\} + [K]\{x\} = -[M]\{J\}\ddot{y}_s(t) \tag{5.24}$$

where $\ddot{y}_s(t)$ = ground acceleration time history; and $\{J\}$ is a vector, and its components are the cosines of angles between a direction of ground motion and directions of corresponding DOF in a vector {x}. If given DOF is rotation, the corresponding component of a vector $\{J\}$ is zero. In that specific case, when the displacements on all DOF are linear and parallel to ground motion direction, $\{J\} = \{I\}$, i.e., vector composed from ones on the diagonal.

and [M] = mass matrix; [C] = matrix of viscous damping coefficients; [K] = stiffness matrix for the mass points of the linear elastic structure; $\{x\}$ = displacement vector; $\{\dot{x}\}$ = velocity vector; and $\{\ddot{x}\}$ = acceleration vector.

In all typical linear matrix formulations, matrices **M, C,** and **K** are symmetric and positive definite.

Direct integration or time-step integration is a numerical method directly applied to the solution of the differential equations of motion of linear structural system, Eq. (5.24), in a step-by-step manner. No uncoupling procedure is necessary to compute the response and the damping matrix need not be proportional.

There are many acceptable schemes available for numerical integration of the equations of motion, such as Newmark β-method [11], Houbolt method [12] and Wilson method [13].

Using the matrix formulation by Chan, Cox and Benfield [14] and Newmark β-method, Eq. (5.23) may be transformed into a finite-difference equation in recurrence form involving displacement:

$$[D]\{x_n+1\} = [B]\{x_n\} - [F]\{x_n-1\} + \beta h^2 \tag{5.26}$$

where $[D] = [M] + (h/2)[C] + \beta h^2[K]$; $[B] = 2[M] - (1-2\beta)h^2[K]$; $[F] = [M] - (h/2)[C] + \beta h^2[K]$; h = length of the time interval used in the direct integration procedure; β = parameter of Newmark β-method in reflecting the type of acceleration function assumed between two time steps and numerical stability of the procedure; and n = the number of time intervals lapsed.

The dynamic responses of a structure at any time may be calculated from the incremental time history of excitations and the previous structural responses.

The direct integration method has the advantage of simplicity in concept and elimination of the eigen value (mode shapes) computational effort. This method, unlike the modal analysis formulation method, may also be extended to deal with nonlinear problems where, for example, stiffness is no longer a constant. For practical solutions to real problems commercially available computer program are used.

It should be understood that unlike the response spectrum method in the time history methods, the time-dependent motion relationship is maintained. To get results the full strong motion duration time history of the earthquake must be used. Appendix N of *ASME B&PVC* Section III [10] specifies strong motion duration of 6.0 seconds preceded by a ramp-up duration of 4 seconds. This duration is characteristic of a moment magnitude 6.0 earthquake. Higher magnitude earthquakes typically have longer durations. A moment magnitude 7.0 might have a moment magnitude strong motion duration of 10–15 seconds, a 7.5 moment magnitude strong motion duration of 15–20 seconds and a moment magnitude of 8.0 with a strong motion duration of 30–35 seconds, with the total durations of shaking three to four times longer than the strong motion duration. However, it should be understood that fault rupture distances for 7.5 and higher magnitude typically exceed the constant isointensity geographic dimensions, which suggest that the fault rupture distance to the site is no longer effective as a constant.

The time history time steps are typically taken at 0.001 seconds intervals. Therefore, the time history response to a 13-second strong motion shaking for a magnitude 7 earthquake would be 13000, which would generate 13000 inertia time-dependent accelerations applied to the masses of the model. In general there is no way to determine which time step would result in the highest + and – stresses or internal force or moment resultants, except to analyze the structural system through each of the 13000 time steps. As a result, for practical application the maximum accelerations for each mass point in each direction during the strong motion duration of the earthquake are determined and conservatively applied simultaneously to get resultant earthquake stresses or internal forces or moments in the system.

Modal Analysis

The response of a linear or elastic SSC is always some combination of its normal modes. In the modal analysis method, the mode shapes are used as principal coordinates to reduce the equations of motion to a set of uncoupled differential equations that describe the motion of each mode n. These equations may be written as:

$$\ddot{A}_n + 2\omega_n p_n \dot{A}_n + \omega_n^2 A_n = -\Gamma_n \ddot{y}_s \tag{5.26}$$

where the modal displacement or rotation, A_n, is related to the displacement or rotation of mass point r in mode n, x_{rn}, by the equation:

$$x_{rn} = A_n \phi_{rn} \tag{5.27}$$

where ω_n = natural frequency of mode n (in radians per second); p_n = critical damping ratio of mode n; and Γ_n = modal participation factor of mode n given by:

$$\Gamma_n = \frac{\overset{n}{\Sigma} m_r \phi^1{}_{rn}}{\overset{n}{\Sigma} m_r \phi^2{}_{rn}} \tag{5.28}$$

and $\phi^1{}_{rn}$ = component of ϕ_{rn} in the direction of the earthquake.

The essence of the modal analysis lies in the fact that Eq. (5.26) is analogous to the equation of motion developed in Eq. (5.24). Multiplying Eq. (5.24) by $[\mathbf{M}]^{-1}$ gives:

$$\{\ddot{x}\} + [M]^{-1}[C]\{\dot{x}\} + [M]^{-1}[K]\{x\} = [M]^{-1}\{f\} \tag{5.29}$$

As the matrices $[M]^{-1}[C]$ and $[M]^{-1}[K]$ are not diagonal, pre-multiplying by $[M]^{-1}$ does not lead in itself to decomposition of the coupled set of equations in the independent ones. The routine is somewhat more complicated. It is based on the modal vectors orthogonality conditions with matrices $[M]$ and $[K]$:

$$\{\Phi_i\}^T[M]\{\Phi_n\} = \begin{cases} 0 \text{ if } i \neq n \\ \{\Phi_i\}^T[M]\{\Phi_i\} \neq 0 \text{ if } i = n \end{cases} \tag{5.30}$$

$$\{\Phi_i\}^T[K]\{\Phi_n\} = \begin{cases} 0 \text{ if } i \neq n \\ \omega_i^2\{\Phi_i\}^T[M]\{\Phi_i\} \neq 0 \text{ if } i = n \end{cases} \tag{5.31}$$

If a set of coupled equations does not contains the damping matrix $[C]$, then the two conditions are enough to get a set of the independent equations. But if a system is damped, i.e., there is $[C]$ in the given set of equations, then their decomposition is possible only if matrix $[C]$ also satisfies the orthogonality condition.

$$\{\Phi_i\}^T[C]\{\Phi_n\} = \begin{cases} 0 \text{ if } i \neq n \\ 2\omega_i p_i\{\Phi_i\}^T[M]\{\Phi_i\} \neq 0 \text{ if } i = n \end{cases} \tag{5.32}$$

When seismic excitation is considered, it is preferable to use equation of motion in the form defined here. The routine of decomposition consists of the following steps.

First one should make the substitution

$$\{x\} = \sum_{n=1}^{m}\{\Phi_n\}A_n(t) \tag{5.33}$$

,

where m = a total number of DOF; $\{\Phi_n\}$ = n mode vector; and A_n = n generalized coordinate dependent on time t. Eq. (5.26) is the r member of this vector. After this substitution the set of equations becomes

$$[M]\sum_{n=1}^{m}\{\Phi_n\}\ddot{A}_n + [C]\sum_{n=1}^{m}\{\Phi_n\}\dot{A}_n + [K]\sum_{n=1}^{m}\{\Phi_n\}A_n = -[M]\{J\}\ddot{y}_s(t) \tag{5.34}$$

Then there is pre-multiplication of this equation by the transposed i mode vector:

$$\begin{aligned}&\{\Phi_i\}^T[M]\sum_{n=1}^{m}\{\Phi_n\}\ddot{A}_n + \{\Phi_i\}^T[C]\sum_{n=1}^{m}\{\Phi_n\}\dot{A}_n + \{\Phi_i\}^T[K]\sum_{n=1}^{m}\{\Phi_n\}A_n = \\ &= -\{\Phi_i\}^T[M]\{J\}\ddot{y}_s(t)\end{aligned} \tag{5.35}$$

Due to the orthogonality conditions (1) to (3) all terms in the sums become zeroes except $i = n$:

$$\begin{aligned}&\{\Phi_i\}^T[M]\{\Phi_i\}\ddot{A}_i + 2\omega_i p_i\{\Phi_i\}^T[M]\{\Phi_i\}\dot{A}_i + \omega_i^2\{\Phi_i\}^T[M]\{\Phi_i\}A_i = \\ &= -\{\Phi_i\}^T[M]\{J\}\ddot{y}_s(t)\end{aligned} \tag{5.36}$$

Dividing by $\{\Phi_i\}^T[M]\{\Phi_i\}$ gives

$$\ddot{A}_i + 2\omega_i p_i\dot{A}_i + \omega_i^2 A_i = -\frac{\{\Phi_i\}^T[M]\{J\}}{\{\Phi_i\}^T[M]\{\Phi_i\}}\ddot{y}_s(t) \tag{5.37}$$

Write Γ_i for the participation factor:

$$\Gamma_i = \frac{\{\Phi_i\}^T[M]\{J\}}{\{\Phi_i\}^T[M]\{\Phi_i\}} \tag{5.38}$$

Thus we come to Eq. (5.26):

$$\ddot{A}_i + 2\omega_i p_i \dot{A}_i + \omega_i^2 A_i = -\Gamma_i \ddot{y}_s(t) \tag{5.39}$$

The critical damping ratio of a single-degree-of-freedom (SDOF) system, *p,* is defined by the equation:

$$p \equiv \frac{c}{c_c} \tag{5.40}$$

where the critical damping coefficient is given by the expression:

$$c_c = 2m\omega \tag{5.41}$$

Substituting Eq. (5.41) into Eq. (5.40) and solving for *c/m* gives:

$$\frac{c}{m} = 2\omega p \tag{5.42}$$

Substituting this expression and the expression for $\omega = k/m$ gives:

$$\ddot{x} + 2\omega p\dot{x} + \omega^2 x = -\ddot{y}_s \tag{5.43}$$

Note the similarity of Eq. (5.26) and (5.39). Thus each mode may be analyzed as though it were an SDOF system and all modes are independent of each other; thus in a time history, analysis can be combined based on an arithmetic sum to give the resultant time history response. By this method a fraction of critical damping, i.e., c/c_C, may be assigned to each mode and it is not necessary to identify or evaluate individual damping coefficients.

5.4 RESPONSE SPECTRA

The response spectra used in seismic analysis are of two types: ground, which have the characteristics of the earthquake motion at the site usually defined at the free-field ground or foundation surface applied to a structure, and floor or in-structure spectra, where the filtering and increased acceleration effect of the building is included.

5.4.1 Ground

Nuclear

Characteristic seismic ground response spectra historically used in hazardous facility design (nuclear) are shown in Figures 5.9 through 5.14. Earthquakes are generally assumed to occur in three orthogonal directions—two horizontal and one vertical. The horizontal ground response spectra are usually assumed to have the same shape and amplitude in the two horizontal directions and a somewhat different shape and amplitude in the vertical direction.

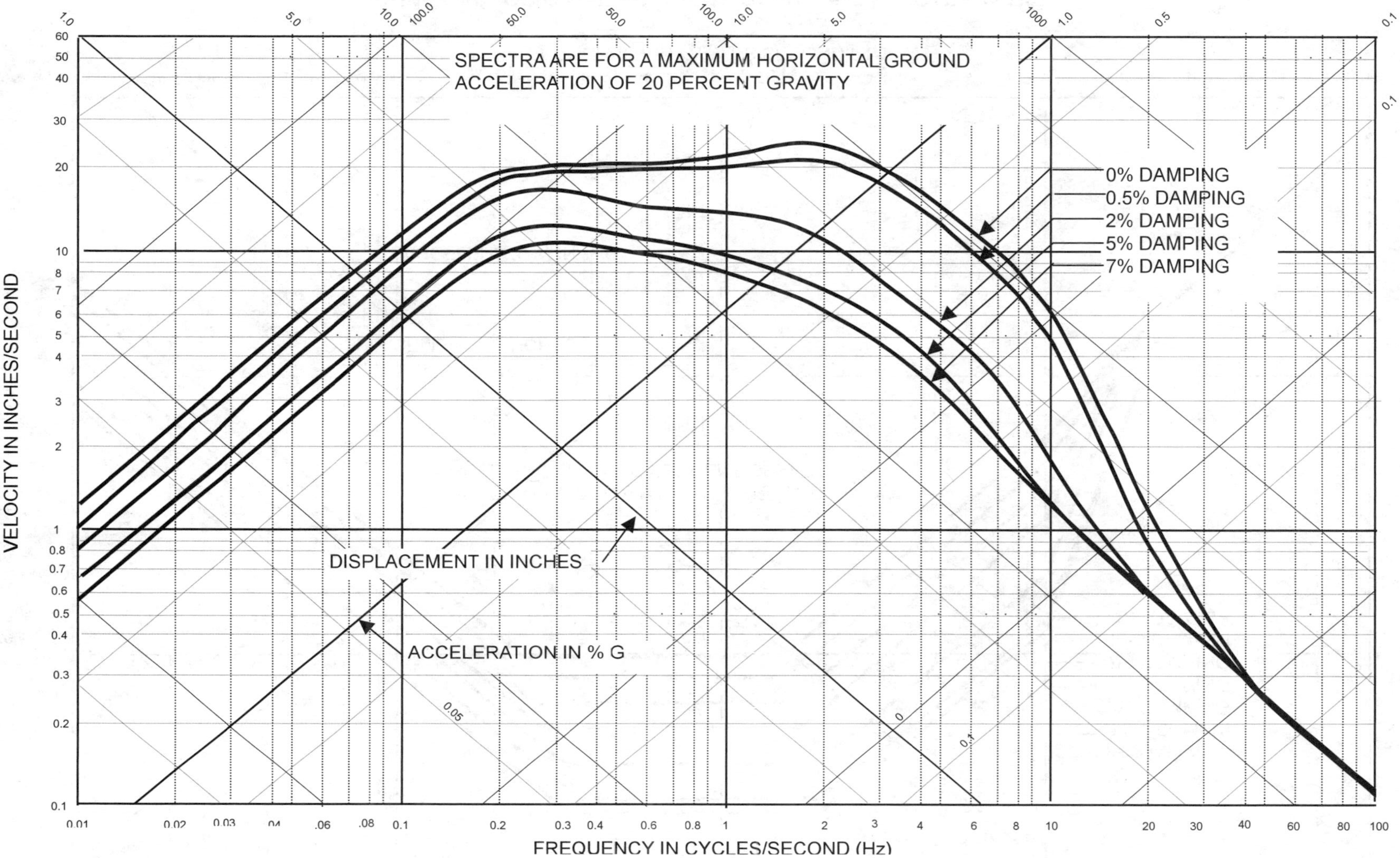

FIGURE 5.9. HOUSNEN-TYPE HORIZONTAL GROUND RESPONSE SPECTRA PLOTTED ON TRIPARTATE PAPER NORMAL TO 20% GRAVITY PEAK GROUND ACCELERATION

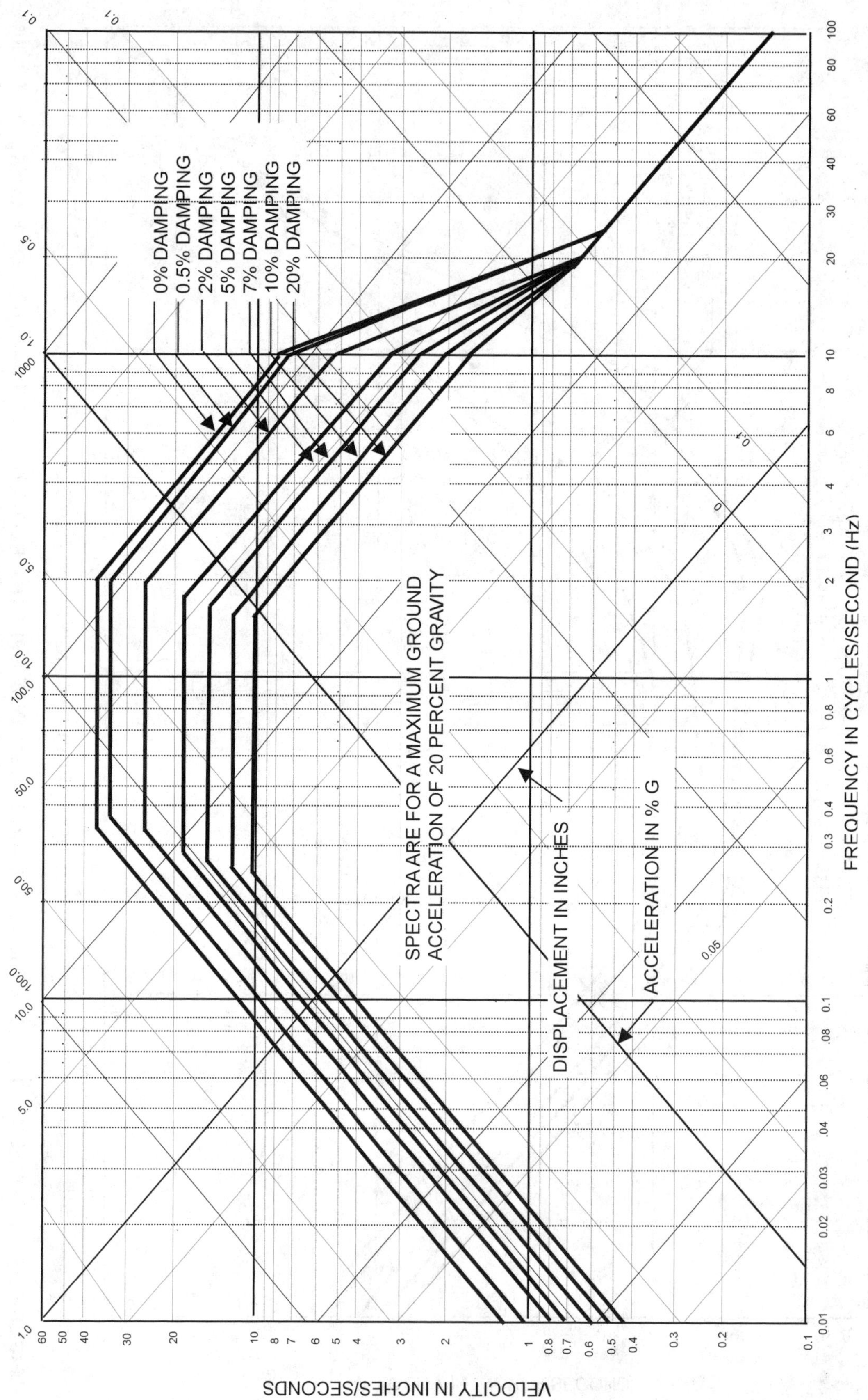

FIGURE 5.10. ORIGINAL NEWMARK RESPONSE SPECTRA (1969)

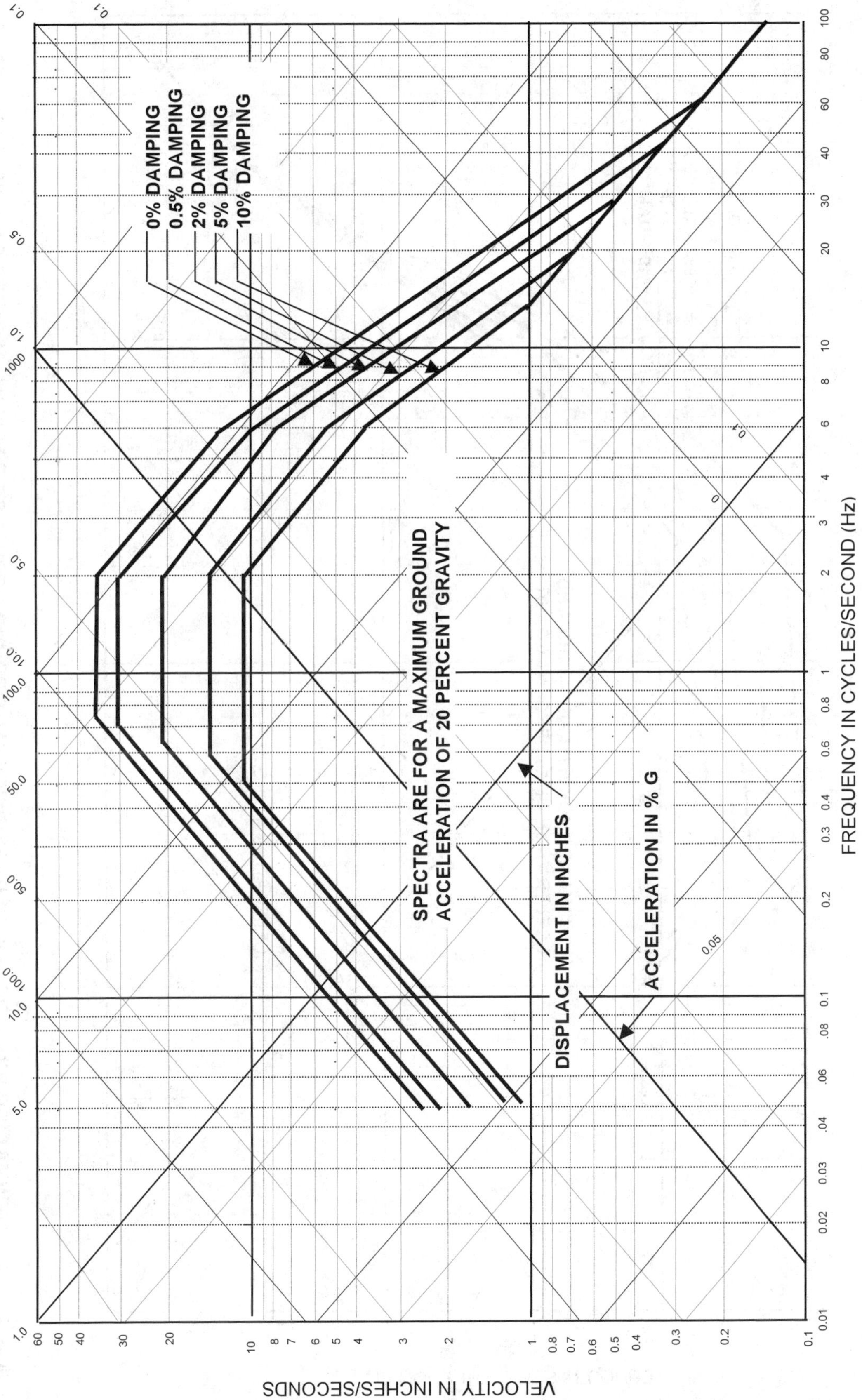

FIGURE 5.11. LATER NEWMARK RESPONSE SPECTRA (1969)

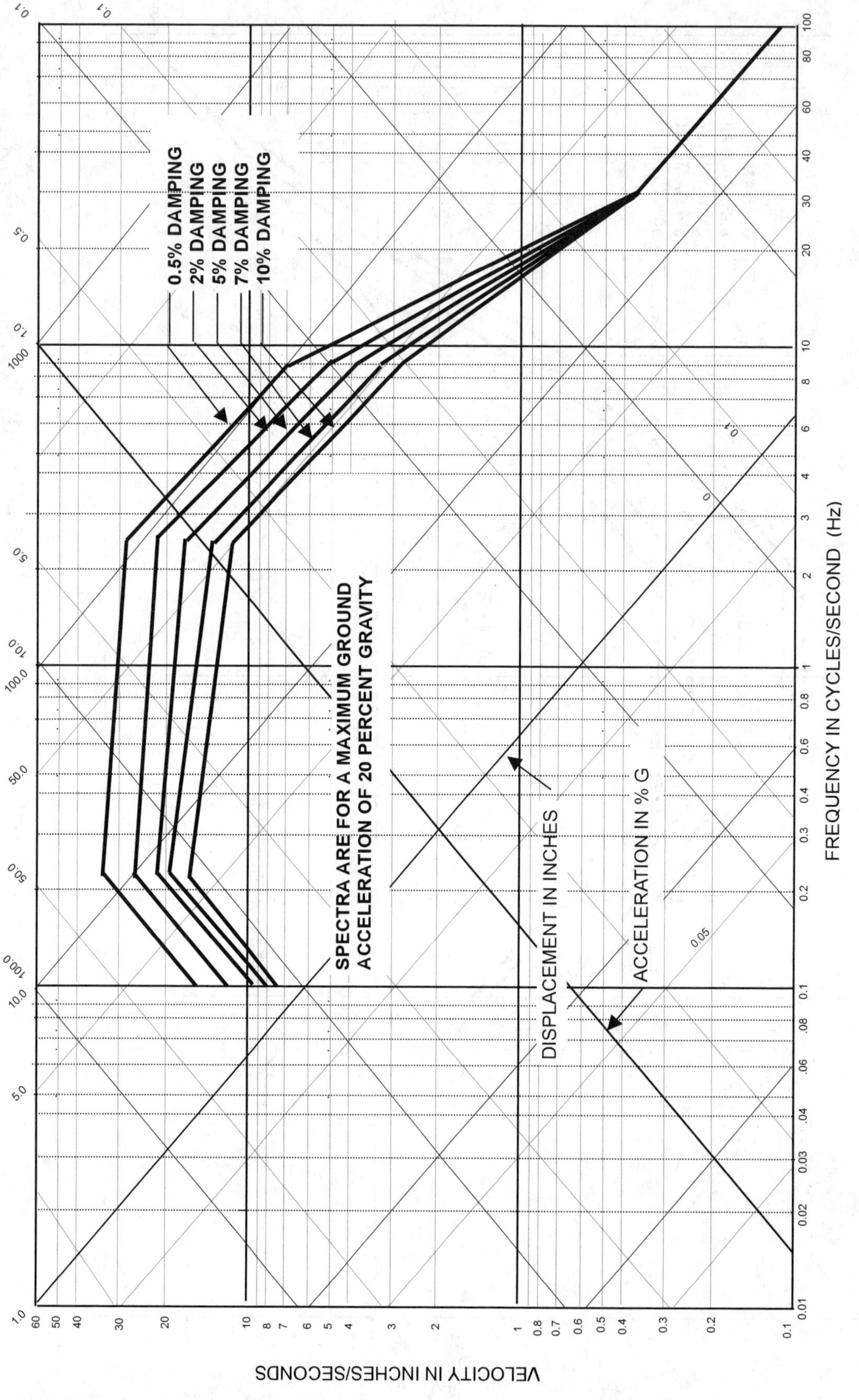

FIGURE 5.12. NEWMARK-BLUME-KAPPOR-TYPE HORIZONTAL GROUND RESPONSE SPECTRA PLOTTED ON TRIPARTATE PAPER

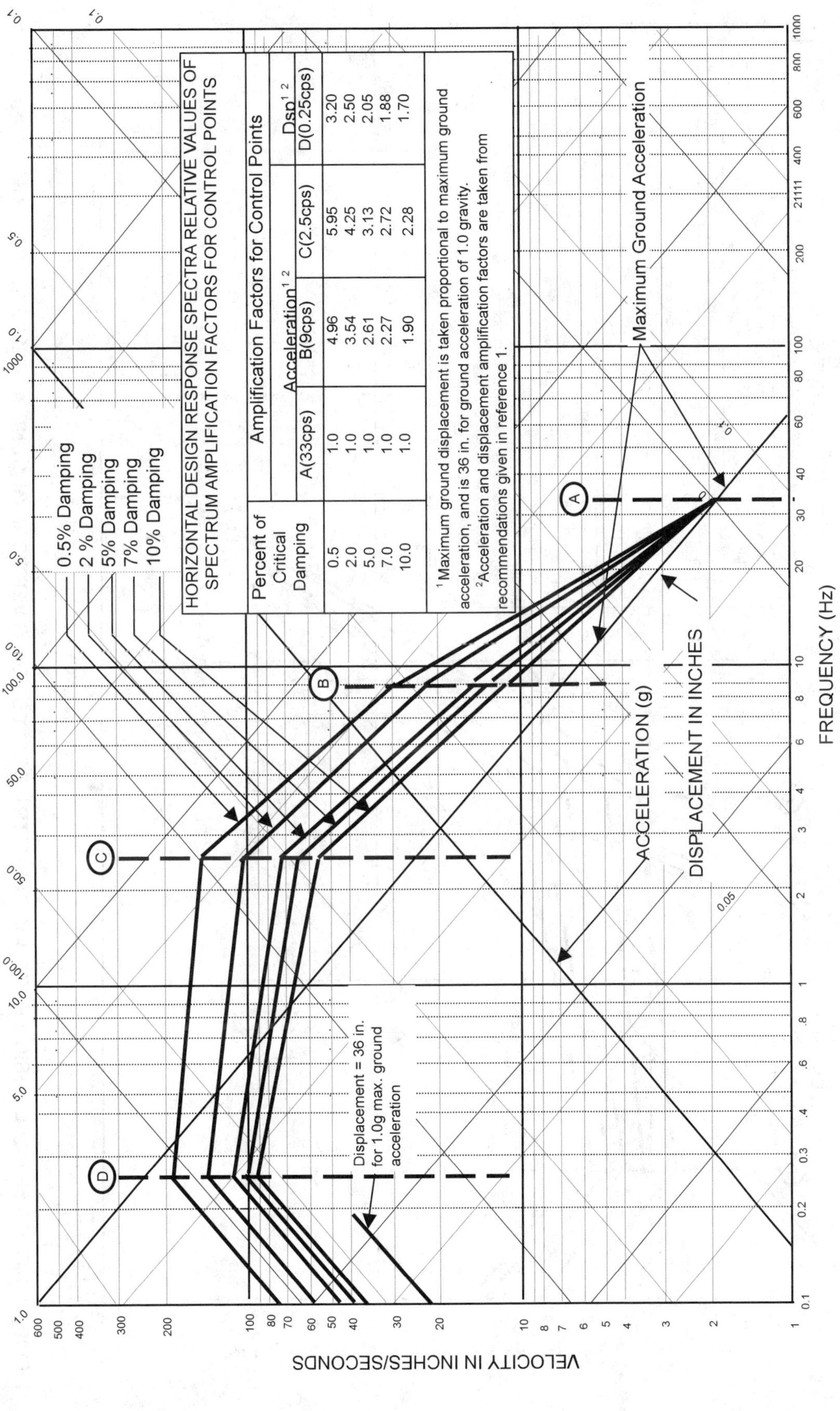

Percent of Critical Damping	Amplification Factors for Control Points			
	Acceleration[1 2]			Dsp[1 2]
	A(33cps)	B(9cps)	C(2.5cps)	D(0.25cps)
0.5	1.0	4.96	5.95	3.20
2.0	1.0	3.54	4.25	2.50
5.0	1.0	2.61	3.13	2.05
7.0	1.0	2.27	2.72	1.88
10.0	1.0	1.90	2.28	1.70

[1] Maximum ground displacement is taken proportional to maximum ground acceleration, and is 36 in. for ground acceleration of 1.0 gravity.

[2] Acceleration and displacement amplification factors are taken from recommendations given in reference 1.

FIGURE 5.13(A). HORIZONTAL DESIGN RESPONSE SPECTRA SCALED TO 1 G HORIZONTAL GROUND ACCELERATION FROM USNRC RG 1.60

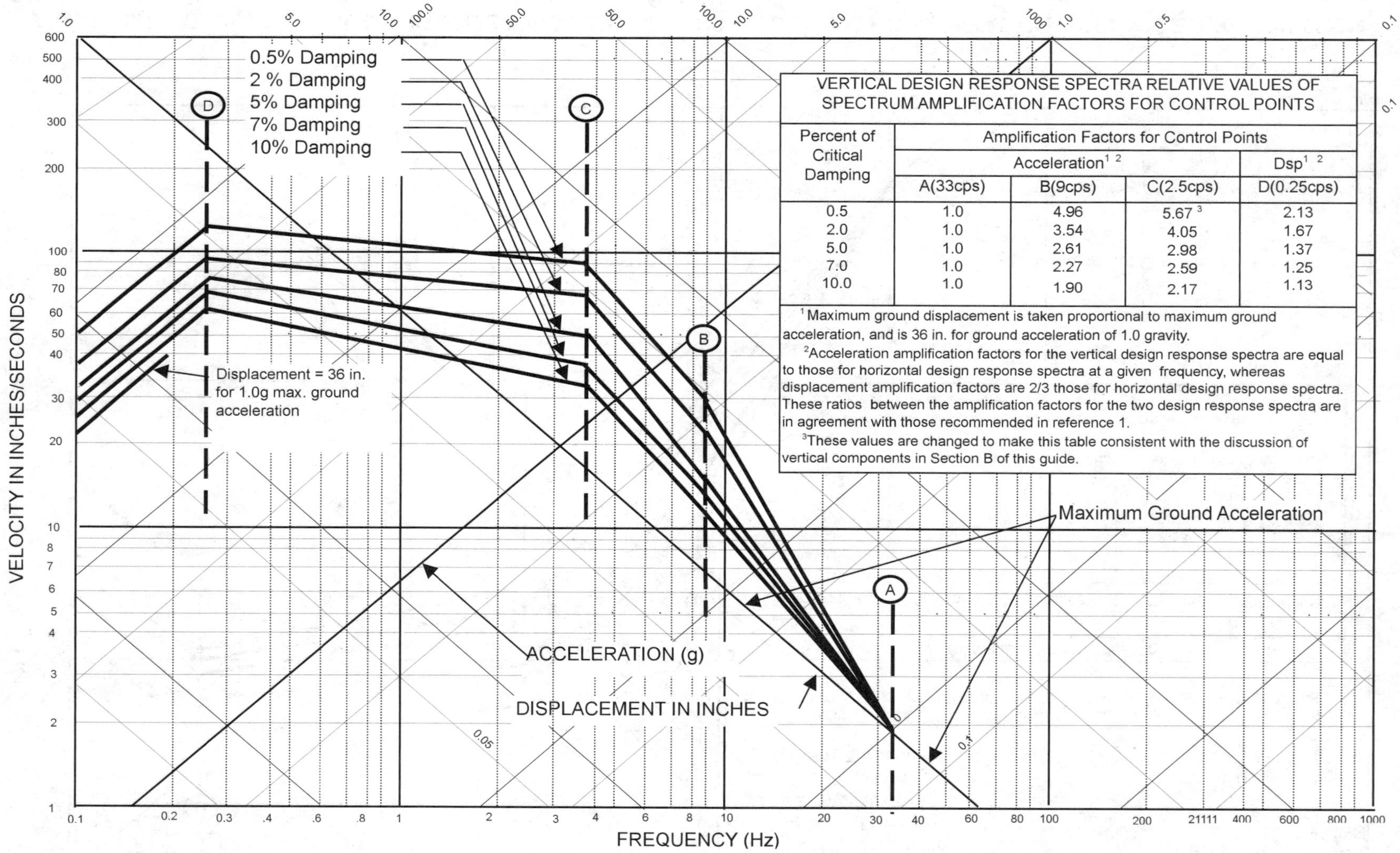

Percent of Critical Damping	Amplification Factors for Control Points			
	Acceleration[1,2]			Dsp[1,2]
	A(33cps)	B(9cps)	C(2.5cps)	D(0.25cps)
0.5	1.0	4.96	5.67[3]	2.13
2.0	1.0	3.54	4.05	1.67
5.0	1.0	2.61	2.98	1.37
7.0	1.0	2.27	2.59	1.25
10.0	1.0	1.90	2.17	1.13

[1] Maximum ground displacement is taken proportional to maximum ground acceleration, and is 36 in. for ground acceleration of 1.0 gravity.

[2] Acceleration amplification factors for the vertical design response spectra are equal to those for horizontal design response spectra at a given frequency, whereas displacement amplification factors are 2/3 those for horizontal design response spectra. These ratios between the amplification factors for the two design response spectra are in agreement with those recommended in reference 1.

[3] These values are changed to make this table consistent with the discussion of vertical components in Section B of this guide.

FIGURE 5.13(B). VERTICAL DESIGN RESPONSE SPECTRA SCALED TO 1 G VERTICAL GROUND ACCELERATION FROM USNRC R.G. 1.6

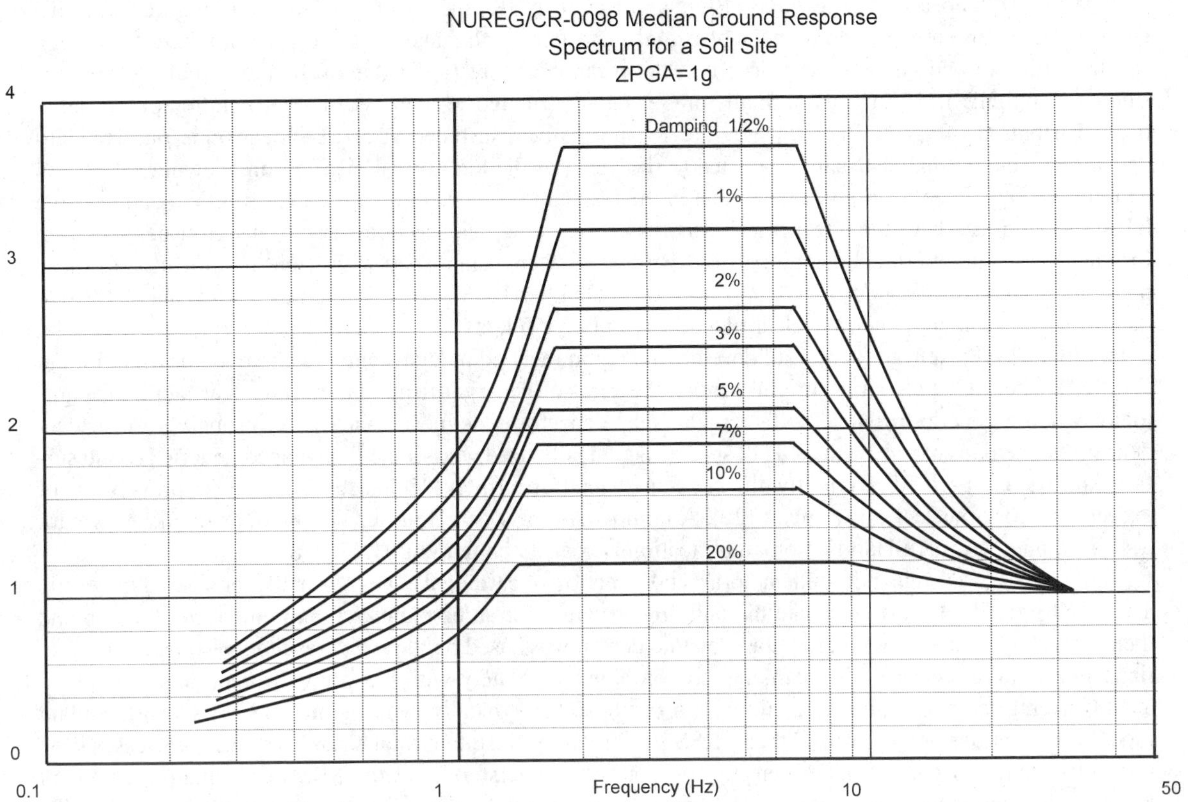

FIGURE 5.14. NEWMARK-HALL MEDIAN-SHAPED SPECTRA FROM NUREG/CR-0098

A typical ground response spectrum results if the series of simulated pendulums are assumed supported on the ground and are subject to free-field ground surface seismic time history input motion.

The time history input can be in the form of acceleration, velocity or displacement. By the use of tripartite paper, as shown in figures 5.9 through 5.13, the acceleration, velocity and displacement can be plotted on a single figure. The response spectra are generated by plotting the maximum ordinate of acceleration, velocity or displacement response for each pendulum frequency during the time history response. A series of plots are generated as a function of the assumed critical viscous damping applied to the series of pendulums.

During the period 1964 through 1969, the Housner [15] ground spectrum shown in Figure 5.9 had been specified for the seismic design of category I SSC in nuclear power plants, and a typical actual recorded time history earthquake motion, such as Taft 1952 (rock site) or El Centro 1940 (soil site), was used to generate floor response spectra. The use of a single time history was generally acceptable provided the response spectrum generated from such an earthquake time history had an amplification higher than the ground response spectrum specified for the site in the frequency range between 1 and 25 Hz.

From 1969 until April 1973, there was a gradual movement away from the Housner curve of Figure 5.9 toward the original Newmark curve [16] shown in Figure 5.10, then toward the later Newmark spectra [17] as shown in Figure 5.11; this later Newmark curve was modified slightly as discussed in Reference [18]. The Newmark spectra are based on the enveloping of individual response spectra rather than the weighted averaging of the individual spectral values, as was done for the Housner spectra. Note: The response spectra curves for each value of damping are parallel lines in the region above 10 Hz for the later Newmark curve. This is not characteristic of most spectra, which converge in the high-frequency range to peak ground response spectra independent of damping. The extension of the response spectra into the rigid region of response from 20 to 33 Hz is also a result of Newmark's work.

In April 1973 appeared the Newmark, Blum and Kapur (NBK) spectra [19], as shown in Figure 5.12. This curve is based on an expanded number of actual measured earthquake response spectra where the design spectra represent one standard deviation from the mean of the average value of the individual spectra considered, assuming the spectra are normally or Gausain distributed. A more recent interpretation of the statistical distribution of spectra assume they are log-normally distributed such that mean plus one standard deviation of the normal distribution is equal to the mean of the log-normal distribution. Newmark-Hall can be found in Table 5.8. It should be noted that in the resonant response range for 2 to 5% critical damping the NBK spectrum can have the effect of a 1.5 increase in seismic loads as compared to the Housner spectrum for the same value of critical damping. The NBK curves were established with slight modification as those presented in the NRC *Regulatory Guide 1.60* [20] in October 1973, shown in Figures 5.13(a) and 5.13(b)—they form the basis for generic nuclear plant design after 1974.

Finally in 1978, Newmark and Hall developed a recommended median centered curve as shown in Figure 5.13 [21]. The *R.G. 1.60* curve currently forms the basis of generic seismic input to nuclear facility designs today, when a generic free-field surface mean shape spectra is defined for a 10^{-4}/yr probability of exceedence. The median shape Newmark-Hall spectra is typically used when a median-shaped generic ground surface spectra is specified. Currently the median-shaped spectra is defined at the 10^{-5}/yr probability of exceedence for a nuclear power plant [22]. A comparison between Housner, original Newmark, Newmark *R.G. 1.60* and Newmark-Hall motion amplification ratios can be found in Table 5.8.

Until late 1972 a single independent horizontal direction of earthquake was normally considered in design. The usual procedure was to evaluate the building structure for an earthquake in a nominal north-south and then separately in an east-west direction. Seismic design was based on the stress resultants from either direction of earthquake considered separately and combined absolutely with the vertical direction stress resultants. Currently combination of seismic inertia response has typically been on an SRSS basis for resultant modal accelerations or stresses and then SRSS for direction. Resultant total earthquake inertia stresses have been given either a plus or minus sensing. It should be understood that the SRSS combination of modal forces or accelerations results in all modal forces or accelerations being given a positive sensing. This distorts mode shapes and does not correctly represent the modal loading of the structure. However, because of the difficulty in applying modal forces to complex structures such combinations are generally allowed. In addition, with the possible exception of piping system continuous over several supports, it results in conservative seismic load resultants.

TABLE 5.8. COMPARISON OF ONE-DEGREE-OF-FREEDOM MAXIMUM AMPLIFICATION FACTOR FOR HORIZONTAL[1] SEISMIC RESPONSE TO PEAK GROUND MOTION FOR VARIOUS DAMPING VALUES

	Weighted average Housner		R.G. 1.60 horizontal median plus one[2] standard deviation		NUREG/CR-0098 horizontal median		NUREG CR-0098 horizontal median plus one standard deviation[2]	
Critical damp (%)	Acc.	Disp.	Acc.	Disp.	Acc.	Disp.	Acc.	Disp.
0.5	6.2	1.4	5.95	3.2	3.68	2.01	5.10	3.04
1	4.6	—	—	—	3.21	1.82	4.38	2.73
2	3.1	—	4.25	2.5	2.74	1.63	3.66	2.42
5	2.3	1.2	3.13	2.0	2.12	1.39	2.71	2.01
7	1.5	1.0	2.72	1.8	1.89	1.29	2.36	1.85
10	—	—	2.28	1.7	1.64	1.20	1.99	1.69
20	—	—	—	—	1.17	1.01	1.26	1.38

Note: Based on a standard relative earthquake maximum values of ground motion: acceleration = 0.10g; displacement = 3.6 in.

1. For a vertical motion slightly different amplification values are sometimes defined.
2. Since a normal distribution was assumed, the median equals the mean. For a log normal distribution, normal median plus one standard deviation is approximately equal to the log normal mean.

Petrochemical and Biomedical

Ground spectra used in the design of petrochemical and biomedical safety-related SSC can be found in reference [7] as shown in Figure 5.15 for 5% damping. The spectra is given as a function of period with spectral acceleration constant at the S_{DS} level between periods T_o and T_s, where $T_o = 0.2\ S_D1/S_{DS}$; $T_S = S_{D1}/S_{DS}$; S_{DS} = the design spectral response acceleration at 0.2 second period and $S_{DS} = 0.67\ S_{MS}$; and S_{MS} = 5% acceleration for the site spectra acceleration adjusted for site foundation effects as defined in Table 9.4.1.24a of [7].

where S_S = 0.2-second 5% spectral acceleration taken from Map Figure 9.4.1.1.a of [7]; and S_D1 = design spectra response acceleration at 1.0-second period determined in a manner similar to S_{DS} for a 0.2-second period

5.4.2 Floor or In-Structure

The first seismic ground response spectra were developed in the mid 1950s [15]

In the mid 1960s it was recognized that building structures tend to modify earthquake ground motion as it is transmitted through the structures. The first attempt to include this effect was to increase the equipment damping level ground spectra by the ratio of the peak floor accelerations to the peak ground accelerations. In general, in shear-wall-type structures this results in an increase of $(1 + 2\ z/h)$

Where z = height of the building where the equipment or system are supported; and h = total height of building.

For concrete-frame structures the increase is $(1 + 3\ z/h)$ and for steel-frame structures the increase is typically $(1 + 4\ z/h)$.

By the late 1960s the initial more-or-less accurate generation of seismic floor or amplified response spectra was to take a time history recorded from a strong motion earthquake and scale it, usually by increasing

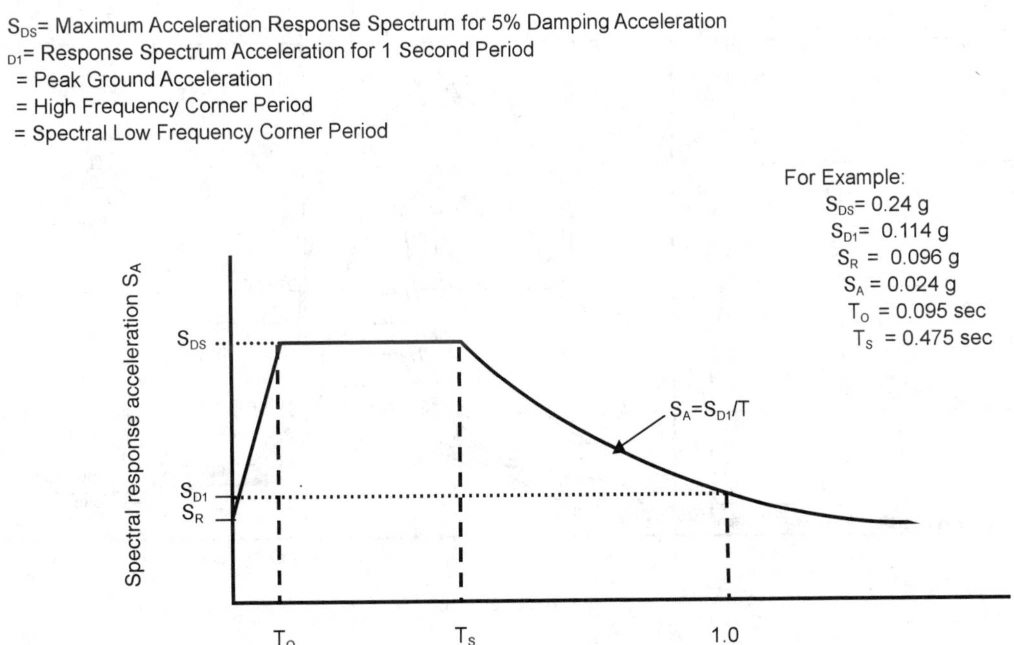

FIGURE 5.15. DESIGN RESPONSE SPECTRUM FROM ASCE-7-05

the peak ground acceleration so that its response spectra enveloped a referenced ground response spectra. This time history was applied at the base of the structure, and the resultant time histories recorded at various elevations of the building were applied to a series of one DOF pendulums in order to generate floor or amplified response spectra. The time history matching of a ground spectra was usually done at 2 or 5% critical damping. It was also necessary to define percent critical damping levels for the building (typically 4% if the building was assumed to remain elastic and 7% of critical damping if the building was responding at or near yield in the reinforcing or structural steel). The floor spectra applied to the seismic design or evaluation of equipment (secondary mass) are typically defined at 2, 3, 5, 7, and 10% critical damping. Typical current in-structure or floor response spectra are shown in Figure 5.16.

It is obvious that the shape of the resultant floor or amplified spectra are dramatically different from that of the ground spectra used as its basis. The building tends to respond at its dominate natural frequency modes of vibration. This has the effect of regularizing or making the seismic motion applied to the building floors more sinusoidal in nature. Whereas the ground motion of the earthquake had been more random in nature, such that the median amplified spectra response acceleration over the peak ground acceleration in the 2–10 Hz frequency range for 5% critical damping was a factor of about 2.5 and had a mean value of about 3.1, the more regular sinusoidal motion within the building has the theoretical upper bound value:

$$A_f = 1/2\beta \tag{5.45a}$$

where A_f = amplification factor; and β = decimal value of critical damping.

In the case where $\beta = 0.05$, 2β is equal to 0.1.
$A_f = 1/0.1 = 10$ as compared to a ground median amplification factor of 2.5 for 5% damping.

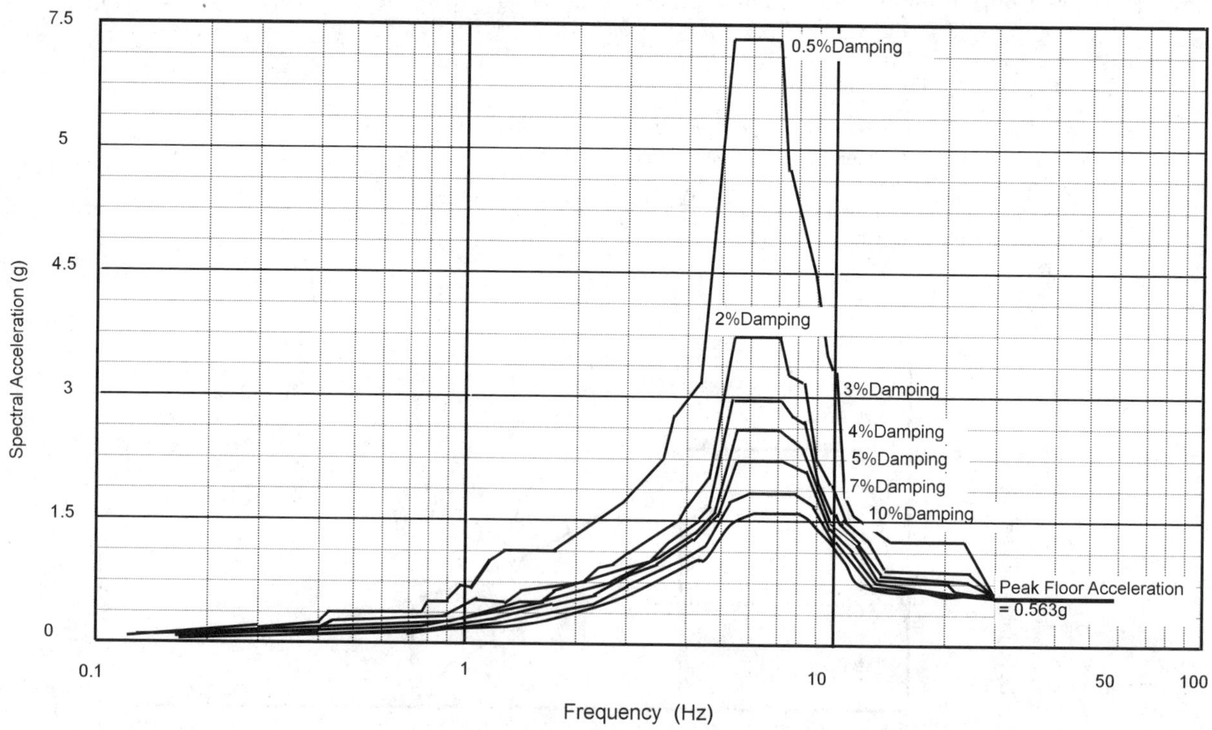

FIGURE 5.16. TYPICAL IN-STRUCTURE MIDHEIGHT ACCELERATION RESPONSE SPECTRA FOR A FOUR-STORY BUILDING—7% BUILDING DAMPING AND EQUIPMENT DAMPING AS SHOW

The resultant floor or amplified spectra are typically the uncoupled seismic input to systems or components (secondary system) supported by a building (primary system). By including the stiffness and mass of the secondary system or component model into the primary or building model, and thereby generating a coupled model, it is possible to significantly reduce the seismic input to the secondary system.

This effect is typically termed the mass ratio effect. In [21] the amplification factor, A_f, is defined as follows:

$$A_f = 1/(\beta_e + \beta_s + \sqrt{\sigma}) \tag{5.45}$$

where β_e = critical damping ration for the system or component (secondary system), and typically ranges from 0.02 to 0.10; β_s = critical damping ratio for the building (primary system), typically taken as 0.04 or 0.07; and σ = mass ratio between the secondary and primary mass, for as little as a 1% mass ratio would result in an effective increase in critical damping of 3%.

Most industrial structures that contain hazardous materials are typically shear-wall-type structures. Such structural types are generally more effective at confining such material or providing security or radiation protection from hazardous material stored or processed within such structures.

Once a ground response spectra is defined as well as a model of the structure, an artificial time history input can be generated where response spectrum just envelopes the reference ground response spectra at 2 or 5% damping and is used to generate amplified or floor response spectra. Alternatively, such spectra can also be generated by a direct method, which requires only a dynamic model of the structure and ground response spectra [24].

Fixed-base building models are representative of buildings founded on stiff soil or rock with foundation shear-wave velocities greater than about 2,500 fps. For structures founded on softer soil, the soil response characteristic frequencies can have a profound effect in defining the shape and amplitude of the resultant floor or amplified response spectra. Dominant floor or in-structure spectra peaks on soil foundation with shear-wave velocities below 2,500 fps range between 2 and 5 Hz. Fixed-based structures' dominant frequencies are typically between 4 and 9 Hz, with the 4 Hz peak response typical of frame-type structures and 9 Hz associated with shear-wall reinforced-concrete structures.

The effect of a soil foundation is to amplify seismic motion at the foundation rock-soil interface as the seismic waves travel from the rock surface through the soil to the foundation of the building, while at the same time the effect of high radiation damping in the foundation soil is to reduce the response of the structure.

Relatively recently computational algorithms have been developed [24] where it is possible to include variable damping as a function of soil strains as well as to include layered soil pressure on embedded structure walls.

5.4.3 Equipment

In equipment response spectra are sometimes used to qualify small parts or devices attached to equipment located at a given floor level or elevation. The procedure used is the same as that to generate floor response spectra. A time history generated on the floor is input into a lumped parameter or finite-element model of the equipment and a resultant time history is generated at various equipment locations. These time histories are then applied to a series of one DOF systems to generate spectra. As in the case of direct ground to building models, it is also possible to directly generate equipment spectra from a dynamic model of the equipment and the applicable floor spectra without application of time history input.

It should be noted that most equipment is modeled assuming a fixed base. This is correct assuming the applied lateral load does not cause uplift of the equipment base. When this happens the stiffness of the anchorage system comes into play. Normally such stiffness is much less than if the component were fixed at its base and the resultant inertia load on the component were typically much greater, since the tendency is to shift the dominant response mode to be more in resonance with the earthquake amplified frequencies. This is illustrated in section 5.5.

5.5 DETERMINATION OF SEISMIC LOADS ON STRUCTURAL SYSTEMS

5.5.1 Introduction

Presented here are simple examples that illustrate the application of the seismic inertia load and resultant seismic stresses to two basic structural systems: vertical cantilever and beams on multiple vertical and lateral supports typically used to model mechanical structures, distribution systems and components.

5.5.2 Cantilever Systems

This modeling is typical of mechanical structures such as ground-supported vertical tanks and towers and base-supported mechanical components such as pumps and vessels. The shear and bending stiffness of these components is typically combined into a cantilever stick model as shown in Figure 5.4. In its simplistic form this would be a single mass node supported by a beam located at the center of gravity of the component. To develop the significant higher modes of response at least a three-mass stick model is required. When there is lack of symmetry in the component, such that the center of shear or rigidity does not coincide with the center of mass, torsional response will result.

Illustrative Problem

Given a carbon-steel-type ASTMA 516 vertical steel cylindrical vessel containing air as shown in Figure 5.17:

Length, $L = 10.0$ feet
Inside diameter, ID = 2.0 feet
Shell thickness, $t = 0.25$ inches
Design pressure, $P_o = 150$ psi
Flat plate top and bottom end closures, $t_2 = 1.0$ inches
R_1 = radius of top closure plate = 12.5 inches
R_2 = radius of bottom closure plate = 16.25 inches
R_3 = radius of bottom plate anchor bolts
Anchorage = Three 1/2-inch-diameter-type ASME A307 carbon steel poured in place of/with anchor bolts spaced 120° apart, as shown in Figure 5.17.

Compute weight of vessel: Weight of steel = $\pi D t_1 L + \pi R_1{}^2 t_2 + \pi\, R_2{}^2 t_2$

where $\pi = 3.14$; $D_1 = 24.25$ inches; D_2 = top PL = 24.5 inches; D_3 = Bottom PL = 32.5 inches; $t_1 = 0.25$ inches; $t_2 = 1.0$ inches; $L = 120$ inches; $R_1 = 12.25$ inches; and $R_2 = 16.25$ inches.

Weight of steel = 490 lbs/cu.ft.
Cylinder steel volume = $3.14 \times 24.25 \times 0.25 \times 120 = 2284$ in^3
2,284 in^3/1,728 cu. in./cu. ft. = 1.32 cu. ft.
490 lbs/cu. ft. $\times$ 1.32 cu. ft. = 647 lbs.

End plates:

Top plate = $3.14 \times 12.5 \times 1$in. = 491 cu. in.
Bottom plate = $3.14 \times 16.25^2 \times 1$ in. = 829 cu. in.
491 + 829 = 1320 cu. in.
1,320/1,728 = 0.764 cu. ft.
$490 \times 0.764 = 374$ lbs
Total weight of vessel = 647 + 374 = 1,021 lbs.
(Say 1,050 lbs. to account for weld material.)

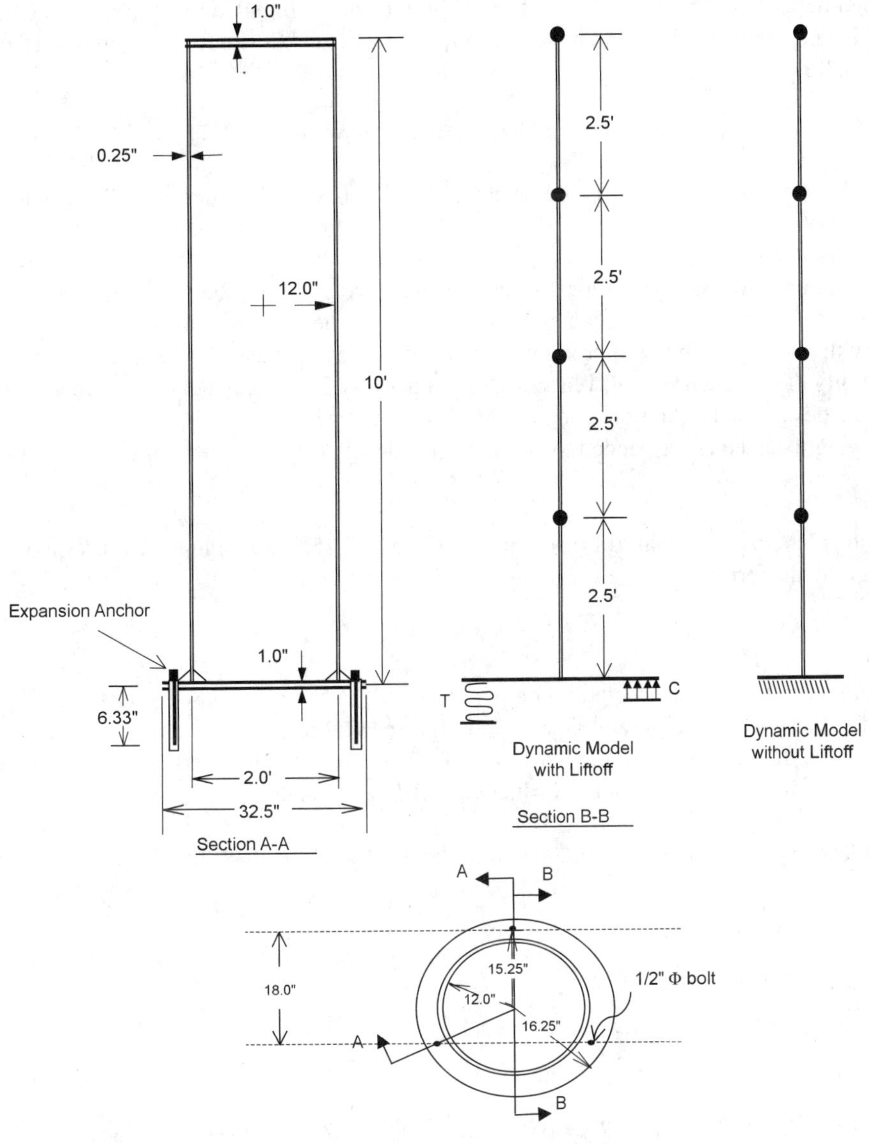

FIGURE 5.17. TYPICAL VERTICAL AIR TANK ANCHORED AT BASE

ASCE 7-05 Computations of Seismic Forces

Determine F_p based on Eq. (5.3):
where a_p = 2.5 from Table 5.3 (Table 9.6.3.2 of [7]); R_p = 2.5 from Table 5.3 (Table 9.6.3.2 of [7]); W_p = 1,050 lbs; and I_p = 1.5 importance factor (Table 9.1.4 of [7]).

With vessel located at the mid point of the supporting building structure, z/h = 0.5; z = height of building to base of component support; and h = total height of supporting building structure.
Determine S_{MS}

$$S_{MS} = F_a S_s \tag{5.46}$$

where S_{MS} = maximum considered earthquake, 5% damped, spectral response acceleration at short periods adjusted for site class effects as defined in Section 9.4.1.2 of [7]; F_a = 1.2, acceleration-based site coefficient (at 0.2-second period) in Table 9.4.1.2.4 (a) from [7] for a class C foundation site; $S_S = 0.3$ g spectral acceleration at 5% damping at a 0.2-second period at a typical Eastern U.S. site from Figure 9.4.1.1(a) of [7]; and $S_{MS} = 1.2 \times 0.3$ g = 0.36 g.

$$S_{DS} = 2/3\ S_{MS} = 2/3 \cdot 36g = 0.24g \text{ from Section 9.4.1.2.5 of [7]} \tag{5.47}$$

where S_{DS} = 5% damped, spectral response design acceleration for a short period (0.2 seconds) as defined in Section 9.4.1.2 of [7].

There does not appear to be any technical basis for multiplying S_{MS} by 2/3 to obtain a design acceleration other than to match the acceleration, which was determined when a 10% probability of exceedence (*UBC-*1997) [25] with a return period of approximately 500 years was the basis for design, to that associated with a 2% probability of exceedence in 50 years or a return period of approximately 2,000 years is now the basis for earthquake design load definition using the [7] design procedures.

The following parameters are needed to define the [7] design response spectra for a particular application of response spectra.

S_a = for periods less than or equal to T_O, as shown in Figure 5.15, the design spectral response acceleration shall be given by the formula:

$$S_a = S_{DS}\ (0.4 + 0.6\mathrm{T/T_O}) \tag{5.48}$$

S_{DS} = design spectral response acceleration at short period (0.2 second) based on S_{MS}, F_a, S_S from Figure 9.4.1.1(a) of [7] as shown in Figure 5.13; S_D1 = design spectral response acceleration at 1.0-second period as a function of S_M1, F_V, S_1 from Figure 9.4.1.1(b) of [7]; T = fundamental period of the structure (seconds); T_O = period 0.2 S_D1/S_{DS} (seconds); and T_S = period S_D1/S_{DS} (seconds).

For the cantilever tank example: $S_{DS} = 0.24$ g from Figure 9.4.1.1(a) of [7] for:

$$S_S = 0.3\ g,\ F_a = 1.2,\ S_{MS} = 0.36\ g$$

$S_D1 = 0.114$ g from Figure 9.4.1.1(b) of [7] for:

$$S_1 = 0.1\ g,\ F_V = 1.7,\ S_M1 = 1.7 \times 0.1 = 0.17\ g$$

then $T_O = 0.2 \times 0.114/0.24$ seconds; $T_O = 0.095$ seconds; $T_S = 0.114/0.24$ seconds; and $T_S = 0.475$ seconds.

From Eq. (5.37) the peak ground acceleration = 0.24 [(0.4 + 0.6 × (0/0.095)] = 0.096 g. The design response spectra curve contained in [7] as Figure 9.4.1.2.6, as shown in Figure 5.15, appears to be median-shaped with an amplification factor of about 2.5 over peak ground acceleration at 0.2-second period and 5% damping.

Then, the total lateral seismic inertia force, F_p, is determined from Eq. (5.3.).

$$F_p = \frac{0.4x2.5x0.24gx1050\ \text{lbs}}{2.5/1.5}(1+(2x0.5)) \tag{5.49a}$$

$$F_p = \frac{1.0x0.24x1050}{1.67}(2.0) = 302\ \text{lbs.} \tag{5.49b}$$

To obtain the bending moment associated with this force applied to the center of mass of the tank, the usual practice for a building structure cantilever beam element is to apply the force distributed as an inverted triangle to the building structure represented as a vertical cantilever beam. The inverted triangle is taken as a reasonable approximation of the dominant first mode response of a cantilever beam as shown in Figure

5.5(b). However, it is also common practice to apply the resultant lateral load, F_p, to the center of gravity (c.g.) of a component. This gives the same base shear but a less conservative overturning moment.

Seismic lateral load = 301 lbs.
Seismic moment at base with lateral load applied as an inverted triangle = 302 × 2/3 × 120 = 24160 in-lbs.
Seismic moment at base if lateral load is applied at c.g. = 302 × 1/2 × 120 = 18120 in-lbs.

This moment applied to the tank cross section is then divided by the section modulus of the tank base to determine the seismic-induced longitudinal membrane stress in the tank for the earthquake applied in one lateral direction. In rotationally symmetric components such as tanks, stacks or vessels, it does not matter if a horizontal single direction or horizontal resultant of two horizontal earthquake components of equal magnitude are applied, since the stress resultants in the rotationally symmetric cylindrical shell is the same. However, for square, rectangle or other cross-sectional component, cross-section shapes or their anchorage system, except in a circular pattern, stress resultants from the two horizontal earthquake components as compared to a resultant will not give the same seismic resultant.

For other than rotationally symmetric structures and anchor bolt patterns, the preferred approach would be to apply each of the two horizontal components of earthquake motion, F_p, separately to compute resultant forces, moments or stresses. The resultant force and moment stresses—as a result of the fact that earthquake motions are applied independently with respect to direction—would then be combined on an SRSS basis. These earthquake stresses are then combined with other applicable load-induced stresses to be compared against acceptance criteria. In any case, earthquake resultant stresses would be given both a plus and minus sensing.

When using *ASCE-7* [7] design procedures, vertical earthquake motions are considered as defined in Section 9.5.2.7 and 9.5.2.4 of [7], where the vertical component of earthquake load is taken as:

$$E_V = 0.2\ S_{DS} D \tag{5.50}$$

For the vertical tank $E_V = 0.2 \times 0.24\text{g} \times 1050$ lbs; and $E_V = 50.4$ lbs.

The stress resultant from the application of this force is given a ± sensing.

Equivalent Static Analysis

In the equivalent static load procedure applied to mechanical SSC, rather than using statically applied lateral seismic load as shown in Eq. (5.3), use K•ma, where *a* is the peak spectral acceleration from the applicable response spectra defined at the support of the component and applied at the c.g. of the mechanical SSC. *K* is a coefficient used to account for higher modes of response and the distortion of the mode shapes resulting from defining all modal displacements as positive. The acceleration times the mass of the SSC are all assumed to act in the same direction; hence, the actual mode shapes, which can have + and – modal coordinates, are lost. This results in external seismic forces being applied—all with the same sense of direction. The resultant seismic stress in the SSC component or anchorage would be given both a (+) and (–) sensing.

When evaluating *K* as a function of the ratio of inertia stresses determined by dynamic analysis to equivalent static stress resultants, values less than 1.0 are usually determined. However, there are systems such as beams continuous over several supports where the value of *K* can exceed 1.0 [8, 9]. For this reason, to provide additional margin a *K* value of 1.5 has often been selected for design purposes [26] for earthquake loads.

The equivalent static design approach is similar to the *ASCE 7-05* approach, except that the F_p term of Eq. (5.3) for horizontal load would be typically replaced as follows:

$$F_p = 1.5 S_{a_{max}} m \tag{5.51}$$

$$F_p = 1.5 \times 0.48\ \text{g} \times 1{,}050 = 756 \text{ lbs}$$

$$M_o = F_p \ell_{cG}$$

$$M_o = 756 \times 60 = 45{,}360 \text{ in-lbs}$$

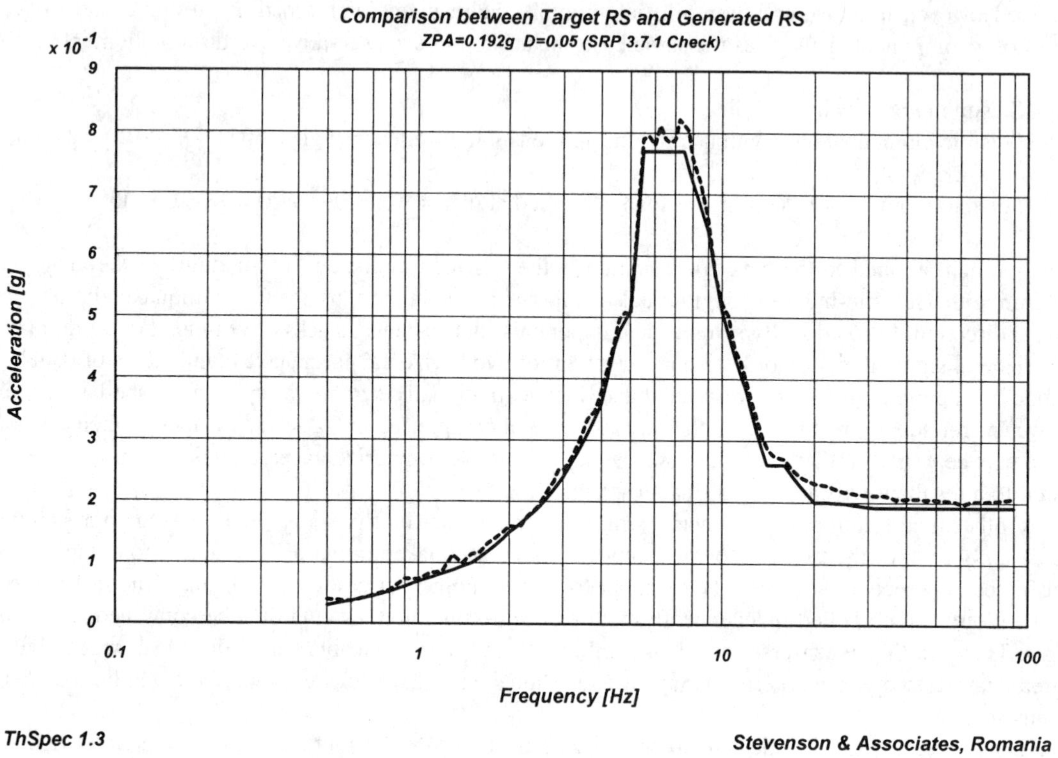

FIGURE 5.18. COMPARISON BETWEEN TARGET IN-STRUCTURE DESIGN TIME HISTORY AND GENERATED RESPONSE SPECTRUM

where $S\alpha_{max}$ = maximum acceleration of the applicable response spectra for the appropriate percent critical damping. In order to permit direct comparison with the *ASCE 7-05* method, the response spectra contained in Figure 5.15 from [7] with a peak design acceleration, S_{DS}, of 0.24 g multiplier by the height adjustment factor $(1 + 2z/h)$ or 0.48 g. Alternatively, the in-structure (Figure 5.18) response spectra could be used. In the vertical direction no building height amplification is used; m = unit masses or lumped mass at the center of gravity; and ℓ_{cG} = center of gravity of component above the support 60 inches.

Values of K less than 1.5 may be used if they can be justified by equivalent static comparison with valid dynamic analysis for the geometry under consideration [6, 26].

In the vertical direction, if the fundamental mode is less than 50 Hz, then Eq. (5.38) is applied to the applicable vertical response spectrum to obtain the vertical earthquake force. If the component has a fundamental frequency above 50 Hz, then the K factor of 1.5 is reduced to 1.0.

For this method and the response spectrum modal analysis methods, when the vertical component of earthquake is considered, its effect may also be added on a SRSS basis. Alternatively a 100, 40, 40% or 40, 100, 40% or 40, 40, 100% directional combination may also be used. The use of the SRSS basis is preferred because it gives rise to fewer load combinations to be considered in design.

For components located inside buildings, the horizontal in-structure response spectrum shown in Figure 5.16 for 5% equipment damping is modified by multiplying all spectral values by the ratio of peak floor acceleration at the support point of the component to the peak floor accelerations of Figure 5.16 for 5% equipment damping. This assumes that the spectral shape of Figure 5.16 is characteristic of all in-structure response spectra shapes as a function of 7% building damping for various equipment damping. Peak floor accelerations, A_p, are considered a function of the type of building structure and height of the component support as follows:

For concrete shear wall construction: $A_p = S_R\ (1+2\ z/h)$
For concrete moment frame construction: $A_p = S_R\ (1+3\ z/h)$
For structural steel moment frame construction: $A_p = S_R\ (1+4\ z/h)$

where S_R = zero period ground acceleration; z = height of component support point in the building; and h = height of the building.

Applying this modification to the tank example with the tank supported at the midheight of a concrete shear wall structure: $A_p = 0.096$ g (1+2•0.5/1.0) = 0.192 g

The peak floor acceleration given in Figure 5.16 is 0.563 g. Therefore, the applicable in-structure response spectrum for 5% equipment damping and 7% building damping as shown in Figure 5.18 and Table 5.9 is determined by multiplying the 5% damped curve of Figure 5.16 by the ratio 0.192/0.563 = 0.341.

Applying the Figure 5.18 spectrum to the equivalent static solution of the tank example results in a shear of: $F_p = 1.5 \times 0.77\text{g} \times 1{,}050 = 1{,}212.8$ lbs

and moment: $M_o = 1{,}212.8 \times 60 = 72{,}765$ in-lbs

Vertical ground response spectra accelerations are typically taken as 2/3 times the horizontal ground spectra. However, the vertical in-structure response spectra are typically taken as having the ground spectra shape (i.e., building is considered rigid in the vertical direction).

This is a valid assumption for structures founded on rock (class A or B as defined in [7] foundation media), but may be unconservative for soil founded buildings (class C and D as defined in [7] foundation media). For soil-founded buildings vertical in-structure response spectra, it is recommended that the vertical ground response spectra be used multiplied by the height correction factor used for horizontal spectra which is a function of the type of building construction.

For the tank example, the horizontal peak ground response spectra value has a peak spectral value of S_{DS} = 0.24 g so a vertical spectra would be 2/3 or this value of 0.16 g. The resultant vertical force on the structure applied at the center of gravity is:

$$0.24\text{g} \times 2/3 \times 1050 = 168 \text{ lbs (rock)}$$
$$2 \times 0.24\text{g} \times 2/3 \times 1050 = 336 \text{ lbs (soil)}$$

Of course, both the horizontal and vertical in-structure response spectra can be generated by applying a time history motion at the base of the building structure, which generates a response time history motion in the structure from which an in-structure response spectra can be generated. The ground motion time history used is developed such that it just envelopes the ground response spectra typically for 5% damping. It is also possible, knowing the shape of the ground response spectrum and the dynamic characteristics of the building, to directly generate in-structure response spectra.

TABLE 5.9. FLOOR RESPONSE SPECTRAL ACCELERATION VALUES (IN G'S) TAKEN FROM THE SHAPE OF THE FIGURE 5.18 SPECTRUM FOR 5% EQUIPMENT DAMPING NORMALIZED TO A PEAK FLOOR ACCELERATION OF 0.192G

Hz															
Acceleration in % gravity															
0.5	1.0	1.5	2.0	2.5	3.0	3.5	4.0	4.5	5.0	.5	6.0	6.5	7.0	7.5	8.0
0.03	0.07	0.10	0.15	0.19	0.24	0.30	0.36	0.47	0.51	0.77	0.77	0.77	0.77	0.77	0.72
HZ															
Acceleration in % gravity															
8.5	9.0	10.0	12.0	14.0	16.0	20.0	24.0	28.0	30.0	∞	—	—	—	—	—
0.68	0.64	0.51	0.38	0.26	0.26	0.2	0.2	0.195	0.192	0.192	—	—	—	—	—

Response Spectrum Modal Analysis

In the response spectrum modal analysis (RSMA) procedure, loads are determined in each mode of response in each direction and should be applied as shown in Section 5.3.3 and Eq. (5.11). To perform the RSMA procedure it is necessary to define response spectrum for design purposes. The design spectra used should be defined at the support point(s) of the component being analyzed. To properly analyze a tank supported halfway up the height of a structure, it is necessary to generate a floor or amplified spectra similar in shape to that shown in Figure 5.16 for 5.0% damping. Using the spectral shapes of Figure 5.16 as a typical floor response spectra it will be necessary to normalize this spectra to a midheight peak floor acceleration used in the tank example of $2 \times 0.096 = 0.192$ g. This means all spectral values of Figure 5.16 would be multiplied by $0.192/0.563 = 0.341$ to obtain the design floor or in-structure response spectrum of Figure 5.18. In actual design applications it would be necessary to perform a dynamic analysis of the building to develop design-basis floor or amplified response spectra.

In evaluating the air tank dynamically, using modal analysis and response spectra methodology, two different boundary (support) conditions are considered. The simplest support assumption is when the base of the tank is rigidly attached to the foundation. This condition occurs when the seismic lateral inertia loads are insufficient to cause tension or liftoff of the base of the component.

When there is no tension or liftoff of the base the base of the tank can be assumed fixed. When tension is developed in the tank foundation anchor bolts, the stiffness of the anchor bolts will begin to affect the restraint of the tank base. This stiffness is defined by:

$$K_{\text{bolt}} = \text{AE/L}$$
$$K = 0.2 \times 30 \times 10^6/6.33$$

where K_{bolt} = bolt stiffness, 9.47×10^5 lbs/in; A = area of bolt, 0.5 in^2; L = effective length of bolt 6.33 inches as defined by expansion anchor manufacturer; and E = bolt modulus of elasticity, 30×10^6 psi.

The effective L for expansion anchor is as defined by the manufacturer. In cases where this is not specified the total length of the anchor assembly may be used. In the instance where cast-in-place bolts are used, the effective length as a function of type of anchor used is given in Tables 5.10 or 5.11.

When liftoff occurs (the reaction to the inertia overturning moment caused by the inertia lateral load, plus the dead weight falls outside the middle third of the base diameter), the simplified rotational stiffness in this example is reduced to that of a single bolt with rotation about a line connecting the two other bolts, as in the case of the tank in Figure 5.17.

Another more conservative method of determining rotation stiffness would be to assume rotation about the centroidal axis of the tank base.

Using the geometric, mass and stiffness properties of the tank and its anchorage, and sketched in Figure 5.17, a computer program eigen value solution can be performed with the following dynamic modal characteristic results (only the first four modes have been considered in the analysis):

For $K_{\text{bolt}} = 9.47 \times 10^5$ lb/in and the rocking stiffness, K_{rocking}, is K_{bolt} times the 18.0 inch lever arm shown in Figure 5.17, $K_{\text{roking}} = 1.7 \times 10^7$ lb/in/rad

a. Flexible Base Analysis

Frequency Analysis (With Flexible Base Stiffness)

Mode	Frequency (Hz)	Modal Mass (% of total mass)
First-mode horizontal direction	5.58	84
First-mode vertical direction	341.9	89

TABLE 5.10. NOMINAL ALLOWABLE CAPACITIES FOR CAST-IN-PLACE BOLTS AND HEADED STUDS ($F'_C \geq 3{,}500$ PSI)[1]

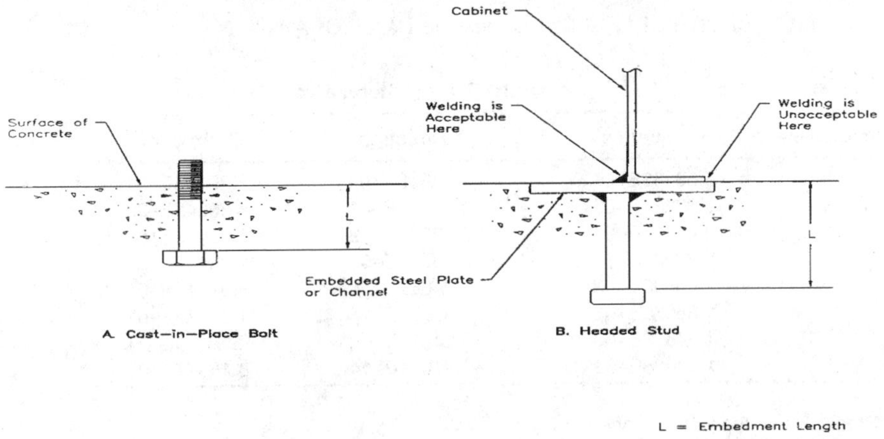

Bolt/Stud Diameter (D, in.)	Pullout Capacity (P_{nom}, kip)	Shear Capacity (V_{nom}, kip)	Minimum Embedment (L_{min}, in.)
3/8	3.74	1.87	3-3/4
1/2	6.66	3.33	5
5/8	10.44	5.22	6-1/4
3/4	15.03	7.51	7-1/2
7/8	20.44	10.22	8-3/4
1	26.69	13.35	10
1–1/8	33.80	16.90	11-1/4
1–1/4	41.72	20.86	12-1/2
1–3/8	50.40	25.25	13-3/4

[1] The pullout and shear capacities are from ASTM A-307 bolts installed in sound, uncracked concrete (i.e., no cracks passing through the anchor bolt installation) with a compressive strength of 3,500 psi or greater.

TABLE 5.11. NOMINAL ALLOWABLE CAPACITIES FOR J-BOLTS CAST-IN-PLACE ($F'_C \geq 3{,}500$ PSI)[1,2,3]

Bar Diameter (*D* in.)	Pullout Capacity (P_{nom}, kip)	Shear Capacity (V_{nom}, kip)
3/8	3.74	1.87
1/2	6.66	3.33
5/8	10.44	5.22
3/4	15.03	7.51
7/8	20.44	10.22
1	26.69	13.35
1–1/8	33.80	16.90
1–1/4	41.72	20.86
1–3/8	50.40	25.25

1. The pullout and shear capacities shown here are from ASTM A-307 J-Bolts installed in sound, uncracked concrete with a compressive strength (f'_c) of at least 3,500 psi.
2. The minimum total embedment length to develop the full tensile capacity of the anchor is $L' = P_{nom}/0.12\ f'_c\ \pi\ D$ where L' is the total length of the anchor including the hooks; D is the nominal diameter of the bolt where *l*oad *r*esistance *f*actor *d*esign is used. When *a*llowable *s*tress *d*esign is used the 0.12 coefficient is reduced to 0.07.
3. The effective embedment length, L', may be used to determine anchor stiffness.

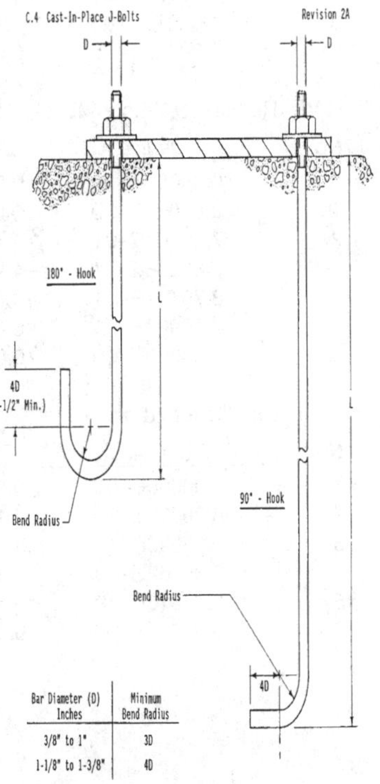

The fundamental vertical (y-direction) frequency is 341.9 Hz, which is well into the rigid range. The resultant vertical acceleration of the vessel would be equal to the vertical acceleration of the floor.

Seismic analysis of vertical vessel with Flexible Base; RS Analysis

		Modal Participation Factors	
Frequencies Mode	**Frequency**	***X*-Direction**	***Z*-Direction**
1	5.58463E+00	1.24618E+00	−2.29015E−31
2	2.32231E+02	−4.94465E−01	−8.56240E−28
3	3.41949E+02	3.32532E−19	1.28476E+00
4	7.89929E+02	2.18624E−01	−1.28187E−25
5	1.00606E+03	5.14067E−15	3.97814E−01
6	1.58076E+03	−2.37479E−14	1.94772E−01
7	1.61533E+03	9.34675E−02	−2.59110E−21
8	1.97268E+03	−2.97631E−14	8.48677E−02

Mass Properties for the Whole model:
Based on translational masses, including masses at the base:

Summation				**Center of Mass**	
***X*-Direction**	***Y*-Direction**	***Z*-Direction**	***X*-Coord.**	***Y*-Coord.**	***Z*-Coord.**
2.7170E+00	—	2.7170E+00	0.0000E+00	0.0000E+00	5.3596E+01

Based on translational masses excluding masses at the base.

Summation		
***X*-Direction**	***Y*-Direction**	***Z*-Direction**
1.8540E+00		1.8540E+00

Seismic analysis of vertical vessel with flexible base; RS analysis

Nodal Results for Whole Model Modal Results Mode = 1

Node	*X*-Dir Displ.	*Z*-Dir Displ.	*Y*-Rot Displ.
1	0.00000E+00	0.00000E+00	8.45609E−03
2	2.55021E−01	−5.62589E−32	8.53999E−03
3	5.12094E−01	−1.06615E−31	8.59331E−03
4	7.70356E−01	−1.45786E-31	8.62024E−03
5	1.02911E+00	−1.69664E−31	8.62781E−03
6	0.00000E+00	0.00000E+00	0.00000E+00

Nodal Results for Whole Model Modal Results Mode = 2

Node	*X*-Dir Displ.	*Z*-Dir Displ.	*Y*-Rot Displ.
1	0.00000E+00	0.00000E+00	8.45609E−03
1	0.00000E+00	0.00000E+00	−2.91283E−02
2	−7.69267E−01	−2.41356E−28	−1.84980E−02
3	−9.63298E−01	−4.22844E−28	6.79064E−03
4	−3.79441E−01	−4.47455E−28	3.03181E−02
5	7.10927E−01	−6.65904E−28	3.93593E−02
6	0.00000E+00	0.00000E+00	0.00000E+00

Nodal Results for Whole Model Modal Results Mode = 3

Node	*X*-Dir Displ.	*Z*-Dir Displ.	*Y*-Rot Displ.
1	0.00000E+00	0.00000E+00	2.25643E−21
2	6.80501E−20	3.15596E−01	2.27882E−21
3	1.36648E−19	5.98089E−01	2.29305E−21
4	2.05562E−19	8.17848E−01	2.30023E−21
5	2.74610E−19	9.51822E−01	2.30225E−21
6	0.00000E+00	0.00000E+00	0.00000E+00

Nodal Results for Whole Model Modal Results Mode = 4

Node	*X*-Dir Displ.	*Z*-Dir Displ.	*Y*-Rot Displ.
1	0.00000E+00	0.00000E+00	5.37450E–02
2	1.06046E+00	–3.89219E–25	–1.76103E–03
3	–6.56819E–02	–1.43682E–25	–5.32690E–02
4	–1.00492E+00	1.53212E–25	8.29890E–03
5	3.89097E–01	5.92333E–26	6.55513E–02
6	0.00000E+00	0.00000E+00	0.00000E+00

b. Rigid Base Analysis

Frequency Analysis (With Rigid Base)

Mode	Frequency (Hz)	Modal Mass (% of total mass)
First-mode horizontal direction	48.88	71
First-mode vertical direction	341.6	89

Seismic Analysis of Vertical Vessel with Rigid Base; RS Analysis

	Modal Participation Factors		
Frequencies Modal	**Frequency**	***X*-Direction**	***Z*-Direction**
1	4.89020E+01	1.14778E+00	5.13618E–30
2	3.36728E+02	6.11301E–01	6.63548E–15
3	3.41949E+02	–8.36630E–15	1.28476E+00
4	9.73930E+02	3.48501E–01	4.90308E–13
5	1.00606E+03	–5.38701E–13	3.97814E–01
6	1.58076E+03	–2.05757E–14	1.94772E–01
7	1.76991E+03	–2.03609E–01	–3.14315E–13
8	1.97268E+03	–7.66567E–18	–8.48677E–02

Nodal Results for Whole Model Modal Results Mode = 1

Node	*X*-Dir Displ.	*Z*-Dir Displ.	*Y*-Rot Displ.
1	0.00000E+00	0.00000E+00	6.40066E–21
2	1.02132E–01	1.46457E–29	6.43226E–03
3	3.63773E–01	6.36664E–30	1.06483E–02
4	7.20824E–01	–5.23398E–30	1.28438E–02
5	1.11876E+00	–2.79441E–30	1.34747E–02
6	0.00000E+00	0.00000E+00	0.00000E+00

Nodal Results for Whole Model Modal Results Mode = 2

Node	*X*-Dir Displ.	*Z*-Dir Displ.	*Y*-Rot Displ.
1	0.00000E+00	0.00000E+00	4.57893E–20
2	5.26198E–01	1.62998E–15	2.55717E–02
3	1.06866E+00	3.08899E–15	4.57877E–03
4	6.53047E–01	4.22399E–15	–3.12019E–02
5	–6.08294E–01	4.91594E–15	–4.74661E–02
6	0.00000E+00	0.00000E+00	0.00000E+00

Nodal Results for Whole Model Modal Results Mode = 3

Node	*X*-Dir Displ.	*Z*-Dir Displ.	*Y*-Rot Displ.
1	0.00000E+00	0.00000E+00	–6.26675E–34
2	–7.20158E–15	3.15596E–01	–3.49976E–16
3	–1.46257E–14	5.98089E–01	–6.26653E–17
4	–8.93764E–15	8.17848E–01	4.27032E–16

5	8.32516E–15	9.51822E–01	6.49624E–16
6	0.00000E+00	0.00000E+00	0.00000E+00

Nodal Results for Whole Model Modal Results Mode = 4

Node	*X*-Dir Displ.	*Z*-Dir Displ.	*Y*-Rot Displ.
1	0.00000E+00	0.00000E+00	1.34559E–19
2	1.02416E+00	1.04257E–12	2.29313E–02
3	3.62189E–01	1.13853E–12	–5.55062E–02
4	–1.00910E+00	2.00769E–13	–4.23255E–03
5	3.27071E–01	–9.19283E–13	6.89246E–02
6	0.00000E+00	0.00000E+00	0.00000E+00

Ignoring the potential for liftoff and resultant rocking seismic response will often underestimate horizontal input response motion. Most mechanical components with their bases assumed rigidly attached to the supporting structure will have dominant response frequencies well above the amplified portion of the response spectrum (i.e., $f_p \geq 16$ Hz).

Employing the procedures in Section 5.3.3:

Rigid base modal analysis results: Base shear = 101.99 lbs; vertical reaction force = 81.64 lbs; and overturning moment = 9847.17 lb-in.

Flexible base modal analysis results: Base shear = 462.42 lbs; vertical reaction = 81.64 lbs; and overturning moment = 43289.10 lb-in.

Due to the fact that the base node is fixed the total mass considered in the analysis correspond to a weight of 716 lb. The remaining mass is attached to the base node (fixed DOFs). The vertical response of the tank is rigid. The vertical reaction can be evaluated by hand calculations:

Vertical reaction = $716 \times \left(\frac{2}{3}\right) \times 0.192 = 91.64$ lb ; the value obtained by RS analysis is 81.64 lb.

The difference represents the contribution of the higher modes. Only the first four modes have been considered. Modes with frequencies greater than 1,000 Hz have been ignored. Typically higher modes are neglected above 50 Hz frequency in a response spectrum analysis. However, care must be taken to check modal effective mass to ensure that at least 80% percent of the mass has participated in the modes considered. If not, a missing mass correction factor is needed.

Dynamic Time History Modal Analysis

The dynamic time history load: In this procedure the time history analysis can be solved by either direct integration of the equations of motion or by time step modal analysis, as described earlier. The time history input to the tank example problem is selected so that the time history equivalent 5% damped response spectra just envelopes the floor response spectral values given in Figure 5.18. The resultant time history is shown in Figure 5.19.

Dynamic time history modal results (only first four modes considered):

Rigid base TH analysis results: Base shear = 142.37 lbs; vertical reaction force = 84.10 lbs; and overturning moment = 11,473.0 lb-in.

Flexible base TH analysis results: Base shear = 468.87 lbs; vertical reaction = 84.10 lbs; and overturning moment = 44,129.4 lb-in.

Dynamic Time History Direct Integration Analysis

As in the case of practical response spectrum and modal time history dynamic analysis, a computer program is required to perform the direct integration time history dynamic analysis. Figures 5.20 to 5.27 show the time history of the base shear and overturning moment for both rigid-base and flexible-base assumptions.

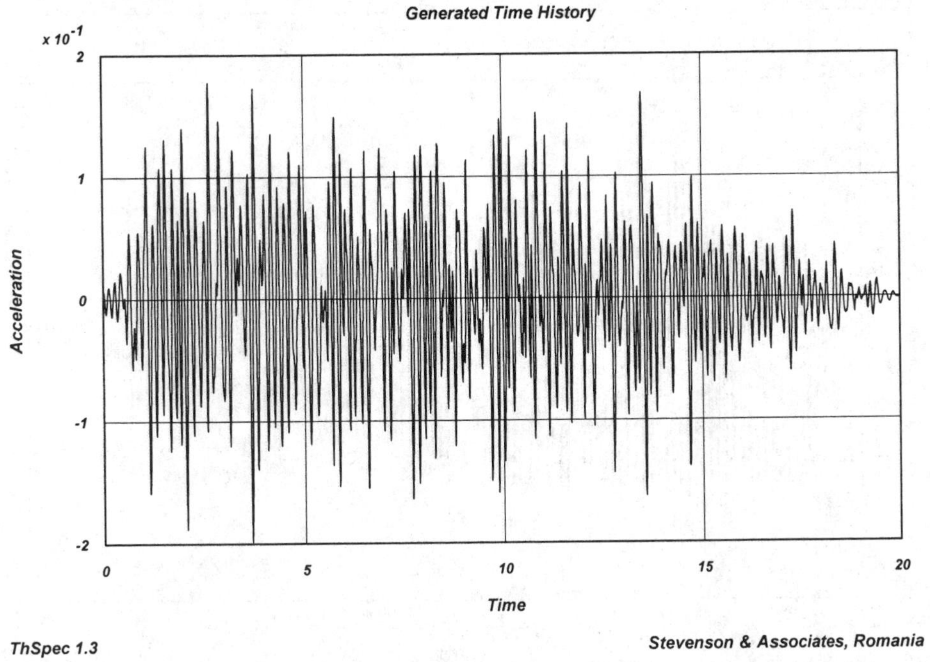

FIGURE 5.19. GENERATED TIME HISTORY COMPATIBLE WITH THE TARGET RESPONSE SPECTRUM OF FIGURE 5.18

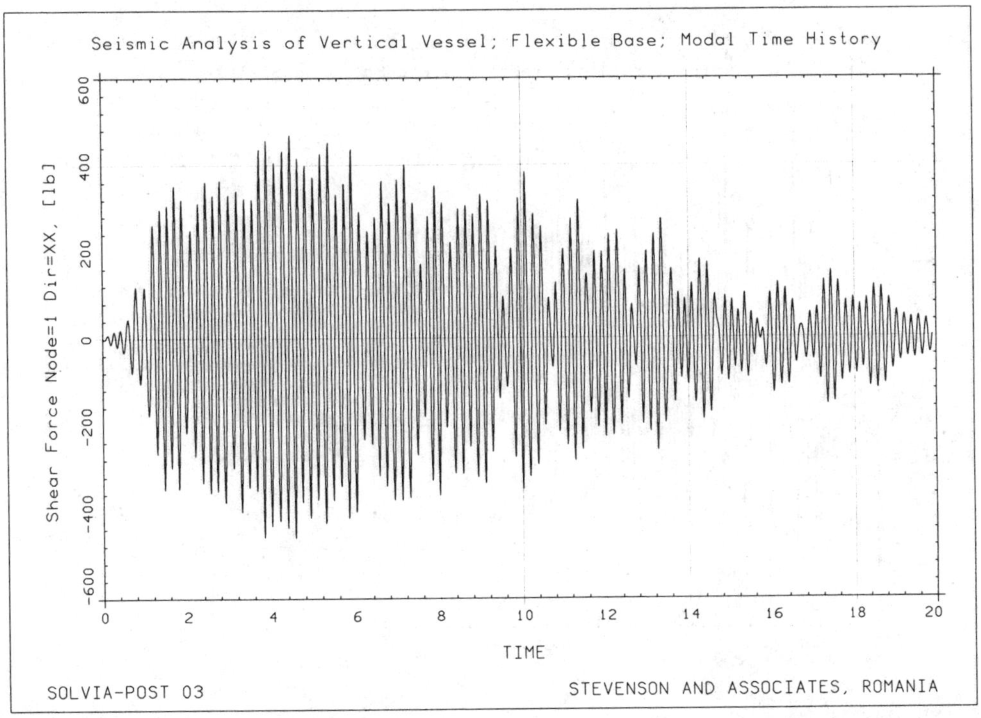

FIGURE 5.20. FLEXIBLE-BASE MODEL: MODAL TIME HISTORY ANALYSIS – BASE SHEAR FORCE

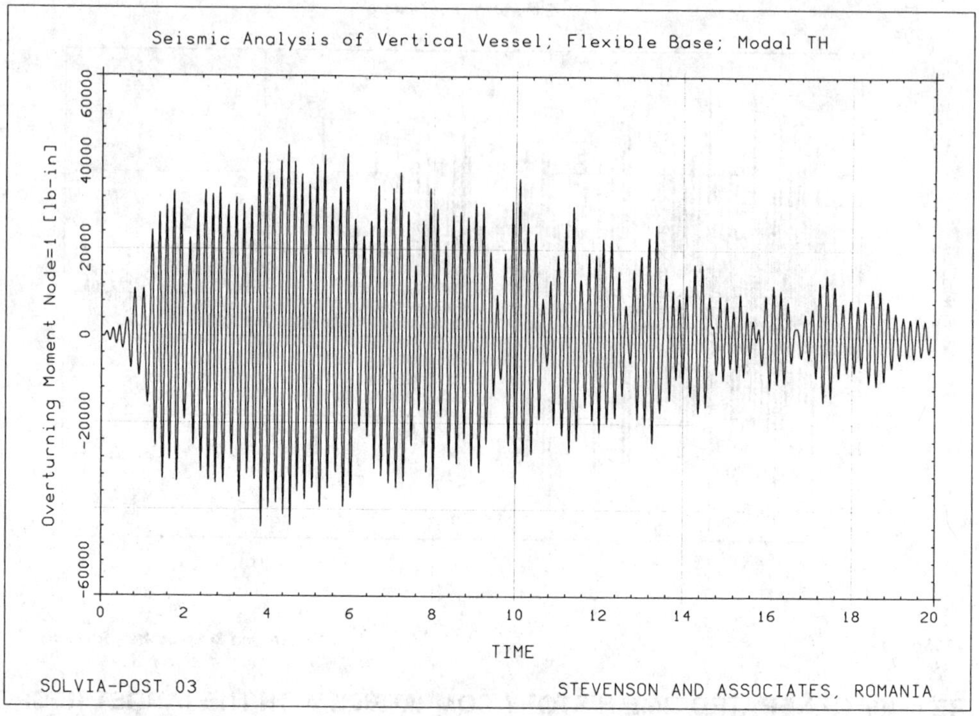

FIGURE 5.21. FLEXIBLE-BASE MODEL: MODAL TIME HISTORY ANALYSIS—OVERTURNING MOMENT (LB-IN)\P

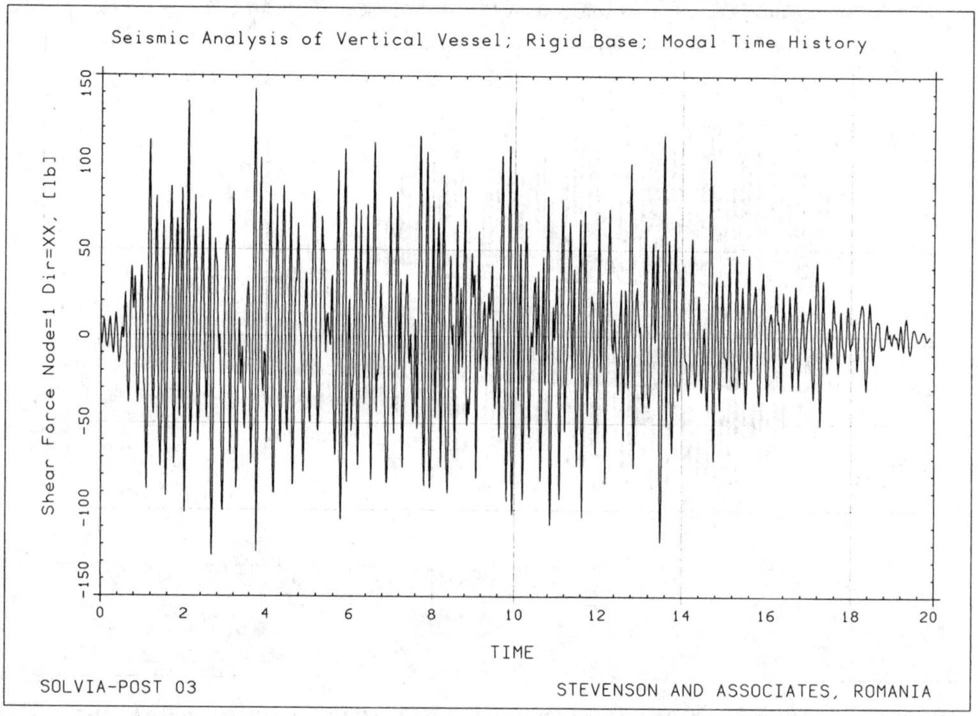

FIGURE 5.22. RIGID BASE-MODEL: MODAL TIME HISTORY ANALYSIS – BASE SHEAR FORCE (LB)

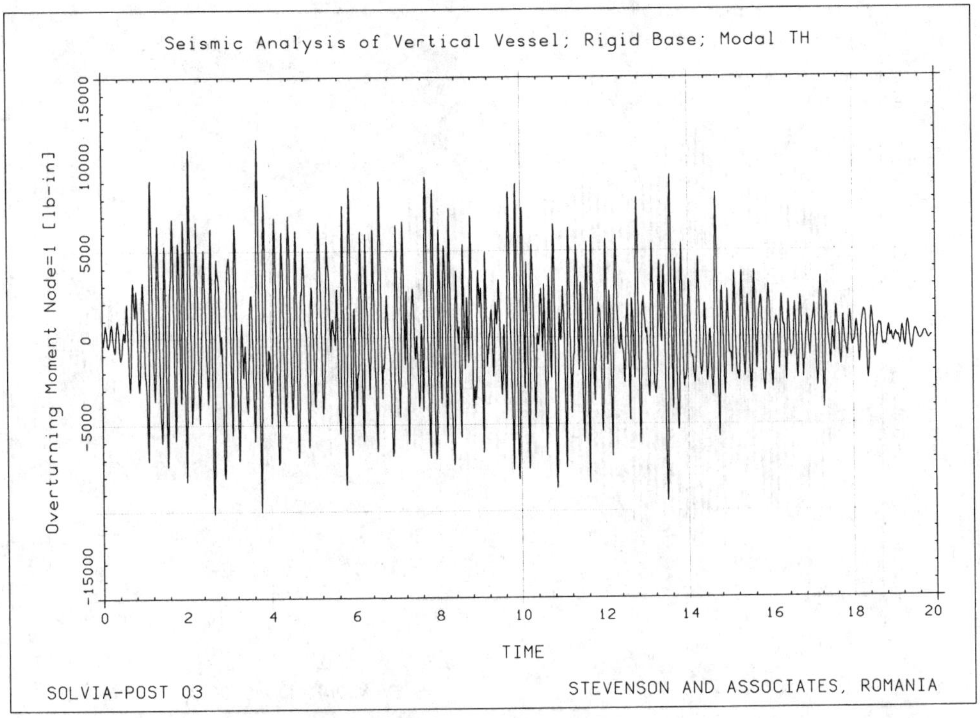

FIGURE 5.23. RIGID-BASE MODEL: MODAL TIME HISTORY ANALYSIS – OVERTURNING MOMENT (LB-IN)

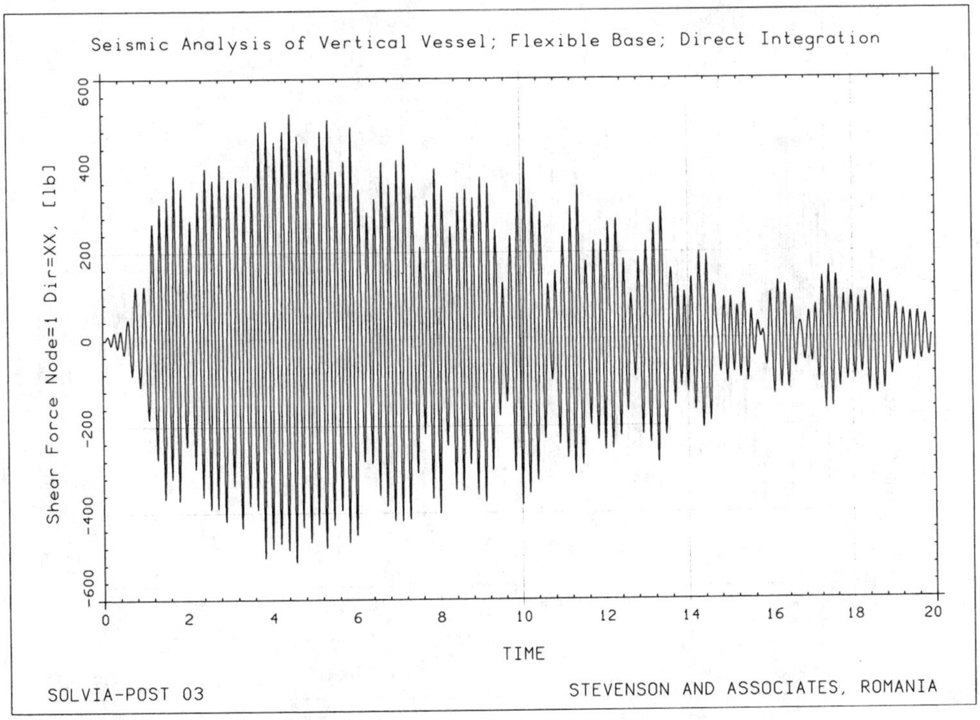

FIGURE 5.24. FLEXIBLE-BASE MODEL: DIRECT INTEGRATION ANALYSIS – BASE SHEAR FORCE (LB)

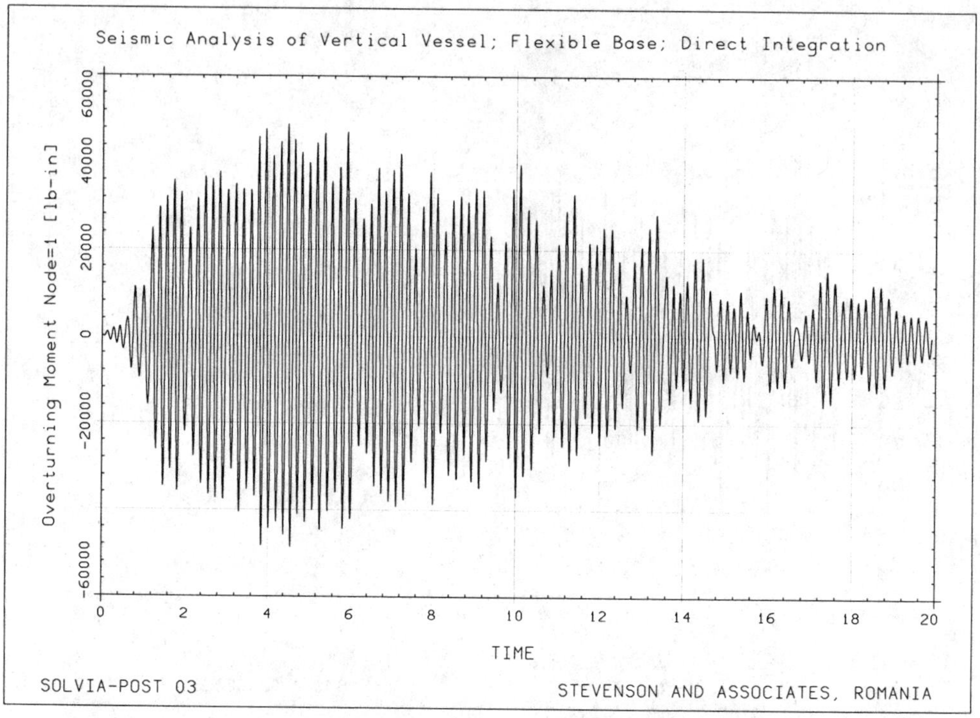

FIGURE 5.25. FLEXIBLE-BASE MODEL: DIRECT INTEGRATION ANALYSIS—OVERTURNING MOMENT (LB-IN)

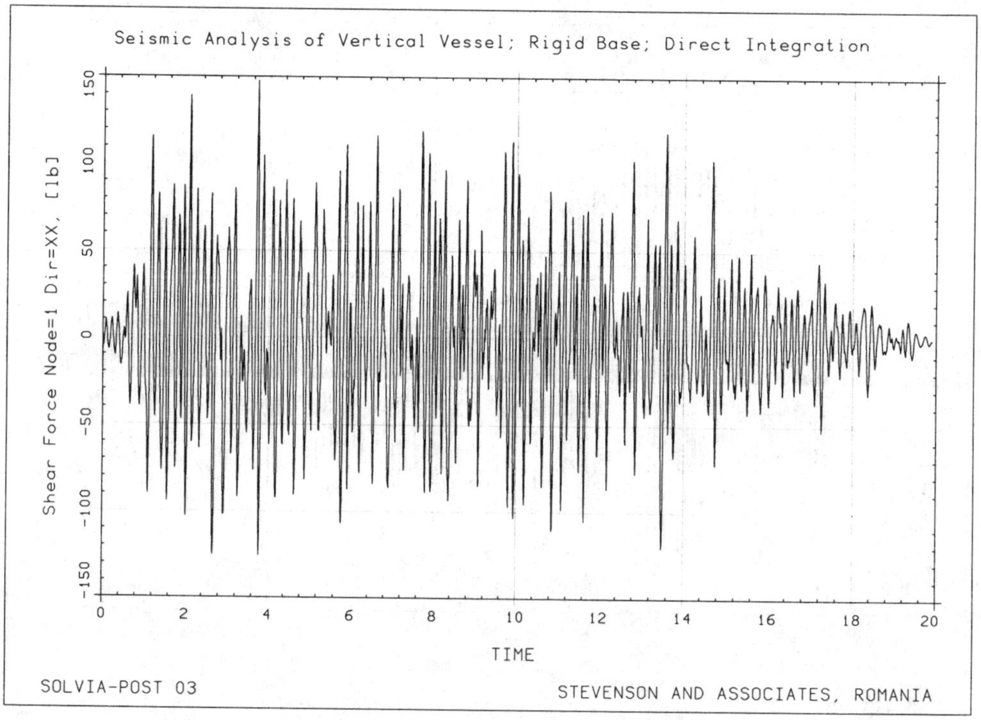

FIGURE 5.26. RIGID-BASE MODEL: DIRECT INTEGRATION ANALYSIS – BASE SHEAR FORCE (LB)

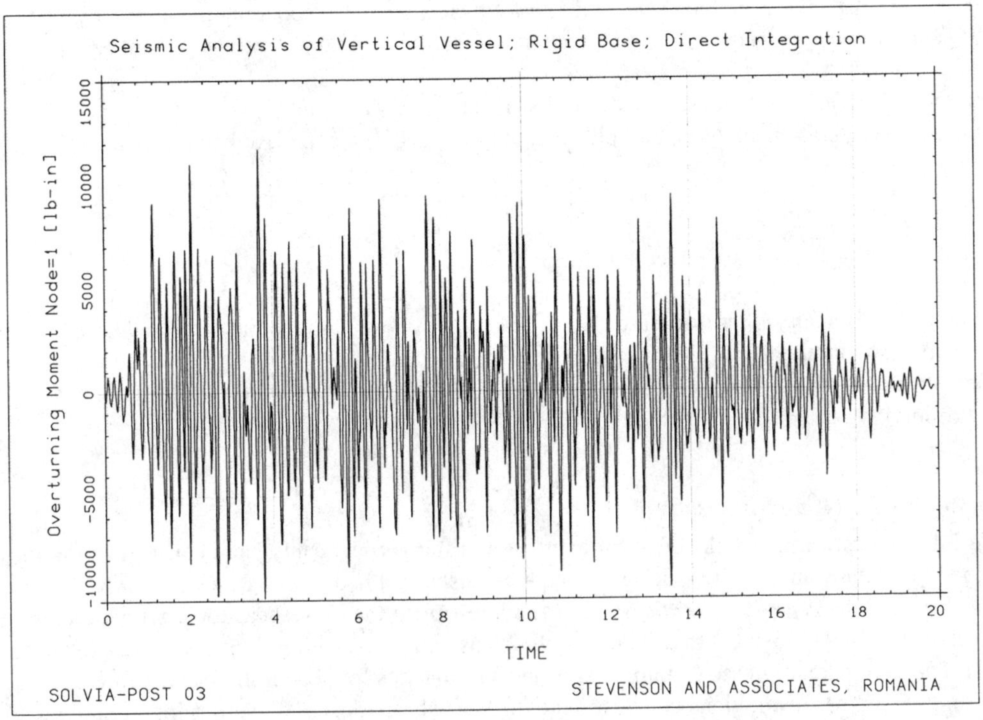

FIGURE 5.27. RIGID-BASE MODEL: DIRECT INTEGRATION ANALYSIS–OVERTURNING MOMENT (LB-IN)

One important difference between modal TH and direct integration is the way of defining the damping properties. In case of modal TH the damping properties are given as modal damping coefficients for each mode. Direct integration requires the construction of a damping matrix:

$$[C] = a[M] + \beta[K] \tag{5.52}$$

where **M** and **K** = mass and stiffness matrices; α and β = Rayleish coefficients; and C = damping matrix represented as a linear combination between **M** and **K** matrices.

According to Appendix N "Dynamic Analysis Methods" [10], the Rayleigh coefficients α and β can be calculated as a function of damping ratio for a frequency range as following:

$$a = \frac{2\omega_r\omega_s(\zeta_s\omega_r - \zeta_r\omega_s)}{\omega_r^2 - \omega_s^2} \tag{5.53}$$

$$\beta = \frac{2(\zeta_r\omega_r - \zeta_s\omega_s)}{\omega_r^2 - \omega_s^2} \tag{5.54}$$

and the frequency dependent damping ratio is:

$$\zeta_i = \frac{1}{\omega_r^2 - \omega_s^2}\left[\frac{\omega_r\omega_s}{\omega_i}(\zeta_s\omega_r - \zeta_r\omega_s) + \omega_i(\zeta_r\omega_r - \zeta_s\omega_s)\right] \tag{5.55}$$

where α and β = Rayleigh coefficients; ω_s and ζ_s = frequency and corresponding damping ratio in frequency point "*s*"; and ω_r and ζ_r = frequency and corresponding damping ratio in frequency point "*r*".

Between ω_s and ω_r ($\omega_s < \omega_r$) the damping ratio has a parabolic variation.

For the direct integration analysis, the following Rayleigh coefficients have been calculated:

$$\alpha = 3.029 \qquad \beta = 4.588\times10^{-5}$$

Dynamical time history results – direct integration:

Rigid base TH analysis results: Base shear = 147.47 lbs; vertical reaction force = 94.50 lbs; and overturning moment = 11,705.6 lb-in

Flexible base TH analysis results: Base shear = 519.86 lbs; vertical reaction = 94.50 lbs; and overturning moment = 48,907.00 lb-in

Summary of Air Tank Analysis Results

In Table 5.12 are shown the results of seismic analysis of the vertical cantilevered air tank using the values determined from static and dynamic design procedures discussed here.

ASCE 7-05 and the *Equivalent Static Methods* underestimate the base shear force with a factor of about 1.50 as an overturning moment with a factor of 1.8–2.0 as compared with response spectra (RS) or time history (TH) analysis – flexible-base conditions. RS and TH analysis for such a simple structure produced very close results. Fixed-base conditions underestimate the base shear force and overturning moment with a factor of about 4 as compared with flexible base conditions. The difference between modal TH and direct integration analysis results are small. The differences are due to the manner in which damping is considered as well as the missing modes (high frequency and small precipitation factor modes have been ignored).

The values given in Table 5.12 assume the system responds elastically. The *ASCE-7* code permits ductility factor F_p greater than 1.0 for the design of PC-1 and PC-2 systems and components and this effectively values the seismic-induced forces and moments considered in design. For PC-3, PC-4 and PC-5 systems and components a divisor, F_u, as shown in Table 5.13, may be applied to the elastically computed seismic inertia forces and moments using equivalent static, response spectrum and time history analysis techniques.

5.5.3 Piping Continuous over Multiple Support Systems

Given the three-dimensional *Z*-bend pipe run shown in Figure 5.28 determine the resultant seismic inertia moments considered in the pipe and reaction loads on supports using static, equivalent static, dynamic response spectrum modal analysis, dynamic time history modal analysis and dynamic time step integration

TABLE 5.12. COMPARISON OF SEISMIC RESULTANT FORCES AND MOMENTS AS A FUNCTION OF THE ANALYTICAL METHOD USED AND SUPPORT ASSUMPTIONS

No.	Analysis Method	Base Shear (lb)	Vertical (lb)	Overturning Moment (lb-in)
1	ASCE-7-02	302.00	49.0	24,080
2	ES Fig. 5.15 Spect. X (1+2 *Z/h*)	756.03	101.7	45,360
3	RS-Analysis—Rigid Base	101.99	81.64	9,847
4	RS-Analysis—Flexible Rot. Base	462.00	81.64	43,289
5	Modal TH-Analysis—Rigid Base	142.37	84.10	11,473
6	Modal TH-Analysis—Flexible Rot. Base	468.87	84.10	44,129
7	Direct Integration—Rigid Base	147.47	94.50	11,705
8	Direct Integration—Flexible Base	519.86	94.50	48,907

TABLE 5.13. EQUIPMENT AND DISTRIBUTION SYSTEMS INELASTIC ENERGY ABSORPTION FACTOR (F_μ[1]) (FROM REFERENCE 3 FOR LIMIT STATE B)

	Factor, F_u[2]
EQUIPMENT	
Vessel	1.25
Heat exchanger	1.25
Coolers	1.25
Chillers	1.25
Tanks – vertical	1.25
Tanks – horizontal	1.25
Pumps	1.25
Fans	1.25
Valves	1.25
Dampers	1.25
Filters[3]	1.50
Glove boxes[3]	1.50
Electrical boards[3]	1.50
Electrical racks[3]	1.50
Electrical cabinets[3]	1.50
DISTRIBUTION SYSTEMS	
Butt-joined groove-welded pipe	1.50
Socket-welded pipe	1.25
Threaded pipe	1.15
Conduit	1.35
Instrument tubing	1.35
Cable trays	1.35
HVAC duct	1.25
EQUIPMENT SUPPORTS[3]	1.50

1. These inelastic energy absorption factors (F_μ) are applicable to equipment functioning in a passive mode. For active components, which must move or otherwise change state, $F_\mu = 1.0$.
2. Except as discussed in 3 (below), the allowable behavior Limits for passive components are based on the *ASME B&PVC* Section III allowables for service level D. It should be noted that *ASME B&PVC* Section III service level D allowable stresses range from 1.6 to 2.0 σ_y.
3. These components are normally designed to AISC allowables, which are typically limited to 0.8 to 1.0σ_y and hence are allowed somewhat higher inelastic energy absorption factors as compared to ASME code allowables, where allowable stresses can be as high as 2.0$_y$.

analytical procedures. The piping system has the following physical characteristics: $D_O = 2.0$ in; $E = 30 \times 106$ psi; $I = 2.5$ in; and $W = 10$ lbs/in.

where D_o = nominal pipe diameter; E = modulus of elasticity; I = moment of inertia; and W = unit weight of pipe.

Static

Because of the three-dimensional nature of the layout a static finite-element computer analysis is used with results as shown in Table 5.14. Note that demand moments are typically computed at elbows and tees because stress indices or intensification factors defined by the applicable Code [27,28] are used to amplify computed moments at these locations due to the localized curved nature of the pipe, they usually control design. The layout of the Z-bend piping system is shown in Figure 5.28. Note that typically a minimum of three mass points are taken between supports or restraints to ensure that higher bending nodes between supports are considered in the analysis.

TABLE 5.14. Z-BEND RESULT SUMMARY

Analysis Method	Seismic Bending Moment SRSS($M^2xx+M^2yy+M^2zz$) (lb-in) Elbow-1	Elbow-2	Run Node 2 First Support	Run Node 7 Last Support
ASCE-7	1,405	1,404	3,340	3,320
Equivalent static (EQ-2)	3,372	3370	8016	7968
RS analysis	485.81	485.81	877.15	877.15
Modal TH	408.96	342.76	697.27	877.51
Direct integration TH	489.34	356.30	719.29	1032.34

Note: For ASCE-7; Horizontal acc = [0.4 × 2.5 × 0.24 × (1+2 × 0.5)/(2.5/5)] = 0.288g; vertical acc = 0.2 × 0.24 = 0.048g.

Equivalent Static *Sa* = 0.77g where *Sa* is the peak in-structure response spectral acceleration; horizontal acc = 1.5 × 0.77g = 1.155g; and vertical acc = 2/3 × 1.155 = 0.77g.

EQ static (2) *Sa* = 0.24 g where *Sa* is the peak ground response spectral acceleration; horizontal acc = 1.5 × 0.24 (1+2*0.5) g = 0.72g; and vertical acc = 2/3 × 0.72 = 0.48g.

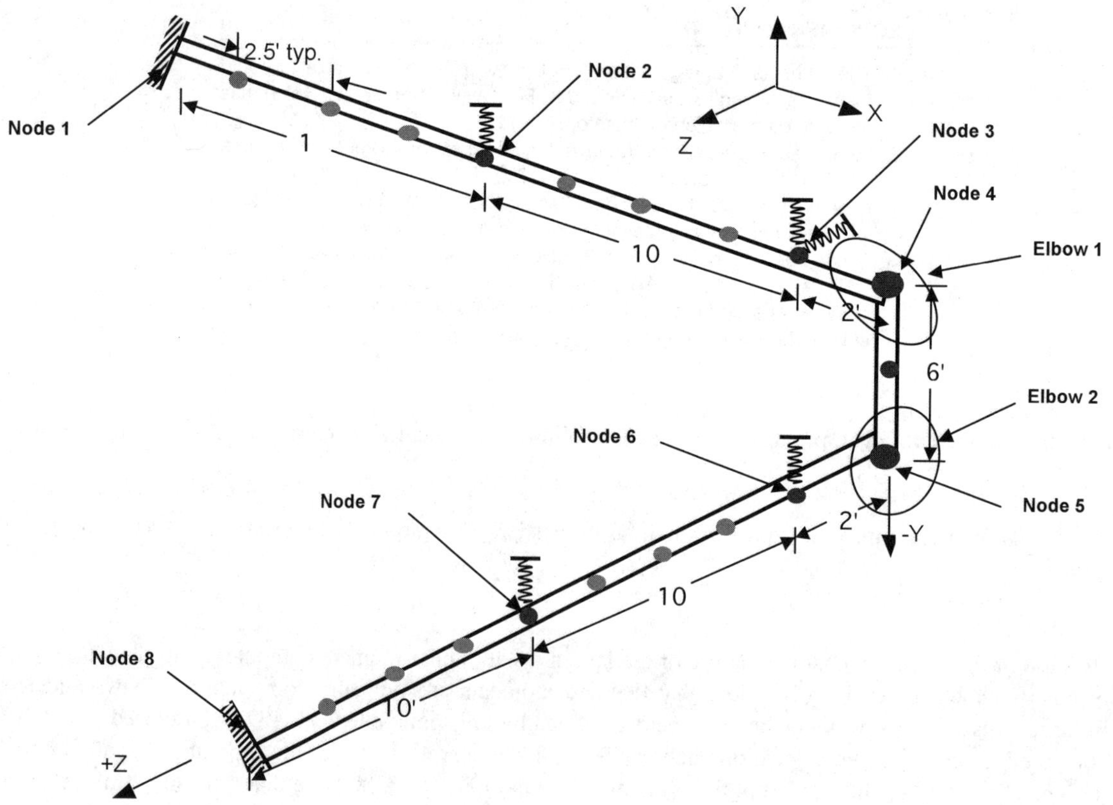

FIGURE 5.28. DYNAMIC FINITE-ELEMENT MODEL OF Z BEND PIPE

ASCE-7 Method of Design

In the analytical procedure the static seismic lateral load, F_P, determined at the support point of the piping system is applied to the lumped mass distribution of the piping system. From Eq. (5.3):

$$F_P = \frac{0.4a_p\, S_{DS}\, W_P\, (1+2z/h)}{(R_P/I_P)} \tag{5.56}$$

where W_P = lumped weight points shown in Figure 5.28.

The W_P at each of the three lumped mass points between supports in Figure 5.28 is typically taken as the weight of the span between support points divided by the number of lumped mass points.

$$F_P = \frac{0.4x2.5x0.24x\,(1{,}200/3)x(1+2\bullet 0.5)}{2.5/1.5} = 105.0 \text{ lbs} \tag{5.57}$$

To obtain the resultant horizontally induced seismic moments in the pipe and reaction at supports and restraints, apply the 105.0-lb load at each of the three mass points between supports perpendicular and parallel to the pipe runs in the horizontal plane. A value of 0.2 S_{DS}D or 0.2 × 0.24 × 1,200/3 equal to 19.2 lbs is applied in the vertical direction.

Equivalent Static

The equivalent static method for piping follows an approach similar to that performed for the tank problem. The $S_{a\,max}$ from the applicable response spectrum is applied to the mass points of the piping system following the Eq. (5.51) approach using the Figure 5.15 spectrum as input, with results given in Table 5.14. Effectively, the inertia force applied to the piping mass points between the supports shown in Figure 5.28 would be:

$$F_P = 1.5x0.24x(1{,}200/3)x(1 + 2z/h) = 288 \text{ lbs} \tag{5.58}$$

The vertical force applied to the piping mass points would be:

$$1.5x2/3\ S_{DS}\ (1{,}200/3) = 96 \text{ lbs} \tag{5.59}$$

If the in-structure response spectrum shown in Figure 5.18 had been used, the resultant bending moments would have been increased in the ratio of 0.77/0.72. In Table 5.14 two equivalent static results are identified. Equivalent static (1) assumes acceleration inputs are available from in-structure or floor response spectrum. The equivalent static (2) results on the ground response spectra are modified horizontally for building height.

Dynamic

Modal Response Spectrum Analysis

Pipeline Z-Bend – RS analysis

Modal Participation Factors

Mode	Frequency	*X*-Direction	*Y*-Direction	*Z*-Direction
1	3.78311E+00	2.85511E–01	–1.26879E–02	2.85511E–01
2	4.34787E+00	–3.02453E–01	–8.03704E–14	3.02453E–01
3	1.20312E+01	1.13443E–01	7.51658E–02	1.13443E–01
4	1.45815E+01	4.69821E–02	7.73353E–12	–4.69821E–02
5	1.46399E+01	–2.76267E–02	1.11110E–01	–2.76267E–02
6	1.58979E+01	–9.27905E–04	–1.33629E–13	9.27905E–04

7	1.91098E+01	−5.37713E−02	2.32344E−01	−5.37713E−02
8	2.40883E+01	1.00612E−01	2.99098E−01	1.00612E−01
9	2.54891E+01	−3.78616E−02	5.72141E−13	3.78616E−02
10	2.71844E+01	8.07886E−02	8.47595E−14	−8.07886E−02
11	3.17685E+01	−1.12463E−02	−5.28919E−01	−1.12463E−02
12	3.74025E+01	−1.81647E−01	1.26952E−01	−1.81647E−01
13	3.82951E+01	−1.97700E−01	4.03818E−13	1.97700E−01
14	5.90692E+01	3.07257E−03	4.94715E−13	−3.07257E−03
15	5.96249E+01	−9.58152E−03	−9.02180E−02	−9.58152E−03
16	6.06052E+01	4.59946E−02	−3.95315E−02	4.59946E−02
17	6.09139E+01	5.47837E−02	−4.38863E−13	−5.47837E−02
18	7.52662E+01	3.68142E−03	−5.20071E−14	−3.68142E−03
19	7.62669E+01	7.41254E−03	2.45385E−03	7.41254E−03
20	8.05030E+01	4.53720E−02	1.11105E−12	−4.53720E−02
21	8.06076E+01	−4.43638E−02	2.28545E−02	−4.43638E−02
22	9.15398E+01	−8.82056E−02	6.85897E−03	−8.82056E−02
23	9.18309E+01	−9.30628E−02	6.68598E−14	9.30628E−02
24	1.18068E+02	−1.33424E−01	2.36328E−02	−1.33424E−01
25	1.18113E+02	1.36479E−01	2.90977E−12	−1.36479E−01
26	1.29159E+02	−3.22726E−02	−2.73288E−13	3.22726E−02
27	1.29698E+02	1.60405E−02	3.17436E−02	1.60405E−02
28	1.32047E+02	1.03543E−01	7.56501E−03	1.03543E−01
29	1.32292E+02	9.51277E−02	−1.29138E−13	−9.51277E−02
30	1.49177E+02	−1.86516E−01	7.08860E−13	1.86516E−01
31	1.50328E+02	−2.15849E−01	1.56273E−01	−2.15849E−01
32	1.54689E+02	2.13593E−01	1.68441E−01	2.13593E−01
33	1.54765E+02	−2.38166E−01	1.85055E−11	2.38166E−01
34	1.76014E+02	−7.84328E−02	−2.45138E−02	−7.84328E−02
35	1.76536E+02	−7.52015E−02	−4.93159E−14	7.52015E−02
36	2.11874E+02	−2.16210E−02	1.33118E−13	2.16210E−02
37	2.17916E+02	−1.78523E−02	6.78020E−02	−1.78523E−02
38	2.25722E+02	−2.11812E−03	2.00387E−02	−2.11812E−03
39	2.26978E+02	−2.07961E−03	−8.90330E−13	2.07961E−03
40	2.38016E+02	6.86948E−03	8.60720E−12	−6.86948E−03
41	2.41995E+02	−1.13662E−02	9.07389E−02	−1.13662E−02
42	2.54340E+02	2.47053E−02	1.30276E−02	2.47053E−02
43	2.56435E+02	1.85297E−02	1.13407E−11	−1.85297E−02
44	2.58172E+02	−2.08004E−03	2.13596E−02	−2.08004E−03
45	2.65610E+02	9.46409E−03	−3.65495E−11	−9.46409E−03
46	2.78455E+02	7.41774E−03	−1.28130E−10	−7.41774E−03
47	2.84801E+02	−1.05006E−02	6.93513E−02	−1.05006E−02
48	3.00876E+02	−4.68707E−02	−1.99378E−02	−4.68707E−02
49	3.05150E+02	4.83721E−02	3.51661E−10	−4.83721E−02
50	3.64974E+02	1.50463E−03	−9.60201E−09	−1.50459E−03

Mass Properties for the Whole Model: Based on Translational masses, including masses at fixed DOF.

Summation *X*-Direction	*Y*-Direction	*Z*-Direction	Center of Mass *X*-Coord.	*Y*-Coord.	*Z*-Coord.
6.5730E−01	6.5730E−01	6.5730E−01	2.1219E+02	−2.4000E+01	6.3809E+01

Based on Translational masses excluding masses at fixed DOF:

Summation *X*-Direction	*Y*-Direction	*Z*-Direction
6.3386E−01	5.9895E−01	6.3386E−01

RS-Dynamic Analysis results

Check Point	Elbow-1	Elbow-2	Run Node 2 First support	Run Node 7 Last Support
Bending moment (lb-in)	485.81	485.81	877.15	877.15

Time History Modal Analysis

Three simultaneous acceleration time history functions have been used as seismic input: two horizontal components and one vertical. The correlation coefficient between any two time history inputs is less then 0.2, so the three input motion can be considered uncorrelated, and hence, independent. The resultant time history moments at selected mass points in the pipe are shown in Figures 5.29 to 5.32.

Modal TH-Dynamic Analysis results

Check Point	**Elbow-1**	**Elbow-2**	**Run Node 2 First support**	**Run Node 7 Last Support**
Bending Moment [lb-in]	408.96	342.76	697.27	877.51

Time History Direct Integration

For direct integration TH dynamic analysis time history outputs of pipe seismic moments at particular mass points are shown in Figures 5.33 to 5.36. The Rayleigh coefficients have been calculated for the following parameters:

$F1 = 3.$ Hz
$F2 = 150$ Hz
$\alpha = 1.81$
$D1 = 5\%$
$D2 = 10\%$
$\beta = 0.2102 \times 10^{-3}$

Direct Integration TH-Dynamic Analysis Results

Check Point	**Elbow-1**	**Elbow-2**	**Run Node 2 First support**	**Run Node 7 Last Support**
Bending Moment (lb-in)	489.34	356.30	719.29	1032.34

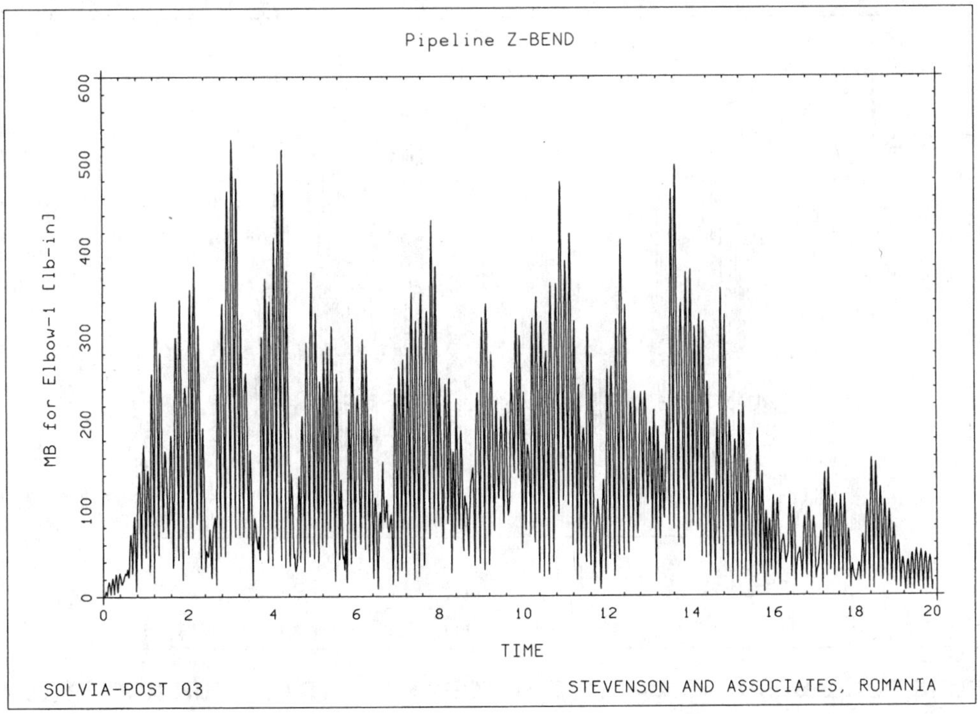

FIGURE 5.29. Z-BEND – MODAL TIME HISTORY RESULTS FOR ELBOW 1 (LB-IN)

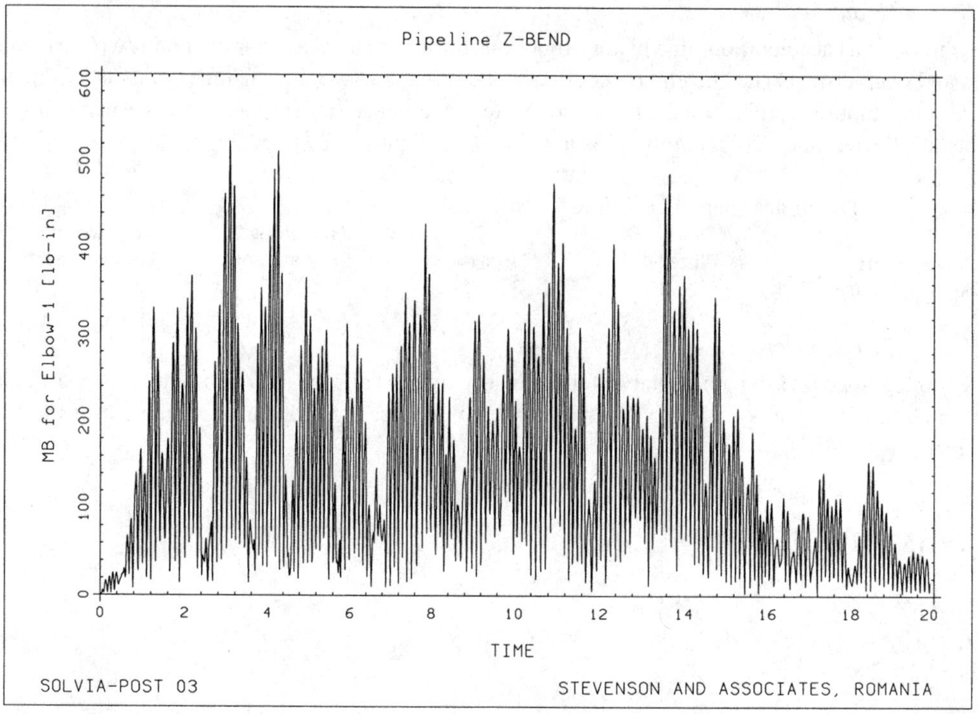

FIGURE 5.30. Z-BEND – MODAL TIME HISTORY RESULTS FOR ELBOW 2 (LB-IN)

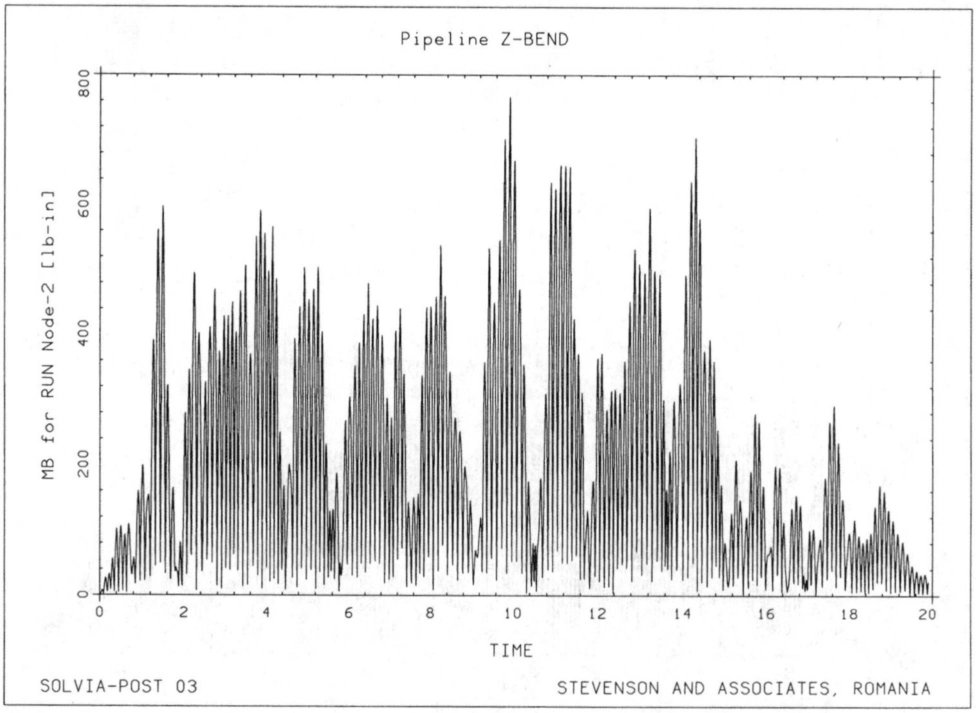

FIGURE 5.31. Z-BEND – MODAL TIME HISTORY RESULTS FOR RUN NODE 2 (LB-IN)

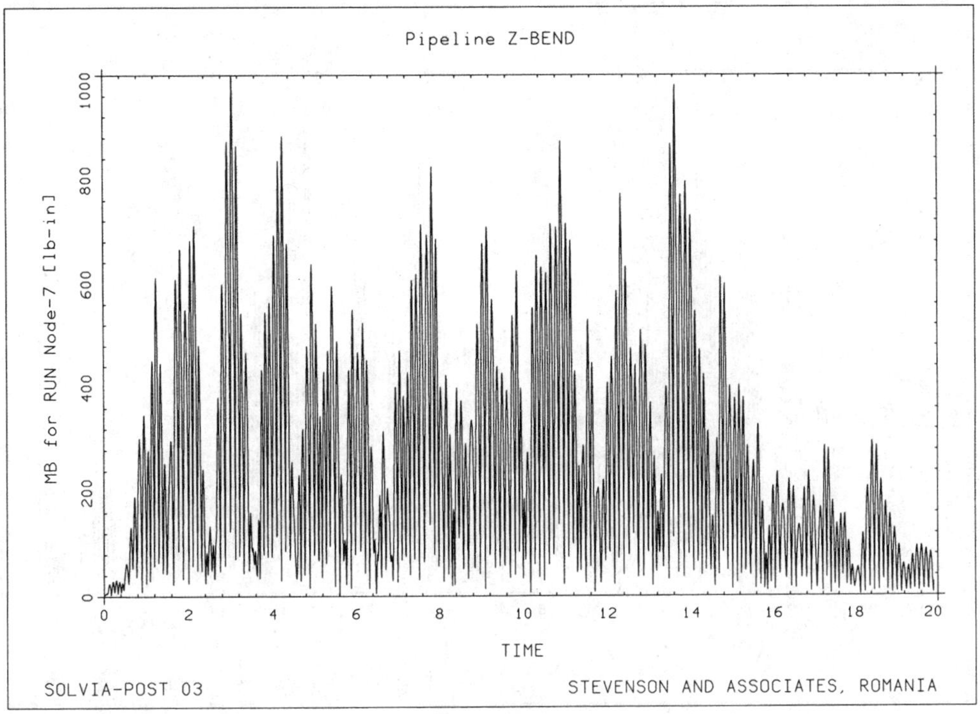

FIGURE 5.32. Z-BEND – MODAL TIME HISTORY RESULTS FOR RUN NODE 7 (LB-IN)

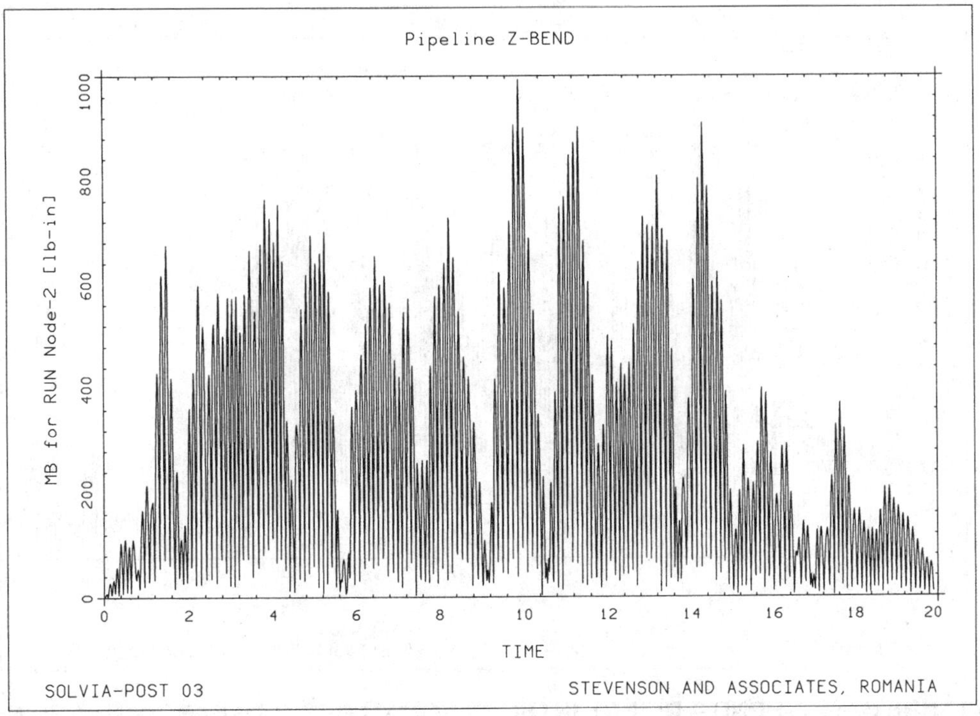

FIGURE 5.33. Z-BEND – DIRECT INTEGRATION TH RESULTS RUN NODE 2 (LB-IN)

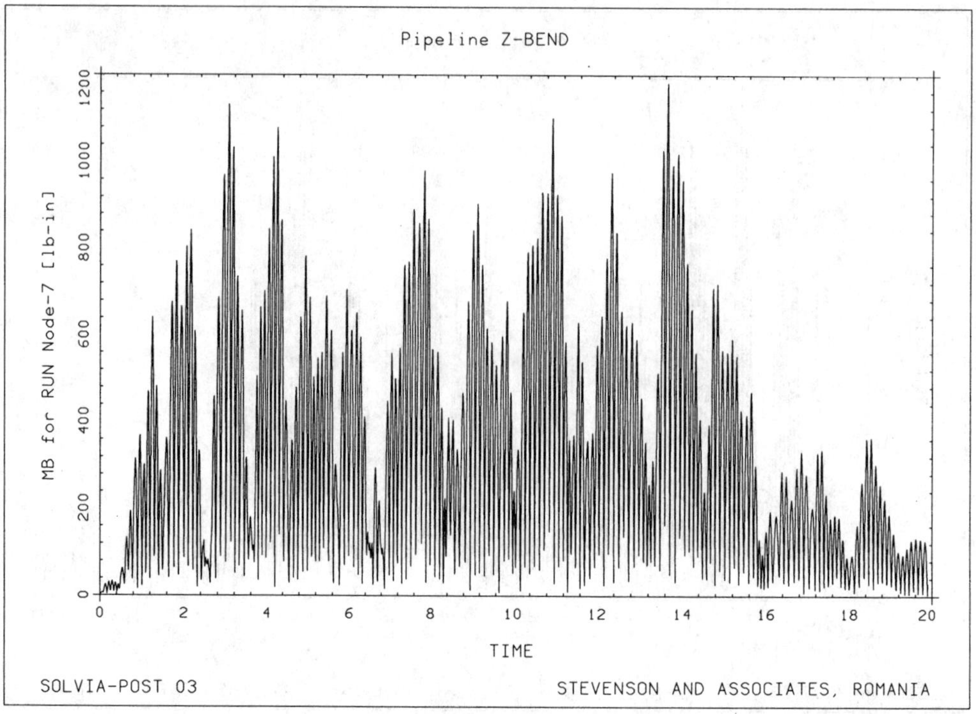

FIGURE 5.34. Z-BEND – DIRECT INTEGRATION TH RESULTS RUN NODE 7 (LB-IN)

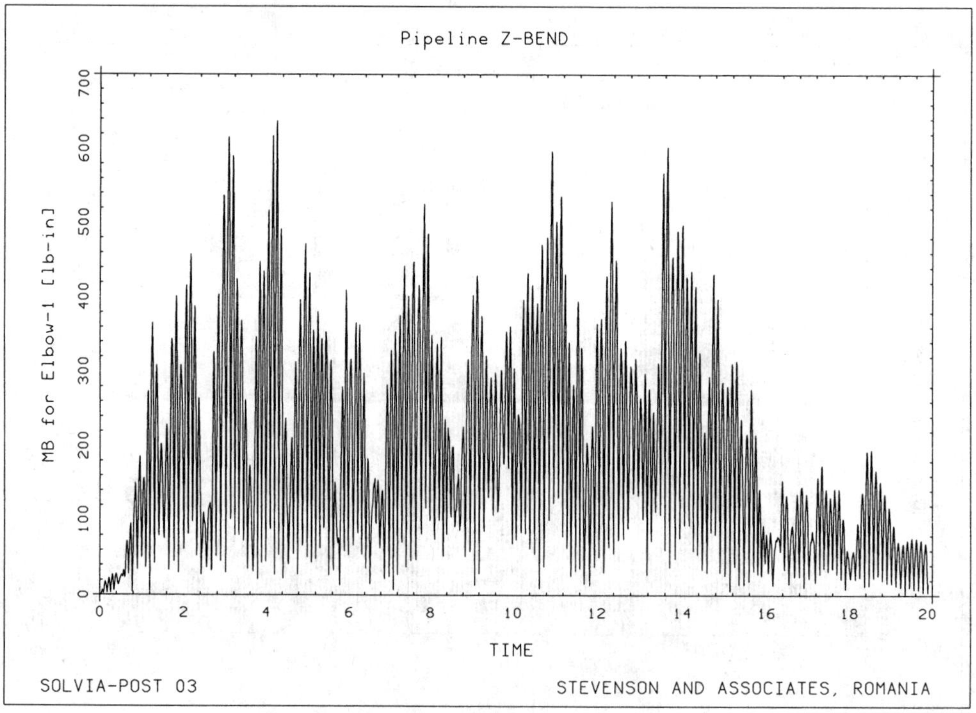

FIGURE 5.35. Z-BEND – DIRECT INTEGRATION TH RESULTS ELBOWS 1 (LB-IN)

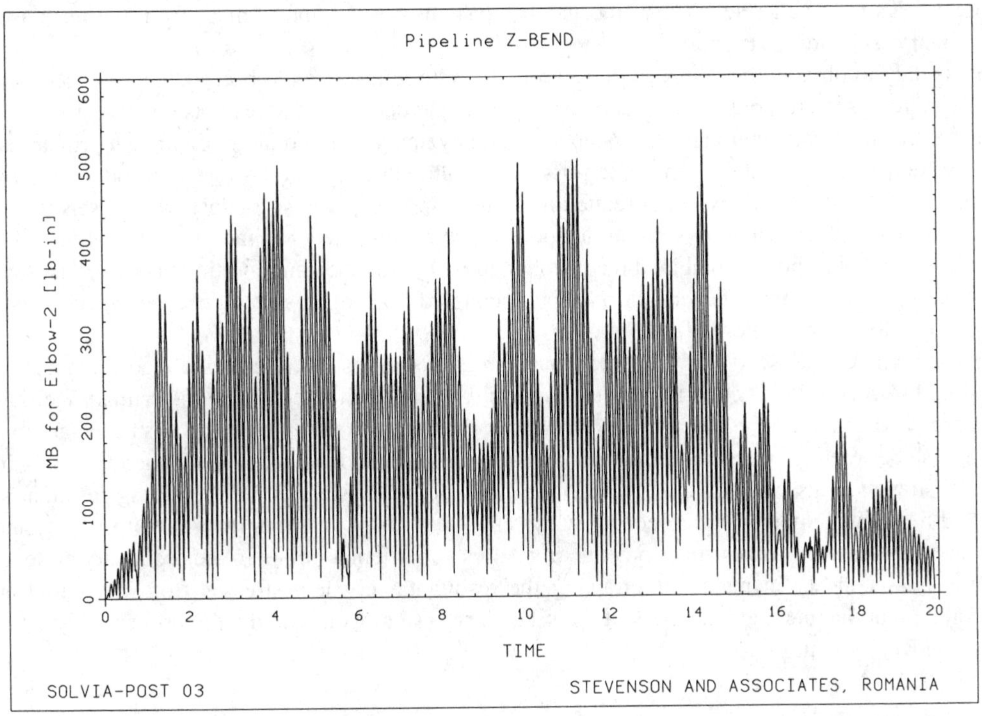

FIGURE 5.36. Z-BEND – DIRECT INTEGRATION TH RESULTS ELBOWS 2 (LB-IN)

Summary

As can be seen by the elastic results from methods other than *ASCE-7,* summarized in Table 5.14 of the piping segment example, there is a significant variation of seismic moments and support reaction resultants for a piping system depending on the analytical method used to compute resultant seismic loads. In general, time history methods are considered the most reliable results. However, because of the many supports and varying gaps between the distribution system and supports the actual response is nonlinear even when stresses in the system and supports are elastic. Procedures described here at best are indices of the seismic stresses induced in a distribution system and its supports. These results can be used for design purposes, but should not be considered an accurate determination of seismic distribution system forces and moments.

The elastically induced seismic inertia moments and support reactions in the mechanical system or component can be reduced by dividing by the appropriate F_u factor given in Table 5.13. In addition, the procedures outlined here assume the supporting building structure is responding elastically. Inelastic response of the primary supporting building structure to earthquake loads would tend to reduce the seismic input to the mechanical secondary components or systems. This phenomenon can be introduced into the analysis by increasing the damping of the supporting structure using the inelastic ductility values of Table 3.4-2 of [3] in Chapter 5 if justified by the resultant inelastic stress levels in the structure (i.e., use of R factors in the structure that exceed 2.5.

5.6 COUPLED VERSUS UNCOUPLED DYNAMIC ANALYSIS OF MECHANICAL (SECONDARY) SYSTEMS

In Section 5.4.2 the use of floor or in-structure response spectra is discussed. Effectively, the use of floor or in-structure response spectra for design of secondary (mechanical) systems in the absence of a mass ratio effect decouples the smaller secondary (mechanical systems or equipment) mass from the response of the larger primary (supporting structure).

One has only to observe the response of a two-mass system in series to observe that this decoupling significantly increases the response of the secondary system and decreases the response on the primary system.

With the advent of computer algorithms capable of analyzing larger and larger combined primary and secondary system dynamic models, it became possible to automatically and correctly introduce the dynamic coupling effect. However, the expected reduction in the response of the secondary system seismic loads as compared to uncoupled floor or in-structure response spectra seismic loads have not been observed. This is because the coupled response combined the primary stress causing seismic inertia loads with the secondary stress resulting from seismic anchor motion. The uncoupled floor or in-structure response spectra analysis considers only the primary inertia load.

It should be noted that secondary seismic anchor motion stresses in the extreme load category in accordance with *ASME B&PVC* Section III service levels C & D are not limited, while the primary inertia caused stresses are. To obtain the benefit of the use of dynamic coupling in seismic analysis it is necessary to be able to distinguish between seismic inertia induced primary stresses and differential static anchor motion induced secondary stresses. This can be done by analyzing the coupled system and determining resultant seismic forces and moments at points in the secondary system due to displacements from the primary system support points. The resultant forces and moments or stresses at particular points in the secondary system in this later case are secondary in nature. Subtracting the resultant secondary stresses from the initial analysis stresses at a point that included the mass of the secondary system would define the primary seismic stresses in the secondary system.

5.7 DEVELOPMENT AND USE OF SPACING TABLES AND CHARTS FOR SEISMIC DESIGN OF PIPING AND OTHER DISTRIBUTION SYSTEMS

5.7.1 Introduction

Seismic design of piping and other distribution systems, by means of MDOF dynamic analysis, is very time-consuming and requires an expensive modeling and analysis procedure. It should be noted that typically there are perhaps four to six buildings in a hazardous facility to be modeled with a combined height typically less than 300 feet. On the other hand there are typically several tens of thousands of feet of pipe and other distribution systems that potentially require similar modeling. In addition, piping and distribution system analyses cannot be completed until all supports that restrain the pipe or distribution system are located. In many cases the resulting analyses are iterative. That is, the analysis results require that additional supports must be provided to reduce stresses in the piping or other distribution systems, and the systems must subsequently be reanalyzed to confirm the adequacy of the system with the new supports. If iterative dynamic analyses are to be avoided, it is necessary to employ simplified methods to locate such supports, whether they are of the movable-, hydraulic- or inertia-(mechanical) snubber or damper-type or of typical fixed U-bolt and hanger-type construction. These simplified, approximate methods employ tables and charts for use in their design or evaluation.

In general the use of spacing tables and charts for the final design of piping systems and their supports are limited to relatively cold piping with design temperatures below 150°F (65°C) or temperature ranges less than 100F° (39°C). For piping systems outside these ranges, thermal indiced stresses can begin to control design and the use of spacing tables and charts can no longer ensure design adequacy and system specific dynamic analysis is required.

5.7.2 Development of Lateral Support Spacing and Inertia Acceleration

Performance category 3 to 5 distribution systems within process facilities typically consist of several thousand feet. This section presents an analytical basis and procedures used for the development of simplified seismic spacing tables and charts. The basic procedure in using tables and charts is to first select a maximum

TABLE 5.15. PIPING FUNDAMENTAL FREQUENCIES AS A FUNCTION OF *ANSI B31.1* SUGGESTED DEAD

Pipe Size (Std) (in.)	Weight (lbs) Water (Ft)	Steam & Air (ft)	L_W (Ft)	L_S (ft)	L_W^3 ×10⁶ (in.³)	L_S^3 ×10⁶ (in.³)	I (in.⁴)	W_W (lbs.)	W_S (lbs.)	f_W (Hz)	f_S (Hz)
1	2.053	1.68	7	9	0.593	1.26	0.0874	15.10	14.05	16.70	11.85
2	5.108	3.66	10	13	1.732	3.80	0.666	51.08	47.60	14.85	10.25
3	10.78	7.59	12	15	2.98	5.83	3.02	129.0	114.0	16.23	11.40
4	16.30	10.8	14	17	4.74	8.49	7.23	228.0	183.7	13.80	11.56
6	31.48	19.0	17	21	8.52	16.00	28.14	535.0	399.0	13.30	11.20
8	50.24	28.6	19	24	11.84	23.89	72.5	955.0	686.0	13.50	11.20
12	98.60	49.6	23	30	21.00	46.66	279.3	2270	1,490	12.95	10.70
16	141.68	62.6	27	35	34.05	74.09	562	3820	2,195	11.13	9.95
20	204.60	78.7	30	39	46.60	102.50	1,114	6,140	3,070	10.70	10.15
24	278.48	94.62	32	42	56.70	128.02	1,943	8,930	3,980	10.40	10.40

1. Frequencies determined = 3.55 $(5W1^3/384\ ELI)^{1/2}$ from Roark[8].
2. $E = 29 \times 10^6$ psi.

length between supports of an "equivalent" straight pipe for the particular pipe size, material, ASME class and applicable seismic acceleration. Deadweight support spans and pipe characteristics are shown in Table 5.15. Table 5.16 gives pipe frequency coefficients as a function of the span ratio selected. The coefficient associated with a partial lateral to vertical support span ratio. (LVSSR) is multiplied by the frequency of the vertical support span given in Table 5.15. This frequency with the applicable floor or in-structure response spectra define the inertia acceleration of the pipe. The basis for the tables and charts is developed in the following paragraphs.

Span Lengths

Seismic support spacing lengths are typically developed for the case of a continuous, straight, horizontal pipe of at least two equal spans and of the same diameter and schedule and material. For these cases, where the seismic load is assumed to act as a uniform load over all spans, the maximum bending moment in a span is determined by the relationship:

$$M = 0.107\ G\ w\ 1_s^2 \tag{5.60}$$

TABLE 5.16. FUNDAMENTAL PIPE FREQUENCIES AS A FUNCTION OF LATERAL TO VERTICAL SUPPORTS SPAN RATIO, LVSSR[2]

Ratio	Fundamental Frequency Ratio Multiplier (Hz)
1	1.0
2	0.34
3	0.16
4	0.09
5	0.06
6	0.04
7	0.03

1. This is the fundamental piping frequency ratio multiplier when the vertical support spacing is taken equal to the recommended dead weight support spacing of *ASME B31.1* or *ASME B&PVC* Section III-NF.
2. For piping system design to resist earthquakes the LVSSR typically ranges between 2 and 4 in order to limit lateral displacements resulting in impacts between adjacent pipe. However, experience with actual earthquakes, with a pga equal to or greater than 0.2 g, ratios as high as 7 have been observed without visible damage to the piping system.

where M = maximum bending moment; G = seismic acceleration in multiples of the acceleration of gravity; w = uniform weight distribution of pipe; and 1_s = length of pipe span between supports.

and the coefficient 0.107 is a dimensionless quantity as a function of the pipe support condition of continuous over several supports.

It is possible to relate this moment to a seismically induced stress in the pipe by the relationship:

$$\sigma_s = \frac{M}{Z} \tag{5.61}$$

where M = moment from Eq. (5.58); Z = elastic section modulus of the pipe; and σ_s = seismic stress in the pipe.

If the stress σ_s is assumed to be a maximum permissible seismic stress which is taken as maximum allowable stress, σ_t minus the dual maximum dead weight (0.1 σ_t) and pressure (0.5 σ_t) stress, then the maximum span length between supports, which is permissible without exceeding this stress, can be obtained by substituting Eq. (5.61) into Eq. (5.60) and solving for 1_s.

$$1_s = \left[\frac{9.35\sigma_s Z}{Gw}\right]^{1/2} \tag{5.62}$$

Thus, if an allowable seismic stress σ_s and the seismic loading (as expressed by acceleration G) are defined, the maximum span length can be obtained from Eq. (5.62). A similar span length can be determined from *ASME B&PVC* Section III class 2 and 3 and *B31.1* or *B31.3* piping can be determined as a function of their total allowable stress.

Allowable Stresses

The allowable seismic stress used in Eq. (5.60) is obtained by subtracting from an allowable total stress as defined by the applicable ASME code allowances for bending deadweight stresses and pressure stresses. For *ASME B31.1* piping, the maximum allowable stress permitted in the pipe is 1.2 S_h and 1.33 S_h for *ASME B31.3* piping for occasional loads such as seismic. A reasonable approximate allowance for the dead weight stress is 0.1 S_h, based on the normally used spacing for deadweight supports of piping systems as expressed in the ASME code as shown in Table 5.15.

The allowance for the pressure stresses was selected as 0.5 S_h based on the assumption that the pipe wall thickness is selected on the basis of pressure stress in the hoop direction by the simple relationship:

$$S_h = \frac{PD_o}{2t_m} \tag{5.63}$$

where P = design pressure in pipe; D_o = outside diameter of pipe; t_m = wall thickness of pipe; and S_h = allowable stress at design temperature.

Since the stress in the longitudinal direction is equal to half the hoop stress, it is assumed that one-half the allowable stress margin in the longitudinal direction is "used: by the pressure and the seismic inertia in the longitudinal direction in the pipe and the remainder is available for lateral seismic stresses."

Thus, using the above allowance for dead load, pressure and longitudinal seismic stress, the maximum permissible seismic lateral stress for *ASME B31.1* piping becomes:

$$\sigma_s = 1.2\ S_h - 0.1\ S_h - 0.5\ S_h = 0.6\ S_h \tag{5.64}$$

A similar approach is taken for *ASME B31.3* piping, except the allowable stress excluding a weld factor is 1.33 S_h.

Seismic Loading

The lateral seismic bending load on the piping system is represented by the term Gw. The accelerations, *G*, is taken as the value specified from the applicable floor or in-structure response spectra.

The value of *G* selected applied in the horizontal directions are conservatively assumed equal. Responses due to the two horizontal and one vertical direction are combined as specified by the applicable ASME code equation.

For example, it was desired to place a lateral support in a 3-inch-diameter water-filled line every third deadweight support, then the lateral frequency of the piping system would be from Tables 5.15 and Table 5.16: $0.16 \times 16.23 = 2.60$ Hz; and the lateral support span length is: $3 \times 12 = 36$ ft from Table 5.15.

From Figure 5.18 for 2.60 Hz, the seismic inertia acceleration applied to the mass distribution of the piping between supports is 0.2*g*.

This approach is applicable to straight runs of pipe. To apply this approach to spans with bends in them it is necessary to develop a figure that considers the particular bend geometry as typically shown in Figure 5.37 and 5.38.

5.7.3 Design Charts for Various Pipe Configurations

The seismic support spacings in Table 5.15 are developed for straight horizontal pipes continuous over multiple supports, and this table can also be used to select maximum span lengths, which will keep the stresses in the pipes within prescribed code limits.

Charts that consider bends in the pipe spans can be developed by iterative static analysis. Specifically, normalized nondimensional seismic design charts are typically prepared for the following three basic configurations:

1. One-bend
2. Two bend, in-plane
3. Two bend, out-of-plane

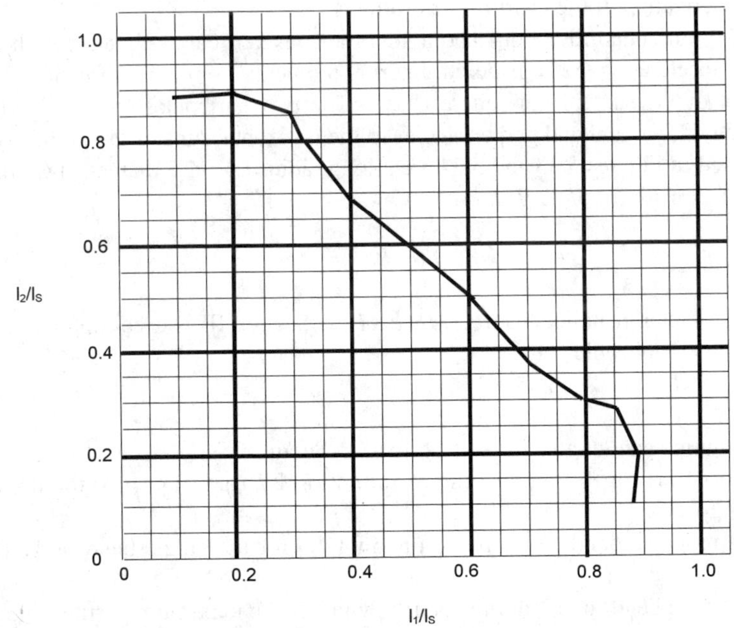

FIGURE 5.37. CONFIGURATION 1 (ONE-BEND) FOR ASME B&PVC SECTION III CLASS 1.

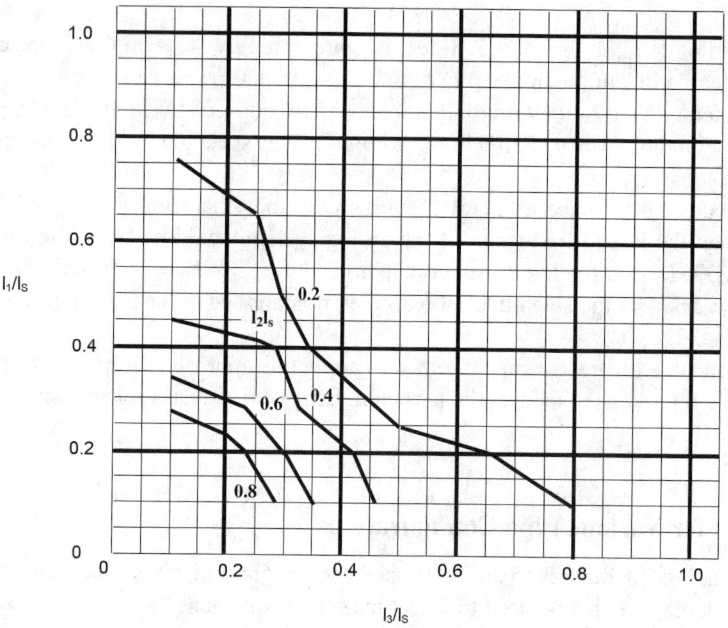

FIGURE 5.38. CONFIGURATION 2 (ONE-BEND, OUT-OF-PLANE) ASME PVC SECTION III CLASS 1 PIPE.

The basic idea of the charts is that they permit selection of the location of the new support if 1_s and the ratios $1_2/1_s$ for one bend or the ratios $1_2/1_s$ and $1_1/1_s$ for the other configurations are known.

Derivation of the Charts

Charts are derived by computer analysis of all span lengths of the selected configurations and orientations. General aspects of the derivation of the charts are as follows.

An earthquake in two horizontal directions and a simultaneous vertical earthquake with an acceleration equal to the horizontal acceleration are considered. Internal moments from these directions have been combined on a square root of the sum of the squares basis. The two horizontal seismic inertia forces were assumed to be oriented both parallel and perpendicular to the horizontal runs of the piping system.

The elbow fittings used in Figures 5.33 and 5.34 are a long radius (R = 1.5 diameter) welding elbow with a stress intensification factor from *ASME B31.1:*

$$i = 0.9/(t_m\ R/r^2)^{2/3} \tag{5.65}$$

The stress intensification factor or indice from *ASME B&PVC* Section III is used to amplify bending stress at fittings as defined by the applicable code.

Use of Charts

a. Find the maximum span length, 1_s, as described in Section 5.7.2.
b. Obtain the ratio $1_1/1_s$ for the single bend or the ratios $1_1/1_s$ and $1_2/1_s$ for the double bend cases.
c. Refer to the applicable bend geometry to obtain $1_2/1_s$ for the single bend or $1_3/1_s$ for the other cases.
d. Calculate 1_2 (single bend) or 1_3 (double bend), which is the maximum permissible distance to the next seismic support measured along the pipe from the last located support.

As an approximation for the effect of concentrated weights, it is suggested that the concentrated weight be replaced by an equivalent weight span length of pipe multiplied by 1.5. For example, if a valve weights 50 lbs in a line having a unit weight of 10 lbs/in, the effective length is 1.5 (50) = 7.5 in. The coefficient 1.5, based on the ratio of maximum moment in a fixed-end beam with load at the center, $Q/8$, to the same beam uniformly loaded, Q $l/12$, where $Q = w1$ and w is the load per unit length of pipe.

The tables and figures provided do not consider any intermediate supports and restraints. In developing the tables and charts, it is assumed that the terminal ends of all support configurations are continuous straight pipes and simply supported. For modeling purposes two spans of straight pipe are used on either side of the geometry in question. If a limiting stress point location is at an elbow or tee, it is necessary to take into account the tee or elbow stress intensification or indice factor defined by the applicable code.

REFERENCES

1. *DOE Std. 1020-2002,* 2002, "Natural Phenomena Hazards Design and Evaluation Criteria for Department of Energy Facilities," U.S. Department of Energy, Washington, D.C.
2. *Draft Regulatory Guide DG-1121,* 2003, "Guidelines for Categorizing Structures, Systems and Components in Nuclear Power Plants According to Their Safety Significance," U.S. Nuclear Regulatory Commission, Washington, D.C.
3. *ASCE Std. 43-05,* 2005, "Seismic Design Criteria for Structures, Systems and Components in Nuclear Facilities and Commentary," American Society of Civil Engineers, Reston, VA.
4. Seed, H.B. and Idriss, I.M., 1982, "Ground Motions and Soil Liquification During Earthquakes," Earthquake Engineering Research Institute,
5. Naeim, F., 1989, *The Seismic Design Handbook,* Van Nustrand Reinhold, New York, NY.
6. *IBC-2003,* 2002, "International Building Code", International Code Council. In Falls Church, VA.
7. *SEI/ASCE 7-02,* 2003, "Minimum Design Loads for Buildings and Other Structures," American Society of Civil Engineers, Reston, VA.
8. Stevenson, J.D. and Lapay, W.S., 1974, "Amplification Factors to Be Used in Simplified Seismic Dynamic Analysis of Piping Systems," *Paper 74-NE-9,* American Society of Mechanical Engineers, New York, NY.
9. Gwinn, J.M. and Goldstein, N.A., 1974, "Equivalent Seismic Load from Amplified Response Curves," American Society of Mechanical Engineers, New York, NY.
10. *ASME B&PVC Section III Division 1 Appendix N,* 2001, "Dynamic Analysis Methods," American Society of Mechanical Engineers, *Boiler and Pressure Vessel Code,* New York, NY.
11. Newmark, N.M., 1959, "A Method of Computation for Structural Dynamics," *Journal of Eng. Mech. Div.,* Vol. 85, No. EM3, pp. 67–94, ASCE, Reston, VA.
12. Houbolt, J.C., 1950, "A Recurrence-Matrix Solution of Dynamic Response of Elastic Aircraft," *J. Aeronautical Sci.,* Vol. 17, pp. 540–555.
13. Bathe, K.J. and Wilson, E.L., 1973, "Stability and Accuracy Analysis of Direct Integration Methods," *Earthquake Engrg. and Struc. Dyn.,* Vol. 1, pp. 283–291.
14. Chan, S.P., Cox, H.L. and Benfield, W.A., 1962, "Transient Analysis of Forced Vibrations of Complex Structural-Mechanical Systems," *J. Royal Aeronautical Soc.,* Vol. 66, July, pp. 457–460,
15. Housner, G.W., Martel, R.R. and Alford, J.L., 1953, "Spectrum Analysis of Strong Motion Earthquake," *Bull. Seismol. Soc. Am.,* Vol. 43.
16. Newmark, N.M., 1967, "Design Criteria for Nuclear Reactors Subjected to Earthquake Hazards," *Proc., IAEA Panel on a Seismic Design and Testing of Nuclear Facilities,*
17. Newmark, N.M., 1971, "Earthquake Response Analysis of Reactor Structures," *Proc., 1st* Int. Conf. on Struc. Mech. in Reactor Technol., Urbana, Ill.,
18. "Preliminary Safety and Analysis Report," 1972, Alabama Power Co., *AEC Docket No. 50-348.*
19. Newmark, N.M., Blume, J.A. and Kapur, K.K., 1973, "Design Response Spectra for Nuclear Power Plants," *Proc., ASCE Nat. Struc. Engrg. Meeting,* Reston, VA.

20. *Regulatory Guide 1.60,* 1973, "Design Response Spectra for Seismic Design of Nuclear Power Plants," Revision 1, U.S. Atomic Energy Commission, Directorate of Regulatory Standards, Washington, D.C.
21. Newmark, N.M. and Hall, W.J., 1978, "Development of Criteria for Seismic Review of Selected Nuclear Power Plants," *NUREG/Cr-0098* Prepared for U.S. Nuclear Regulatory Commission, Washington D.C.
22. *Regulatory Guide 1.165,* 1977, "Identification and Characterization of Seismic Sources and Determination of Safe Shutdown Earthquake Ground Motion," U.S. Nuclear Regulatory Commission, Washington D.C.
23. Stevenson and Associates, 2002, *ThSPEC Program for Digital Processing of Time Series,* Romania.
24. Lysmer, J. et. al., 1981, "SASSI, A System for Analysis of Soil-Structure Interactions," *Report No. UCB 81-02,* U.S. Berkeley.
25. *Uniform Building Code,* 1997, *Structural Engineering Design Provisions, Vol. 2,* International Conference of Building Officials, Oakland, CA.
26. Section 3.7.2, 1981, "Equivalent Static Load Method," *U.S. Standard Review Plan, Rev. 1,*
27. *ASME B&PVC Section III, Subsection NB,* 2004, "Pressure Retaining Components – Class 1," American Society of Mechanical Engineers, New York, NY.
28. *ASME B31.1,* 2004, "Power Piping Code," American Society of Mechanical Engineers, New York, NY.
29. SOLVIA Engineering AB, 1987–2003, *SOLVIA System 03 Computer Program,* Trefasgatan 3, SE-721, Sweden.

CHAPTER

6

Natural Hazard Design and Analysis other than Earthquake

6.1 WIND

The structural loads associated with extreme straight and hurricane wind velocities have usually been considered statically applied and distributed in accordance with the American Society of Civil Engineers (ASCE) *7-05 Standard* for buildings and other structures to include chimneys, tanks and rooftop equipment [1] or *ACI-307* [2] for concrete stacks and *STS-1-2000* [3] for steel stacks.

Tornado wind effects as a design-basis load have generally not been defined or required for consideration in design as a normal or severe load as defined in Chapter 3. They have been prescribed as an extreme load for nuclear safety-related structures. In some cases they have also been considered for petrochemical facilities in tornado-prone regions, but this has not become a standardized design practice associated with hazardous, petrochemical or biomedical facilities. As a result a tornado wind field and the methods used to combine the effect of wind or "dynamic pressure" and the associated reduced atmospheric pressure to include building openings and potential airborne missile effects have not been defined in general industry standards for design purposes, except as applied to the nuclear industry. A rational yet simple combination of the wind field and reduced atmospheric pressure effects for use in analysis and design or evaluation of structures subject to design-basis tornado loadings is presented in this chapter. There is also a suggested procedure for including the effect of postulated, straight, hurricane and tornado-borne missiles in structural design in this chapter as well as in Chapter 8.

Note: For the tornado return periods defined in [4] for design of nuclear facilities, a very low probability design-basis frequency (10^{-7}/yr) (10,000,000-year return period) is assumed for nuclear power plants [5]. Higher frequency, less intense tornadoes are typically considered in design or evaluations of less hazardous nuclear facilities [6]. For these lesser hazard nuclear facilities, a 2×10^{-6} and 2×10^{-5}/yr probability of tornado exceedence are specified.

In, [1], there is a regional map of straight and hurricane wind speeds expressed as a 3-second gust for a 50-year return period (2×10^{-2}/yr probability of exceedence). In Table 6.1 from [7] is a means to convert the 50-year return period wind to other return periods. In Figure 2.11 of Chapter 2 is a recommended regional map of the U.S. to define tornado regions of high, moderate and low intensities, as defined in Figures 2.8 to 2.10 and. At the present time Figures 2.8 to 2.10 of Chapter 2 apply to safety-related SSC in nuclear facilities regulated by the U.S. Department of Energy (DOE), and Figure 6.1 and Table 6.2 apply to existing nuclear power plants regulated by the U.S. Nuclear Regulatory Commission (NRC).

TABLE 6.1. CONVERSION FACTORS FOR OTHER MEAN RECURRENCE INTERVALS

	50-yr Mean Recurrence Interval (MRI); Peak Gust Wind Speed, *V* (mph) (m/s)	
	Continental U.S.	
MRI (years)	*V* = 85–100 (straight winds) (38–45 m/s)	*V* > 100 (hurricane) (44.7 m/s)
500	1.23	1.33
200	1.14	1.21
100	1.07	1.105
50	1.00	1.00
25	0.93	0.89 (84 mph min) (37.5 m/s)
10	0.84	0.73 (76 mph min) (33.9 m/s)
5	0.78	0.52 (70 mph min) (31.3 m/s)

6.2 PHYSICAL PHENOMENA

6.2.1 Straight Winds

For straight winds with a 3-second gust equal to or below 90 mph there is usually no requirement to consider missile loads. In addition, differential pressure loads are also not considered for straight and hurricane winds. Straight winds result from the movement of large weather fronts and squall lines and are characterized, by an observer on the ground, as not having a recognizable rotational component of wind motion.

6.2.2 Hurricanes, Cyclones and Typhoons

Hurricane, cyclones and typhoons are rotating windstorms of tropical origin whose distance to maximum wind speed from the center or eye of the storm is typically measured in miles (km). They have design wind speeds, as shown in [1], applicable to U.S. eastern, southern and western coastlines greater than straight winds from weather fronts, but for design purposes they are treated in the same manner as straight winds but with significantly higher wind speeds.

TABLE 6.2. DESIGN-BASIS TORNADO CHARACTERISTICS FROM NRC R.G. 1.76

Region	Maximum Wind Speed[a] (mph)	Rotational Speed (mph)	Translational Speed (mph)		Radius of Maximum Rotational Speed (feet)	Pressure Drop (psi)	Rate of Pressure Drop (psi/sec)
			Maximum	Minimum[b]			
I	360	290	70	5	150	3.0	2.0
II	300	240	60	5	150	2.25	1.2
III	240	190	50	5	150	1.5	0.6

[a] The maximum wind speed is the sum of the rotational speed component and the maximum translational speed component.

[b] The minimum translational speed, which allows maximum transit time of the tornado across exposed plant features, is to be used whenever low travel speeds (maximum transit time) are a limiting factor in the design of the ultimate heat sink. The ultimate heat sink is that complex of water sources, including associated retaining structures, and any canals or conduits connecting the sources with, but not including, the intake structures of nuclear reactor units. Regulatory Guide 1.27 (Safety Guide 27), "Ultimate Heat Sink," describes a basis that may be used to implement general design criterion 44 of Appendix A to 10CFR50 with regard to the ultimate heat sink.

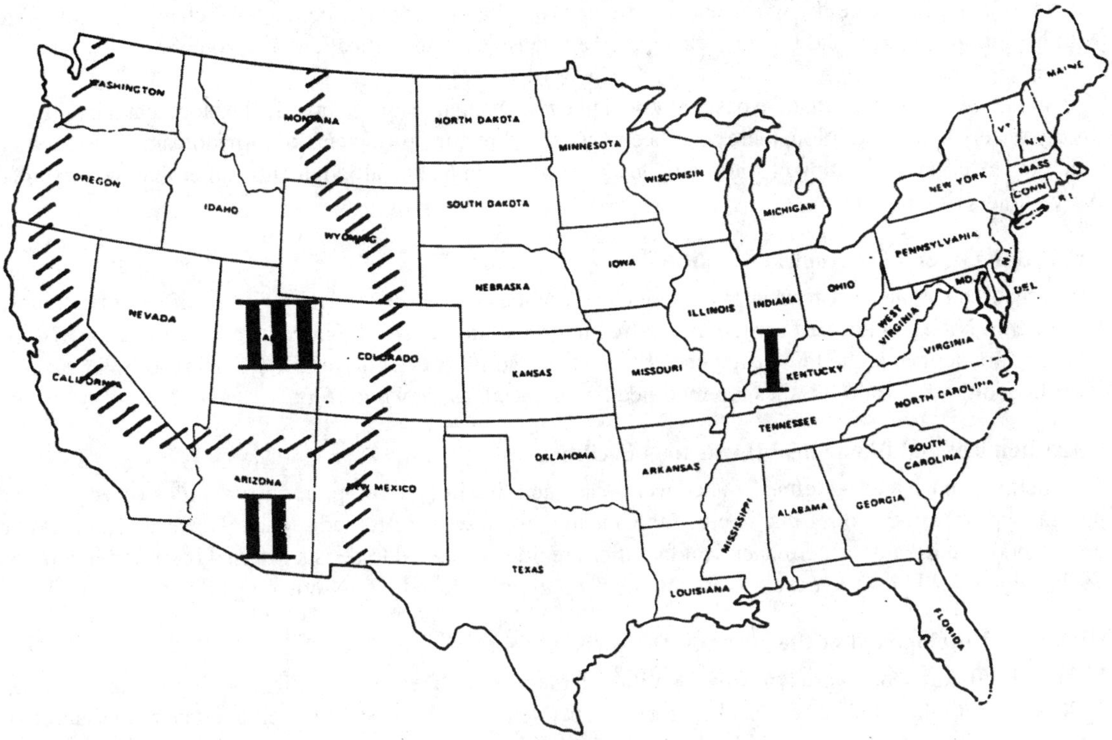

FIGURE 6.1. TORNADO INTENSITY REGIONS

6.2.3 Tornadoes

Applied to Nuclear Power Plants

Tornado wind load effects on safety-related structures have become a general structural design requirement for nuclear power plants located east of the Rocky Mountains since 1967, and since 1971 for plants on the West Coast as well.

Prior to 1994, for nuclear power plants the maximum wind speed was considered on a probabilistic basis at the 10^{-7}/yr probability of exceedence level [5]. Currently there are proposed changes to nuclear power plant criteria which would reduce the maximum wind speed of Table 6.2 by 20 percent. A tornado design requirement was extended to lesser hazard U.S. Department of Energy nuclear facilities in 1994 [6]. The general requirements associated with nuclear power plant design based on ref. 5 have become reasonably well standardized, with the following typical design parameters specified in region I of Figure 6.1 east of the Rocky Mountains:

1. Tangential design wind velocity = 440 fps
2. Translational wind velocity = 88 fps
3. External pressure drop = 3 psi

For West Coast region II sites, tornado wind requirements had not been so well defined, but typical characteristics used were as follows:

1. Tangential design wind velocity = 260 fps
2. Translational wind velocity = 60 fps
3. External pressure drop = 2psi

It should be noted that, in addition to defining rotational and translation wind velocity and pressure drop, Table 6.2. also defines rate of pressure drop and radius to maximum rotational wind. Not defined for design and analysis purposes are the wind fields in the horizontal and vertical directions, the change of wind field with height, gust factors and possibly changes in air densities with vertical and horizontal dimension of the tornado.

It should be understood that the basic physical phenomena that result in tornado development are still very poorly understood. In addition, there have been very few direct measurements of tornado parameters, so that much of the criteria developed for tornado design is based on technical judgments rather than well-defined basic engineering principles.

Applied to Other DOE Nuclear Facilities

The tornado phenomena criteria applied to nuclear facilities owned by the U.S. DOE is different from those for the U.S. NRC as indicated earlier. Two different probability levels of exceedence for a tornado strike on a facility are defined at the PC-4 category of 5×10^{-6}/yr and PC-3 category of 5×10^{-5}/yr for U.S. DOE facilities. In addition, tornado missiles are modified and reduced, as shown in [6].

Petrochemical and Biomedical Hazardous Facilities

In general, wind loads as defined for conventional facilities in [1] are applied to facilities other than the nuclear-type hazardous ones that exclude the specific application of tornado loads. However, it should be understood that hazardous petrochemical facilities are often designed to resists external blast or deflegration loads, which would tend to envelop tornado loads up to Fujita Level F3 as shown in Table 6.3.

Historical Development of the Tornado Design Requirement

Prior to 1870, tornadoes were typically identified in the U.S. at the rate of only five or six per year. Between 1870 and 1880, the number of tornadoes reported per year increased rapidly to an average rate of approximately 200 per year. This rate held fairly steady until 1953, when the rate jumped to approximately 600 per year. Starting about 1970, the rate jumped to approximately 1,000 per year [10] Current tornado occurrence rates are at about 1,300 per year [9]. It should be understood that there does not appear to be any real increase in the number of tornadoes in the continental U.S., but rather, they are reported more diligently. In addition, as population densities increase and weather radar-scanning techniques improve, it would be expected that tornadoes, which would have otherwise gone undetected, can now be identified.

Contrary to popular belief, tornadoes are not strictly a U.S. phenomenon. Tornadoes occur throughout the world; however, elsewhere they are on the average considerably smaller in the areas they affect and in their maximum wind intensities, but they do not necessarily have a reduced occurrence rate per geographic area. Exceptions to this trend, with tornado frequency and intensity approaching that of the U.S., are Western India and Bangladesh.

Until the development of *Regulatory Guide 1.76,* there was little evidence that tornado frequency, intensity and size, which vary substantially as a function of geographic region and topography, were used to define tornado design requirements. Certainly, the definition of a design-basis tornado requirement for specific sites has not received nearly the attention devoted to earthquake definition. It is anticipated that tornado design requirements, which essentially did not exist prior to 1967, will draw ever-increasing public attention and support, since the losses associated with such phenomena are expected to rise with increased population densities and, unlike earthquakes, occur on a fairly consistent basis year after year in the entire U.S. It is interesting that deaths resulting from earthquakes have averaged 13 per year over the past century, while tornado deaths in the U.S. averaged more than 100 per year.

Description of the Tornado Wind Phenomenon

In order to explain tornado wind behavior and to develop analytical tools to predict tornado effects, several models have postulated. Typical of these models are the one-cell vortex, the two-cell vortex and the one-cell vortex with recirculation bubble as shown in Figure 6.2. However, since the basic mechanisms that cause tornadoes are very poorly understood, any model describing wind motion is somewhat conjectural in nature. A simple model, which considers the cyclonic wind velocity as a function of radius and a constant circulation in the outer regions and constant angular velocity in the central core, is typically assumed. Tangential wind velocity as a function of cyclonic radius is presented in Figure 6.3, which also shows the translational component or 70 mph of wind velocity combined with the rotational velocity.

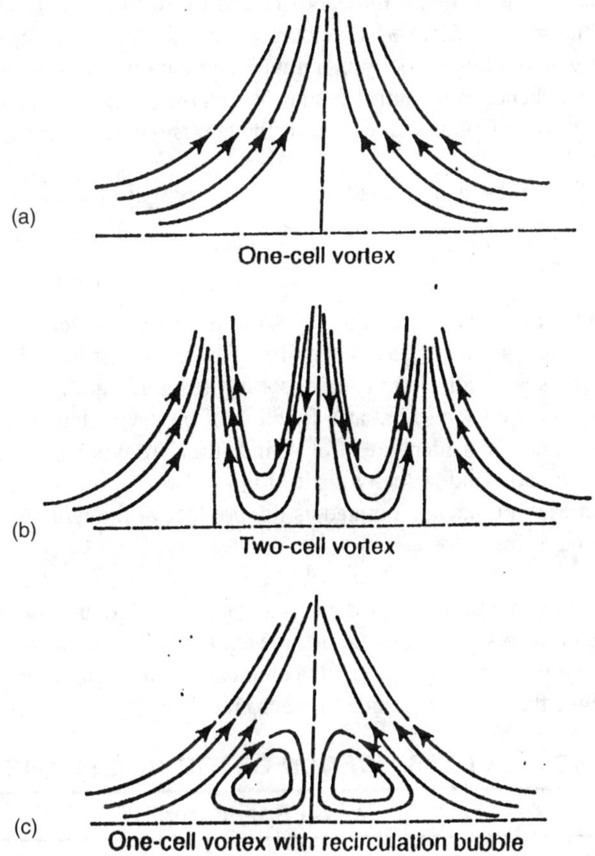

FIGURE 6.2. VERTICAL FLOW PATTERNS

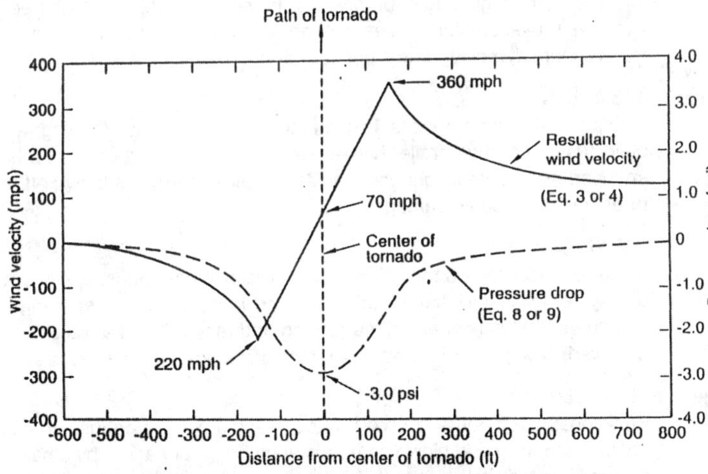

Tornado parameters Region 1	V_c = 290 mph, V_t (max) = 70 mph, V_t (min) = 5 mph R_c = 150 ft, P_{max} = 3.0 psi, max (dP/dt) = 2.0 psi/s

FIGURE 6.3. STANDARD TORNADO, WIND AND PRESSURE FIELD FOR NRC REGION 1 (V_C IS ROTATIONAL SPEED, V_T IS TRANSLATIONAL SPEED, R_C IS RADIUS OF MAXIMUM ROTATIONAL SPEED AND P IS PRESSURE DROP)

Since the basic wind motion mechanism for tornadoes is still not well understood and potentially quite variable, estimates and distributions of vertical wind velocities are also, for the most part, pure conjecture. It is unlikely that vertical wind would attain velocity magnitude comparable to the tangential wind or have a maximum at the same point, and, hence, not control design. Vertical wind speeds used for missile evaluation purposes have typically ranged from one-third to two-thirds the tangential component, with the higher velocities located near the vortex [11].

In two separate papers in [11], Stevenson and Shanahan have suggested more complete tornado wind models.

Tornado Classification

A tornado classification as to the intensity was defined by a Fujita scale in which he categorized five intensity classes by that speeds that vary in the range of 40 mph to 318 mph, as shown in Table 6.3. This Fujita tornado wind damage and wind speed correlation scale is currently undergoing review, with the possible result that wind speeds may be reduced somewhat as a function of observed damage.

Tornado classifications should also consider size effects since they are typically narrow spectrum effects. In addition to developing a tornado wind velocity scale, Fujita also classified tornadoes into categories according to their individual area, and they are defined as the product of the length and mean width of tornado damage [9]. These categories are:

(GI) Giant tornado: A tornado with its individual area equal to or larger than 10 square miles.

(MA) Macro tornado: A tornado with its individual area equal to or larger than 1 square mile, but less than 10 square miles. This category had been identified as regular tornado, but the term "regular" is so vague that it was changed to macro tornado.

TABLE 6.3. FUJITA (F) SCALE CLASSIFICATION OF TORNADOES BASED ON DAMAGE

Damage	Wind Speed (mph)
(F0) Light damage	40 to 72 This speed range corresponds to Beaufort 9 through 11. Some damage to chimneys or TV antennae; breaks branches off trees; pushes over shallow-rooted trees; old trees with hollow insides break or fall; sign boards damaged.
(F1) Moderate damage	73 to 112 73 mph is the beginning of hurricane wind speed or Beaufort 12. Peels surface off roofs; windows broken; trailer houses pushed or overturned; trees on soft ground uprooted; some trees snapped; moving autos pushed off the road.
(F2) Considerable damage	113 to 157 Roofs torn off frame houses, leaving strong upright walls standing; weak structures or outbuildings demolished; trailer houses demolished; railroad boxcars pushed over; large trees snapped or uprooted; light-object missiles generated; cars blown off highway; block structures and walls badly damaged.
(F3) Severe damage	158 to 206 mph Roofs and some walls torn off well-constructed frame houses; some rural buildings completely demolished or flattened; trains overturned; steel-framed hanger-warehouse-type structures torn; cars lifted off the ground and may roll some distance; most trees in a forest uprooted, snapped or leveled; block structures often leveled.
(F4) Devastating damage	207 to 260 Well-constructed frame houses leveled, leaving piles of debris; structures with weak foundation lifted, torn and blown some distance; trees debarked by small flying debris; sand and soil eroded and gravels fly in high winds; cars thrown some distance or rolled considerable distance finally to disintegrate; large missiles generated.
(F5) Incredible damage	261 to 318 Strong frame houses lifted off foundation and carried considerable distance to disintegrate; steel-reinforced concrete structures badly damaged; automobile-size missiles carried a distance of 100 yards or more; trees debarked completely; incredible phenomena can occur.

(ME) Meso tornado: A tornado with its individual area equal to or larger than 0.1 square mile, but less than 1 square mile.

(MI) Micro tornado: A tornado with its individual area equal to or larger than 0.01 square mile, but less than 0.1 square miles.

(DE) Decimicro tornado: A tornado with its individual area equal to or larger than 0.001 square mile, but less than 0.01 square mile.

(TR) Trace tornado: A tornado with its individual area less than 0.001 square mile identified as "trace."

(FU) Funnel cloud: A funnel cloud that does not induce tornadic wind on the ground. If an evidence of tornadic wind damage on the ground is confirmed beneath the path of a funnel cloud, one should classify the storm as a tornado under a reasonable assumption that a vertical column of rotating air extended all the way to the ground.

The above categories are based upon orders of magnitude of tornado areas. The proper selection of the area category can easily be made based upon a rather approximate estimate of damage provided the estimate does not deviate more than one order of magnitude. For example, when a tornado swath extends a distance of 10 miles with damaged houses and trees scattered along a very narrow swath, this storm is classified as a meso tornado. In order to call a storm of the same path length a macro tornado, about a one-city-block width is needed throughout the entire swath. An accurate damage area estimate can be made through detailed surveys, but both newspaper clippings and eyewitness accounts are generally good enough to permit a tornado area assessment with "one-order of magnitude" accuracy.

Fujita also classified a tornado as a function of damage, as shown in Table 6.3.

Tornado Wind Field Simplified Analytical Models

A simple engineering tornado model has been widely used—it considers a two-phase relationship, where at relatively large distances from the center of the vortex the circulation *(C)* is considered constant [12]. Typically tornado wind velocity is assumed not to vary with height and to be based on a 3.0-second wind gust. A maximum rotational wind velocity (V_c) is normally also defined at the distance
R_c as the radius to maximum wind velocity to be determined from Eq. (6.1).

$$C = v(r)r \tag{6.1}$$

where C = circulation constant; $v(r)$ = tangential wind velocity at radius r; and r = radius from center of the vortex.

Alternatively, V_c and R_c can be determined from a physical definition of the tornado for which a design or evaluation is desired. These two parameters can then be used to define C.

Obviously, Eq. (6.1) breaks down as r approaches zero. In the central core region, where r is less than R_c, the angular velocity *(T)* is assumed constant (i.e., solid body rotation is assumed):

$$\omega = \frac{v(r)}{r} = \frac{V_c}{R_c} \tag{6.2}$$

where the radius of cyclonic boundary between the two regions is defined as R_c:
and

$$v(r) = V_C \frac{r}{R_C} \quad \text{for} \frac{r}{R_C} \leq 1 \tag{6.3}$$

$$v(r) = \frac{C}{r} \quad \text{for} \frac{r}{R_C} \geq 1 \tag{6.4}$$

$$V_C = \frac{C}{R_C} \quad \text{for} \frac{r}{R_C} = 1 \tag{6.5}$$

$$v(r) = V_C \frac{R_C}{r} \quad \text{for} \frac{r}{R_C} > 1 \tag{6.6}$$

The resulting pressure distribution is determined by the cyclostropic wind equation

$$\frac{1}{\rho}\frac{dp}{dr} = \frac{v(r)^2}{r} \tag{6.7}$$

where Δ = ambient air density; and p = differential air pressure at a distance r from the center of the cortex. If Eqs. (6.2) and (6.3) are substituted into Eq. (6.7) and the pressure drop is integrated, then:

$$p(r) = -\rho V_c^2 \left[1 - 0.5\left(\frac{r}{R_C}\right)^2\right] \quad \text{for} \frac{r}{R_C} \leq 1 \tag{6.8}$$

$$p(r) = -0.5\rho \left[V_c^2 \left(\frac{R_C}{r}\right)^2\right] \quad \text{for} \frac{r}{R_C} \geq 1 \tag{6.9}$$

The actual wind loading on the structure depends on the geometry of the structure and would require the application of the various structure wind interaction coefficients (C_d) given in national building codes [1]. The differential pressure is maximum at the center of the vortex and half that value where the rotational wind velocity is a maximum.

The wind pressure load on a portion of a structure can be determined by:

$$q(r) = 0.5 C_d \Delta v(r)^2 \tag{6.10}$$

as define in the *ASME STS-1-99 Standard*

where $q(r)$ = dynamic wind pressure load on the structure at a distance r from the center of the vortex; C_d = drag coefficient, depending on the orientation and shape of the structure being considered; Δ = air density of 0.002378 bs mass/cu. ft.; and $v(r)$ = wind velocity at radius r from center of vortex.

The model for determining wind velocity and pressure at any point in the wind and pressure field as just expressed is quite simple in nature when radial and vertical wind components are not considered and there is no variation of tangential velocity with elevation. Although such components obviously exist, they tend to be of much lower magnitude than the combination of tornado translational and tangential wind effects. A typical tornado wind field profile is shown in Figure 6.4.

The differential pressure load $p(r)$ at any point on an SSC would be determined by the application of Eqs. (6.8) or (6.9). In general, an SSC will relieve itself of the differential pressure load of a tornado before it becomes a significant load on the SSC if the surface area of the SSC comprises 5% or more openings. Otherwise, differential pressure loads should also be applied concurrently with the wind pressure load $q(r)$.

6.3 LOAD DEFINITION

6.3.1 Wind Pressure

Wind pressures on SSC are developed from wind impinging of a structural surface. On the windward side of a structure the wind pressure is defined as acting toward the inside of the structure. On the top, sides and lee

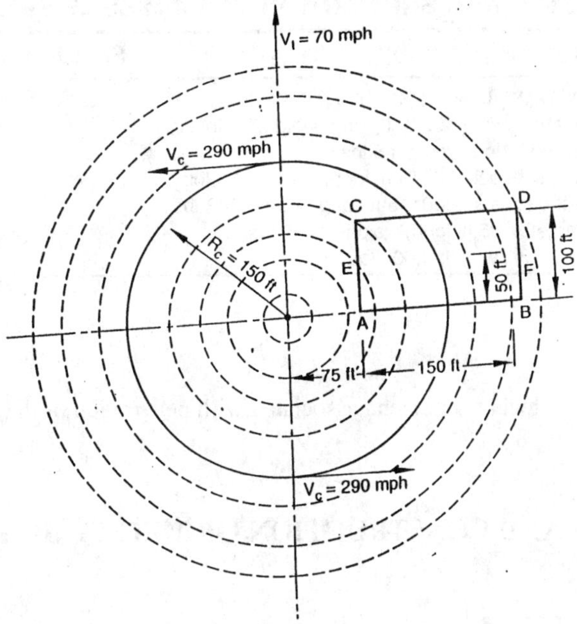

FIGURE 6.4. NRC REGION 1 TORNADO WIND FIELD MAXIUM VELOCITY APPLIED TO THE CENTERLINE OF A RECTANGULAR BUILDING (HEIGHT, *H*, OF THE BUILDING IS 30 FT., AMBIENT AIR DENSITY, P, IS 0.002378 SLUG/FT^3, V_C IS ROTATIONAL SPEED, V_T IS TRANSLATIONAL SPEED, R_C IS RADIUMS MAXIMUM ROTATIONAL SPEED AND *P* IS PRESSURE DROP; AB AND BD ARE SPECIFIED AS WINDWARD WALLS, CD AND AC ARE LEEWARD WALLS, AND EF IS ROOF CENTERLINE)

(back) of the SSC, wind pressure is acting outward away from the SSC. Wind pressure is a function of velocity of wind, wind density, drag coefficient of the wind gust factor and terrain over which the wind is passing, and finally the height of the structures above ground level as defined in reference 1, 2 and 3.

6.3.2 Differential Pressure

Associated with tornado loads is an external pressure drop, which results in the actual pressure acting away from all surfaces of a closed structure or has opening carrying less than 10 percent of its surface area.

6.3.3 Missiles

Associated with wind fields with 3-second gust velocities equal to or less than 90 mph (km) are potential missiles picked up and accelerated by the wind field as a function of the drag and lift coefficient associated with the shape of the potential missile. In [13] are defined typical missiles considered in nuclear power plant (high hazard) design as shown in Table 6.4. Typically there are two limiting types of missiles: penetrating, as characterized by missiles C, D and E of Table 6.4 and impactive, as characterized by missile G of Table 6.4. In [6.6] are missiles and their design parameters associated with other types of (moderate hazard) nuclear facilities. Described in [6.14] and Chapter 8 is the effect of tornado missile impacts associated with impact-type loads on the structure and the local penetration of the structure.

Impact

The impact effect of the missile results from connecting the kinetic energy of the missile to potential energy, which must be carried by the impacted structure as described in Chapter 8.

TABLE 6.4. NRC SRP SECTION 3.5.1.4 MISSILE SPECTRUM A

	Missile Descriptions	Fraction of Total Tornado Velocity
A	Wood plank: 4 in. × 12 in. × 12 ft., weight 200 lbs.	0.8
B	Steel pipe: 3 in. diameter, schedule 40, 10 ft. long, weight 78 lbs.	0.4
C	Steel rod: 1 in. diameter × 3 ft. long, weight 8 lbs.	0.6
D	Steel pipe: 6 in. diameter, schedule 50, 15-ft. long, weight 285 lbs.	0.4
E	Steel pipe: 12 in. diameter, schedule 40, 15 ft. long, weight 743 lbs.	0.4
F	Utility pole: 131/2 in. diameter, 35 ft. long, weight 1,490 lbs.	0.4
G	Automobile, frontal area 20 ft^2, weight 4,000 lbs.	0.2

Penetration

Penetration of missiles is a local phenomena associated with penetration of the missile into the target structure as discussed in Chapter 8.

6.4 WIND DESIGN AND CONSTRUCTION OF MECHANICAL COMPONENTS AND SUPPORTS

6.4.1 Mechanical Structures

The mechanical structure designs normally controlled by wind are steel stacks and towers. In geographical regions of the U.S. of low seismicity and high wind loadings, such as the areas bordering on the Gulf of Mexico, wind may control design of exposed vertical ground supported stacks, towers and tanks. Wind loads may also control design of duct and pipe and filtered openings in structures and components.

6.4.2 Distribution Systems

Mechanical distribution systems whose design is normally controlled by wind are HVAC located on the roof of building structures. Design of outside piping systems in petrochemical facilities may also be controlled by wind loads.

6.4.3 Ventilators and Filters

There are mechanical components typically with relatively large surface areas whose design adequacy may be controlled by wind loads. The design adequacy of these types of components is controlled by the SMACNA [15] and *ASME AG-1* Codes [16].

6.5 IN-LINE WIND AND DIFFERENTIAL PRESSURE LOADING

Straight and hurricane winds load the front, sides, top and back of structures differently as defined in ref 1. Tornado wind loading depends on the relative orientation of the tornado wind field and the structures. The load associated with maximum wind load plus half the maximum differential pressure loads and missile impact is applied to the windward side of a building.

This loading typically generates the maximum positive (external-inward) load on the windward element of the structure or component. The maximum negative (interior-outward) load is on the top, leeward and side elements. Also to be considered is overturning load and uplift and sliding of the overall structure. This load combination for a closed structure having no opening is expressed as follows in Eqs. (6.12) or (6.13):

$$W_t = K_i\, W_w + 0.5\, W_p + W_m \tag{6.11}$$

or

$$W_i = -W_p \tag{6.12}$$

$$W_t = K_i + 0.5\ W_p \tag{6.13}$$

where W_t = total wind load on a structural element; W_w = wind load on a structural element; W_p = tornado differential pressure load, taken as 0 if not a tornado; W_m = tornado missile load usually taken as 0 if design-basis 3-second gust wind is 90 mph or less; and K_i = load coefficient as a function of wind tunnel tests as defined in ref. 1, 2 and 3 or 307.

To design or evaluate structural elements for the limiting tornado loads, the orientation of the tornado would be moved to develop the maximum load consistent with the proximity and orientation of adjacent structures. If there is a small gap between two structures (i.e., less than one-quarter of the radius to maximum wind), it is not considered credible that the tornado could drop into the gap to load the structures; hence tornado wind loads need not be considered in the gap. Differential pressure loads resulting from the passage of the tornado over the gap should be considered in the design of structural elements being loaded from the gap.

6.5.1 Openings

The atmospheric differential pressure load on a building is always negative (outward). Figure 6.3 shows the wind differential pressure load on a closed or essentially closed structure for a 290 mph rotational wind where the pressure drop at maximum wind velocity is one half of the maximum pressure drop.

The orientation of the tornado with maximum wind speed at the centerline of the building results in the maximum lateral wind load on the structure. The orientation of the tornado with the center of rotation at the centerline of the structure results in the maximum uplift load on the structure. Because the maximum wind velocity, the maximum tornado missile load effect on the structure would occur at the point where the centerline of the maximum wind speed is perpendicular o the windward side of the structure.

For structures with openings (i.e., about 10% or more of the surface area of the building wall perimeter surface not designed to be tornado-resistant or louvered to open upon reduced outside pressure), the differential pressure load would be effectively eliminated by equalization of the pressure through the openings. Another criteria used for this effec1.0-ft^2 opening for every 1,000 ft^3 of building volume. The same phenomenon would happen in structures where elements of the structure, such as roof to walls, begin to separate. This is of little value in design, however, because the structure would be greatly weakened by such behavior.

Most residential or commercial and industrial structures have sufficient window or door openings or relatively weak connections between the roof and side walls such that failure of these openings permits an equalization of the differential pressure. However, such equalization takes a finite amount of time so that failure at the wall-to-roof structure connection may occur before the equalization, and the resistance of the walls or roof to wind loads is thereby reduced. In tornado-resistant structures, care must be taken to ensure that connections between the roof and sidewalls can withstand the differential pressure load effects, including the localized effects at eaves and corners described in [6.1], without failure so that the capacity of the sidewalls or roof to withstand wind load effects is not reduced. When building opening areas exceed about 10% of the wall surface area of the building, the effect of differential pressure loads can be ignored.

Openings, particularly on the windward side of the building, can result in significant internal pressurization (the "balloon" effect) when such openings equal or exceed 5% of the wall surface area, and such openings have not been designed to be tornado, wind, pressure or missile load resistant. Because louvered openings are designed to open upon reduced outside pressure, depressurization effects through openings in side or leeward walls may also need to be considered in design.

6.5.2 Dynamic Load Considerations

Straight, hurricane and tornado wind and differential pressure loads perpendicular to the impacted surface have usually been considered as statically applied with a dynamic load factor of 1.0 for design purposes. This is a valid assumption provided the response of the building or building element has a fundamental frequency greater than about 1.0 Hz. This is based on an assumed triangular load pulse with a rise time of about 1.5 seconds. For structures and structural elements with fundamental frequencies less than about 1.0 Hz, it would be generally necessary to evaluate more accurately the potential for dynamic response of the structure or structural element to tornado wind and differential pressure loads as a function of the rate of loading on the structure (i.e., in structures with fundamental frequencies less than about 1.0 Hz, the wind and differential loads normal to the impact surface should be considered as dynamically applied).

Vortex-Shedding Design Methodology

A steel stack or tower response to vortex-induced wind loads is based upon dimensions, modal properties for the vibration mode being considered, the structural damping, β_s, and aerodynamic damping, β_a. The vortex-shedding design methodology is covered in more detail in *ASME STS-1-2000* [3].

For the vortex-shedding response, when V_c is in the vicinity ±30% of V_zcr the aerodynamic damping is negative and shall be included.

Response is highly sensitive to small changes in damping β_s where values of A_2 are near 1.

General Theory

The root-mean-square motion at the point of maximum displacement is given by:

$$\frac{\bar{a}_M}{D} = \frac{C_1 C_M}{m_r \sqrt{\lambda(\beta_s + \beta_a)^{1/2}}} \tag{6.14}$$

where $\beta_a = -\frac{C_2}{m_r}\left[1-4\left(\frac{\bar{a}_m}{D}\right)^2\right]$; $\lambda = H/\bar{D}$; $m_r = m_e/\rho\bar{D}^2 g$; $m_e = \int_o^H m(z)\phi^2(z)dz \Big/ \int_o^H \phi^2(z)dz$; $C_2 = 0.6$ *(for parallel stacks only)*; $C_M = \phi(z_M)\frac{\left(\frac{1}{H}\int_o^H \phi^2(z)dz\right)^{1/2}}{\frac{1}{H}\int_o^H \phi^2(z)dz}$; *$\phi(z_M)$ = value of $\phi(z)$ at* maximum *deflection,* z_M = *H for cantilever* mode; $C_M \cong 2.0$ *for the fundamental* mod *e of vibration;* $C_1 = 0.12$ *for an isolated steel stack;* and $C_1 = \frac{0.12}{(5S)^2}$ *α for a grouped steel stack.*

The equation defining $\tilde{a}_M/\bar{D}$ can also be written as;

$$\frac{\tilde{a}_M}{\bar{D}} = \frac{A_1}{\left\{1-A_2\left[1-4\left(\frac{\tilde{a}_M}{\bar{D}}\right)^2\right]\right\}^{1/2}} \tag{6.15}$$

where $A_1 = C_1 C_M / m_r \sqrt{\beta_s \lambda}$; and $A_2 = C_2/m_r\beta_s$

For $m_r\ \beta_s > 0.8$; $\frac{\tilde{a}_M}{\bar{D}} \cong \frac{A_1}{\left[1-\left(\frac{C_2}{m_r\beta_s}\right)\right]^{1/2}}$

For $m_r\ \beta_s < 0.4$: $\frac{\tilde{a}_M}{\bar{D}} \cong 0.5\left(1-\frac{m_r\beta_s}{C_2}\right)^{1/2}$

Practical Application

The general solution may be reduced to the following formulas of vortex shedding and then used to determine equivalent static loads. For any values of $m_r\beta_s$;

$$\tilde{a}_M = \overline{D}\left(\frac{-(1-A_2)+\sqrt{(1-A_2)^2+16A_1^2A_2}}{8A_2}\right)^{1/2} \tag{6.16}$$

The peak values for vortex-shedding response are:

$$\hat{a} = \hat{g}\tilde{a}_M \text{ and } a_s = g_s\tilde{a}_M \tag{6.17}$$

where $\hat{a}$ = maximum value and is used to calculate peak loads and stress while a_s defines equivalent constant amplitude for fatigue calculations. The values of $\hat{g}$ and g_s are determined from the following:

$$\text{For } m_r\beta_s > 0.8,\ \hat{g} = 4.0 \text{ and } g_s = 2.0 \tag{6.18}$$

$$\text{For } m_r\beta_s < 0.4,\ \hat{g} = 1.6 \text{ and } g_s = 1.5 \tag{6.19}$$

Linear interpolation is used for $0.4 < m_r\beta_s < 0.8$.

Equivalent Static Loads

The equivalent static loads corresponding the displacement, $\hat{a}_M$, are given by:

$$\hat{w}(z) = \hat{a}(2\pi n_1)^2\phi(z)m(z)/g_c \tag{6.20}$$

$$w_s(z) = a_s(2\pi n_1)^2\phi(z)m(z)/g_c \tag{6.21}$$

The number of cycles in 50 years at the equivalent constant amplitude a_s is given by:

$$N/n_1 = 10^{10}\left(\frac{V_c\sqrt{I}}{V_{z_{cr}}}\right)^2 \exp\left[-15\left(\frac{V_c\sqrt{I}}{V_{z_{cr}}}\right)^2\right] \tag{6.22}$$

V_c and V_zcr are evaluated at the same height.

Variable Diameter Stacks or Towers

For variable diameter stacks, the preceding method may be used with the following modifications to account for the range of possible critical diameters. The previous method is used with the following changes in formulas.

The peak response is determined by varying the range of height being considered for any mean diameter, $\overline{D}$, for a portion of the steel stack where the diameter varies 15% from this mean diameter. The peak response is determined by iterations over the full height of the stack.

$$\lambda = \frac{z_2 - z_1}{\overline{D}} \tag{6.23}$$

$$C_M = \phi(z_M)\frac{\left(\frac{1}{H}\int_{z_1}^{z_2}\phi^2(z)dz\right)^{1/2}}{\frac{1}{H}\int_{o}^{H}\phi^2(z)dz} \tag{6.24}$$

$$R = \frac{\left(\int_{z_1}^{z_2} \phi^2(z)dz\right)}{\int_o^H \phi^2(z)dz} \quad (6.25)$$

$R = 1.0$ *for nearly parallel;* and $C_2 = 0.6R$.

The limits change to: $m_r\beta_s > 0.8\ R$; and $m_r\beta_s < 0.4\ R$.
Symbols and definitions:

- A = center-to-center stack spacing for interference effects (ft)
- A_1 = constant
- A_2 = constant
- $\hat{a}$ = maximum value amplitude for static equivalent design loads (ft)
- a_s = maximum value amplitude for static equivalent fatigue loads (ft)
- $\tilde{a}_M$ = root mean square dynamic displacement at z=z_m (ft)
- C_1 = constant for grouped/isolated stacks
- C_2 = constant
- C_M = mode shape constant
- $\bar{D}$ = mean diameter for the segment z_1 to z_2, or for stacks with less than +/–10% variation over the top 1/3 the value of $\bar{D}$ is the average over the top 1/3 (ft)
- H = height of steel stack (ft)
- $\hat{g}$ = constant for maximum static equivalent loads
- g = gravitational acceleration (32.2 ft/sec^2)
- g_c = gravitational constant (32.2 lbm-ft/ft-lb/lbf-sec^2)
- g_s = constant for fatigue static equivalent loads
- m_e = equivalent uniform mass per unit length (lbm/ft)
- $m(z)$ = mass per unit length at height z (lbm/ft)
- m_r = dimensionless mass
- n_1 = natural frequency of mode (Hz)
- R = constant for tapered stacks
- S = Strouhal number
- V_c = critical speed for the segment z_1 to $z_2 = 5n_1\ D$ (ft/sec)
- V_zcr = mean hourly design speed (50-year return period) at the critical height z_{cr} used for evaluating the critical wind velocity (ft/sec)
- z = height z under consideration (ft)
- z_{cr} = ½ (z_1+z_2) or, for stacks with less than +/–10% variation over the top third, z_{cr} = 5/6 H (ft)
- z_M = height at maximum modal shape displacement (H for mode 1) (ft)
- z_1,z_2 = upper and lower limits of a section of the stack over which the diameter changes by 30% (i.e., $\bar{D}$ ±15%) (ft)

For tapered stacks, vortex-shedding loads shall be evaluated over all ranges of height for the distance of +/–15% of mean diameter of that height. For a strongly tapered stack, the range z_1 to z_2, is not necessarily near the top. A larger response can be lower where the diameter is larger.

- a = amplification factor
- β_a = aerodynamic damping
- β_s = structural damping
- λ = aspect ratio
- ρ = air density (0.00238) (lbm-sec^2/ft^4)

$\phi(z)$ = normalized mode shape at height z (ft)
$\phi(z_M)$ = max normalized modal displacement $\phi(z)$ for mode at $z = z_M$, for the first mode $z = H$ (ft)
$\hat{w}(z)$ = equivalent static load (lbf/ft)
$w_s(z)$ = equivalent fatigue load (lbf/ft)

6.6 MISSILE LOADS

6.6.1 Penetrating

Penetrating missiles for design purposes fall into two categories: (1) penetrating missiles; and (2) impact missiles. With respect to nuclear facilities the DOE and the NRC have developed a spectrum of wind and tornado design missiles as discussed in Chapter 8 and shown in Table 6.4 for NRC tornado missiles from [13].

It has been determined that, of the six missiles identified in Table 6.4, the missile with the greatest potential for penetration is the 12.0 in.-diameter pipe traveling horizontally ends at a velocity of 47.0 m/s for NRC region I. The missile with the greatest impact or kinetic energy effect on the structure is the automobile. It has been common practice to limit the automobile missile impact to within 30 feet of grade level.

The impact of the missile may have penetrating effects that reduce the local ability of the target structure to resist simultaneous wind pressure and further penetration effects due to the backside scabbing. This local penetration phenomenon should be considered in the overall resistance of the target structure to resist all missile effects. The potential for scabbing is usually ignored provided the nominal wall thickness is at least three times the computed depth of penetration. (Scabbing is when fragments on the inside wall of the target break off due to the impact of a projectile on the outside of that wall or plate.) If the effect of backside scabbing is considered, including the effect of the generation of secondary missiles, wall thicknesses down to twice the penetration depth are permitted. Dynamic structural impact load should be combined with wind and differential pressure loads for design and evaluation purposes. Scabbing of metal plates due to missile impact can also occur, but generally at impact velocities above 1,500 ft/sec.

6.6.2 Inelastic Structural Response

In response to tornado wind and differential pressure effects, structures and structural elements that are meant to remain effectively leak-tight are expected to remain essentially elastic (i.e., ductility factor $\mu \leq 1.3$). For missile penetration effects, local empirical penetration formulas are used. In regions outside the missile impact and penetration area, the design is controlled by the ability of the structure or structural element to absorb the kinetic energy and rapid application of the missile load. In this instance, local ductility factors well in excess of elastic response are as recommended as defined in Table 8.1, from Chapter 8. For preliminary evaluation purposes, a composite global ductility of 5.0 is often used.

6.7 FLOOD AND PRECIPITATION LOAD

6.7.1 Introduction

Flood and precipitation (snow, rain and ice) loads can in some instances control the design of mechanical SSC. Snow loads can be a limiting design parameter for tank roofs and roof mounted ductwork, particularly when such roofs are flat with perimeter parapets, which may result in significant drifting. Rain loads including ponding effects can control the roof design of mechanical SSC subject to such loads, particularly when it is assumed that the primary roof drain system is plugged. When such conditions exist it is recommended that design be based on increased vertical live load caused by the precipitation. Flood action, and in particular wave or flowing water action, can control the design adequacy of mechanical SSC when subjected to such loads. Flood and precipitation load design procedures are given in Section 5.7 and 8 of *ASCE Std. 7-2005* [1] for flood and precipitation loading phenomena.

6.7.2 Loads During Flooding (by permissions from ASCE)

Hydrostatic Loads

Hydrostatic loads caused by a depth of water to the level of the design flood elevation shall be applied over all surfaces involved both above and below ground level for design; the depth shall be increased by 1 ft.

Reduced uplift and lateral loads on surfaces of enclosed spaces below the design flood elevation shall apply only if provisions are made for entry and exit of floodwater.

Hydrodynamic Loads

Dynamic effects of moving water shall be determined by a detailed analysis utilizing stagnation pressure loads developed when the moving water is stopped by the SSC [1].

Where water velocities do not exceed 10 ft/sec, dynamic effects of moving water may be converted into equivalent hydrostatic loads by increasing the design flood elevation for design purposes by an equivalent surcharge depth, d_h, on the headwater side and only above ground level, equal to:

$$d_h = \frac{aV^2}{2g} \tag{6.26}$$

where V = average velocity of water [in ft/sec (m/s)]; g = acceleration due to gravity, 32.2 [in ft/sec] and a = coefficient of drag or shape factor (not less than 1.25).

The equivalent surcharge depth shall be added to the design flood elevation design depth and the resultant hydrostatic pressures applied to, and uniformly distributed across, the vertical projected area of the structure that is perpendicular to the flow. Surfaces parallel to the flow or surfaces wetted by the tailwater shall be subject to the hydrostatic pressures for depths to the design flood elevation only.

Wave Loads

Wave loads can be determined by one of the following three methods: (1) using the analytical procedures outlined in ref. [1]; (2) more advanced numerical modeling procedures considering fluid structure interaction by finite element modeling; or (3) by laboratory test procedures (physical modeling).

Wave loads are those that result from water waves propagating over the water surface and striking a structure. Design and construction of structures subject to wave loads shall account for the following loads: waves breaking on any portion of the structure; uplift forces caused by shoaling waves beneath the structure, or portion thereof; wave run-up striking any portion of the structure; wave-induced drag and inertia forces; wave-induced scour at the base of the structure or its foundation. Wave loads shall be included for both *V*-zones and *A*-zones. In *V*-zones, waves are 3 ft high, or higher; in coastal floodplains landward of the *V*-Zone, waves are less than 3 ft high.

Breaking waves loads can be calculated using the procedures described, which are taken from Section 5.3 of [1]. Breaking wave heights used in this procedure herein shall be calculated for *V*-zones and coastal *A*-zones using Eqs. (6.27) and (6.28).

$$\mathrm{H_b = 0.78\ d_s} \tag{6.27}$$

where H_b = breaking wave height in ft and d_s = local stillwater depth in ft.

The local stillwater depth shall be calculated using Eq. (6.28), unless more advanced procedures or laboratory tests are used.

$$\mathrm{D_s = 0.65\ (BFE\text{-}G)} \tag{6.28}$$

where BFE = Base Flood Elevation in ft; and G = Ground Elevation in ft.

Breaking Wave Loads on Vertical Walls or Plates

Maximum pressures and net forces resulting from a normally incident breaking wave (depth-limited in size, with $G_b = 0.78d_s$) acting on a rigid vertical wall can be calculated by the following:

$$P_{max} = C_p\gamma_w d_s + 1.2\gamma_w d_s \tag{6.29}$$

and

$$F_t = 11.C_p\gamma_w d^2_s + 2.4\gamma_w d^2_s \tag{6.30}$$

where P_{max} = maximum combined dynamic ($C_p\gamma_w d_s$) and static ($1.2\gamma_w d_s$) wave pressures, also referred to as shock pressures [in lbs/ft^2]; F_t = net breaking wave force per unit length of structure, also referred to as shock, impulse or wave impact force [in lbs/ft], acting near the stillwater elevation; C_p = dynamic pressure = 3.5 for hazardous facilities; γ_w = unit weight of water [in lbs/ft^3 (kN/m^3)] = 62.4 pcf for fresh water and 64.0 pcf (10.05 kN/m^3) for salt water; and d_s = stillwater depth [in ft] at base of building or other structure where the wave breaks.

This procedure assumes the vertical wall causes a reflected or standing wave against the waterward side of the wall with the crest of the wave at a height of 1.2 d_s above the stillwater level.

This procedure also assumes the space behind the vertical wall is dry, with no fluid balancing the static component of the wave force on the outside of the wall. If free water exists behind the wall a portion of the internal hydrostatic pressure counters the hydrostatic component of the wave pressure and force, and the net force shall be computed by Eq. 6.30 [the maximum combined wave pressure is still computed with Eq. (6.29)].

$$F_t = 1.1C_p\gamma_w d_s^2 + 1.9\gamma_w d_s^2 \tag{6.31}$$

where F_t = net breaking wave force per unit length of structure, also referred to as shock, impulse or wave impact force [in lbs/ft], acting near the stillwater elevation; C_p = dynamic pressure coefficient (3.5 for hazardous facilities); γ_w = unit weight of water [in lbs/ft^3] = 62.4 pcf for fresh water and 64.0 pcf for salt water; and d_s = stillwater depth [in ft] at base of building or other structure where the wave breaks.

For more detailed flood and precipitation loads and analyses see [1] and [17].

6.8 VOLCANO EFFECTS

Volcano-induced design loads for hazardous facilities are limited to ash loading on roofs and the potential for clogging air intakes. For more severe volcano loads associated with lava or plastic mud-ash flow, hazardous facilities should not be located where they are credible phenomena.

For more information on volcanoes and siting, see [18].

6.9 LIGHTNING PROTECTION

6.9.1 Introduction

Lightning protection systems are intended to minimize damage to hazardous facilities that may be subject to atmospheric lightning strike. The damage caused by lightning may be thermal (fire), electrical (flashover) and/or mechanical (torsional forces). The energy in a lightning strike can ignite flammable vapors and damage the safety-related equipment and structures by the flow of lightning discharge current through any resistance in its path. In process plants, electrical equipment, all equipment containing hazardous liquids and gases, outdoor metal structures, vessels or other tall equipment subject to lightning, strikes should be grounded.

6.9.2 Protection System Design

Types of Lightning Protection Systems

There are several types of lightning protection systems for hazardous facilities, including overhead wire (catenary), integral and mast. These systems feature air terminals, down conductors and the earth electrode system (ground terminals).

a. Air terminals; An air terminal is the component of a lightning protection system that is intended to intercept lightning strikes.
b. Earth electrode system (ground terminals): The earth electrode system (ground terminals) dissipate the current from a lightning strike to ground. Earth electrode system (ground terminals) may be ground loop conductors, grounding rods, ground plates, a cable immersed in salt water or chemical grounds that are installed for the purpose of providing electrical contact with earth.
    1. A resistance to earth of 25 ohms or less is required. If this resistance cannot be achieved with ground rods alone, a ground loop (counterpoise) system (regularly tested) is acceptable even if the resistance is greater than 25 ohms.
    2. The earth electrode system (ground terminals must meet the requirements of NFPA 78) [21].
c. Down conductors: Down conductors connect any of the air terminals described above to the earth electrode system (ground terminals). Any bend in down conductor wires shall have a minimum radius of 8 inches, and will not result in an angle of less than 90 degrees. These down conductors shall meet the requirements of the NFPA Lightning Protection Code [21].

Striking Distance Requirement

Lightning protection systems designed for hazardous facilities shall be based on 100-foot striking distance arc.

6.9.3 Facility Protection

Side Flash

Definition of and protection from side flash can be obtained by either separation distance or by bonding. The required separation distance can be determined by using the formula in *NFPA 78* [21]. Bonding is used to reduce the possibility of a side flash as well as to ensure that lightning current produces no potential differences. Chapter 3 of *NFPA 78* may be used to determine the minimum bonding requirements.

a. Facility metal items such as radiators, tanks, permanent machinery, stair rails, ventilators, metal doors, air-conditioning ducts, metal columns and metal siding shall be bonded to the lightning protection system if within six feet of the lightning protection system.
b. Exterior metal doors shall have at least one conductive bond (a hinge may suffice) between the door and its frame. The frame must be bonded to the lightning protection system with a separate bond strap or via the structural steel in the building.
c. Fences and railroad tracks located within 6 feet of a structure's lightning protection system shall be bonded to the structure's lightning protection system.

Incoming Utility Requirements

All conductors, including shielded cabling, power cabling, communication lines and conduit shall be buried underground for a minimum of 50 feet prior to entering the structure. Steam, water and air-conditioning lines may run aboveground as long as they are bonded to the structures lightning protection system prior to entering the structure.

REFERENCES

1. *ASCE Standard 7-05,* 2003, "Minimum Design Loads for Buildings and Other Structures," American Society of Civil Engineers, Reston, VA.
2. *ACI-307-98,* 1998, "Design and Construction of Reinforced Concrete Chimneys and Commentary," American Concrete Institute, Detroit, Mich.
3. *ASME STS-1-2000,* 2001, "Steel Stacks," American National Standard and American Society of Mechanical Engineers, New York, NY.
4. Hazard Mitigation Center, 2000, "Development of a Probability Tornado Wind Hazard Model for the Continental United States," Lawrence Livermore National Laboratory,
5. *Regulatory Guide 1.76,* 1974, "Design Basis Tornado for Nuclear Power Plants," U.S. Atomic Energy Commission,
6. *DOE Std. 1020-2002,* 2002, "Natural Phenomenon Hazards Design and Evaluation Criteria for Department of Energy Facilities," Department of Energy, Washington, D.C.
7. *ANS Std. 2.3-1993,* 1993, "Standard for Estimating Tornado and Extreme Wind Characteristics at Nuclear Power Sites," American Nuclear Society (Draft),
9. Fujita, T., 1971, "Proposed Characteristics of Tornadoes and Hurricanes by Area and Intensity," *SMRP Research Paper 91,* Dept. of the Geophys. Sci., University of Chicago, IL.
10. Grazulis, T.P., *Significant Tornadoes 1680–1991, A Chronology and Analysis of Events,* The Tornado Project of Environmental Films,
11. *Proceedings of the Symposium on Tornadoes – Assessment of Knowledge and Implications for Man,* 1976, Texas Tech. University,
12. *Standard Review Plan Section 3.3.2,* 1981, "Tornado Loading," Rev. 2, Nuclear Regulatory Commission,
13. *SRP Section 3.5.1.4,* 1981, "Missiles Generated by National Phenomena," Rev. 2 U.S. Nuclear Regulatory Commission,
14. ASCE Committee on Impact and Impulse Loads, 1980, *Proc. Second ASCE Conf. on Civ. Engrg. and Nuclear Power,* Vol. V, Reston, VA.
15. SMACNA, "HVAC System Duct Design," Sheet Metal and Air Conditioning Contractors National Association, TK, TK
16. *AG-1 Div. II-BA,* "Ventilation Air Cleaning and Air Conditioning Fans and Blowers," American Society of Mechanical Engineers, New York, NY.
17. *SEI/ASCE 24-98,* 1998, "Flood Resistant Design and Construction," American Society of Civil Engineers, Reston, VA.
18. *Provisional Safety Standard Series No. 11,* 1997, "Volcanoes and Associated topics in Relation to Nuclear Power Plant Siting," International Atomic Energy Agency.
19. Center for Chemical Process Safety, 1993, "Guidelines for Engineering Design for Process Safety," American Institute of Chemical Engineers.
20. *EPRI-EL-503G,* 1987, "Grounding and Lightning Protection," Electric Power Research Institute.
21. *NFPA 78-1983,* 1983, "Lightning Protection Code," National Fire Protection Association.
22. ASCE Standard 43-05, "Seismic Design Criteria for Structures, Systems and Components in Nuclear Facilities," American Society of Civil Engineer, Reston, VA.

CHAPTER

7

BLAST DESIGN AND ANALYSIS

This chapter is intended to provide side-on and reflected static pressure loads on structures, systems and components (SSC) from an external explosive blast. These dynamic pressures are then converted to equivalent static loads for evaluation or design purposes. Figures are provided to describe the effect of external equivalent static blast loads as a function of explosive quantity, quality and distance in the 1 to 10,000 lbs range and 25 to 2,000 ft distance. Finally, example problems are presented for blast-equivalent TNT phenomena associated with side-on and reflected pressure and their conversion to equivalent static loads for design or evaluation purposes.

It should be noted that procedures contained here are simplified approximations. To obtain more accurate response of an SSC it is necessary to consider the time-dependent blast wave transversing the SSC. The result load on the SSC involves not only the characteristics of the blast wave and the global and local response frequency of the SSC, but also the geometry of the structure.

7.1 INTRODUCTION

In Table 7.1 is a summary of some common explosive material blast effects relative to a TNT equivalent located on the ground surface. In Table 7.2 are hydrocarbons equivalent to TNT, which can be used for blast wave definition purposes.

Section 7.2 references Figures 7.1 to 7.4 peak pressures and duration that give the magnitude of the impulse loads expressed in pounds at a given range (in feet) to a target as defined in *TM 5-1300* [1].

Section 7.3 describes the global (total load on the SSC) and local (windward, side walls, etc. load on the SSC) to include conversion of the impulse-type loads defined in Section 7.2 to equivalent static loads. These loads are defined as a function of the duration of the impulse, the natural period of the SSC being struck by the impulse blast wave and the amount of ductile global and local response permitted in the SSC being struck.

Table 7.3 in Section 7.4 describes the effect equivalent static loads have on building primary load path structures such as steel or concrete frames and concrete, masonry or brick shear walls globally, as well as architectural elements such as windows, doors, exterior and interior partitions, and both individual reinforced- and unreinforced-masonry infill walls. It also describes the effect blast loads have on mechanical systems and components. Table 7.4 contains a description of the effect blast pressure loads have on occupants of buildings that are directly affected by the blast wave.

Section 7.5 contains example problems, which demonstrate the application of external blast loads to target objects. Evaluation of interior blast load effects involve pressurization due to the gases generated by the explosion or detonation, plus the multiple reflection of blast waves from interior walls and objects. The complexity of this evaluation for practical purposes requires the use of existing computer algorithms [2, 3, 4].

Missile effects on SSC are discussed in Chapter 8.

TABLE 7.1. CHARACTERISTICS OF PRINCIPAL COMMERCIALLY AVAILABLE U.S. EXPLOSIVES

Name	Smallest Cap Required for Detonation	Relative Effectiveness as External Charge	Velocity of Detonation (fps)	Intensity of Poisonous Fumes	Water Resistance	Packaging	
TNT	Special blasting cap, electric or nonelectric	1.00	23,000	Dangerous	Excellent	1 lb, 50 or 56 to box	
Composition C3, M3, M5		1.34	25,000	Dangerous	Good	16 21/4-lb blocks in wooden box	
Composition C4, M112, M5A1		1.34	26,000	Slight	Excellent	24 21/2-lb blocks in wooden box	
Ammonium Nitrate[1] (cratering charge)	Special blasting cap, electric or nonelectric	0.42	14,800	Dangerous	Poor	40 lb charge in metal can	
Straight dynamite (commercial)	No. 6 commercial cap, electric or nonelectric						
40%		0.65	15,000	Dangerous	Poor	102	Sticks per 50-lb box
50%		0.79	18,000		Good	103	
60%		0.83	19,000		Excellent	106	
Ammonia dynamite (commercial)	No. 6 commercial cap, electric or nonelectric						
40%		0.41	8,900	Dangerous	Good	110	Sticks per 5-lb box
50%		0.46	11,000		Good	110	
60%		0.53	12,700		Good	110	
Gelatin dynamite	No. 6 commercial cap, electric or nonelectric						
40%		0.42[1]	8,000	Slight	Good	130	Sticks per 50-lb box
50%		0.47	9,000		Ver Good	120	
60%		0.76	16,000		Very Good	110	
Black powder	NA	0.55	1310 Max. depends on confinement	Dangerous	Poor	Bulk	
Nitrostarch	Special blasting cap electric or nonelectric	0.88	15,000	Dangerous	Satisfactory	1-lb blocks	

1. In combination with fuel oil, relative effectiveness can be increased to 0.70 and with commonly available other admixtures to 1.20 times that of TNT.

TABLE 7.2. EXPLOSIVE CHARACTERISTICS OF HYDROCARBONS

Name	Relative Effectiveness as External Charge	Effective Mass Conversion (%)
Hydrogen, natural gas and hydrocarbons, which are gases at standard conditions.	1.1	20
Gasoline, alcohol, kerosene and hydrocarbon fuels, which are liquid at standard conditions.	1.2	10

1. Explosive effective relative to TNT is greater for most gaseous hydrocarbons but center of detonation is usually airborne, and thus has less effect.
2. Effective detonation mass for hydrocarbons typically varies as a function of confinement, volatility and turbulence and has been observed to vary from 3 to 40%.

7.2 EXPLOSIVE QUANTITY AND DISTANCE AND RESULTANT PRESSURES AND DURATIONS

In Figures 7.1 to 7.4 are given the resultant free field (side-on) and reflected pressures and durations resulting from scaled distance *Z*, from 0.1 to 1,000, where:

$$Z = R/W^{1/3} \tag{7.1}$$

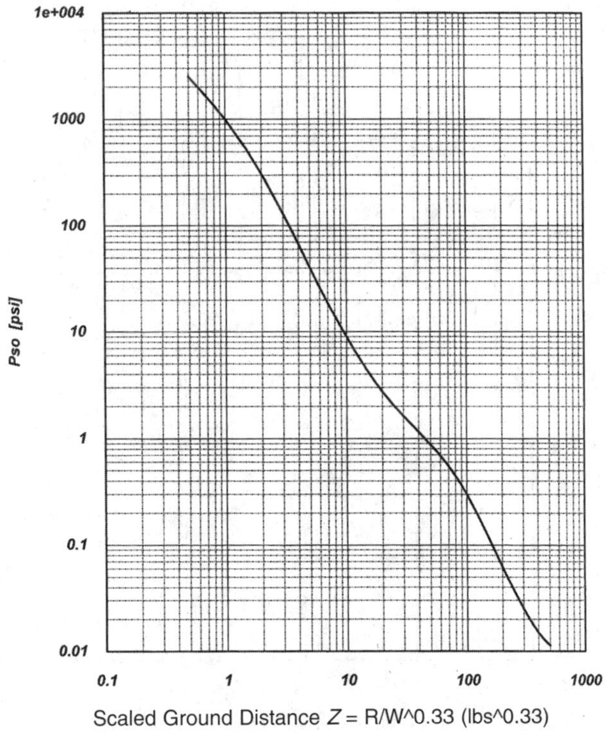

FIGURE 7.1. P_{SO} PEAK POSITIVE SIDE-ON INCIDENT PRESSURE

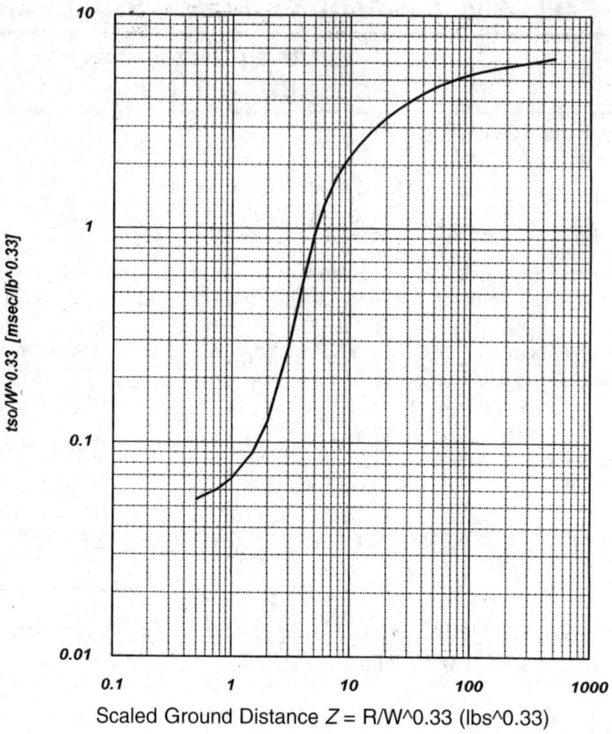

FIGURE 7.2. T_{SO}/W0.33 SCALED POSITIVE DURATION OF POSITION PHASE

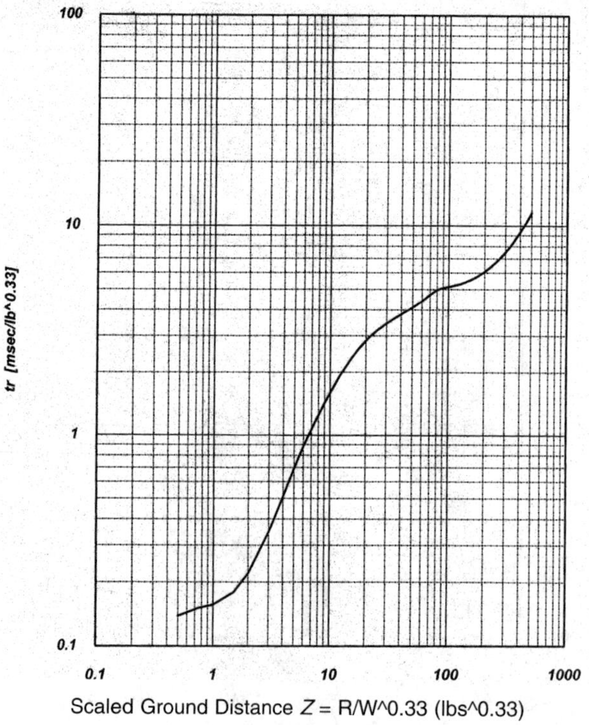

FIGURE 7.3. P_r – PEAK POSITIVE NORMAL REFLECTED PRESSURE

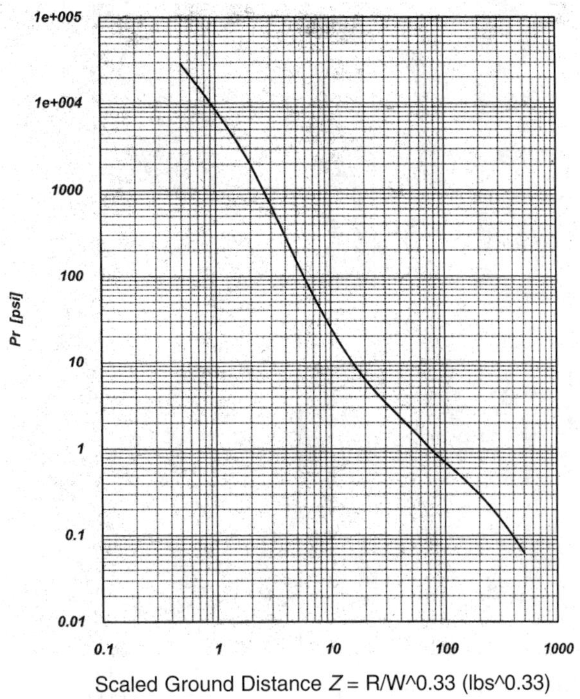

FIGURE 7.4. T_R/W0.33 SCALED POSITIVE DURATION OF NORMAL REFLECTED PRESSURE

and R = distance to the center of the explosion, detonation (feet); and W = weight of TNT (lbs) equivalent quantity of explosive detonated at the ground surface.

The resultant impulsive pressures and durations given in Figures 7.1 to 7.4 are based on Figure 4.12 of *TM 5-1300* published by the U.S. Army [1]. *TM5-1300* does not give the duration of reflected pressure directly. This must be determined from the reflected pressure impulse P_R from Figure 2.3 (Chapter 2) evaluated by the relationship $t_R = 2i_R/P_R$. The resultant duration of the reflected pressure is shown in Figure 7.4.

Peak dynamic pressure levels above about 35 psi (240 KN) will generally result in essentially complete destruction of conventional commercial, industrial and residential buildings, structures, architectural elements, building contents, mechanical and electrical systems and components, and severe injury or death to occupants. Exceptions to destruction might be SSC with very long periods of response and high ductility.

7.3 EQUIVALENT STATIC LOADS (ESL)

The equivalent static load (ESL) on a structure, its contents or occupants determined from an impulsive-type load is a function of the pressure impulse magnitude and duration as well as the natural period and ductility of the target structure contents and occupants.

For explosions inside building, structures multiple reflected pressure pulses occur along with a pressure buildup from the gases released by the explosion. This multiple reflected pressure impulse tends to increase the damage potential of interior explosion as compared to external blasts. Countering this effect is the potential for interior partition walls to fail at relatively low pressure levels, which reduces the effect of reflected pressure pulses. The complexity of interior explosion generally requires they be evaluated by computer analysis [2, 3, 4].

It should be understood that, as a function of the quantity of solid explosive or gaseous or liquid hydrocarbon detonation and range distance, it is quite possible to have the same impulsive pressure magnitude but

TABLE 7.3. FAILURE AND DAMAGE OF STRUCTURE ELEMENTS AND COMPONENTS DUE TO EQUIVALENT STATIC LOAD PRESSURE

Structural Element or Component	Failure Pressure-Equivalent Static Load (psi)		Dynamic Characteristic	Ductility		Failure Mode
	HCLPF	Median	Period (msec)	HCLP (F)	Median	
a. Ordinary window glass	0.2	0.5	25	1.0	1.0	Shatter
b. Doors	0.4	0.75	50	1.0	2.0	Displace
c. Interior plaster board and stud partitions						
i. Unanchored	0.5	1.0	200	1.0	2.0	Displace and overturn shatter
ii. Anchored	1.0	2.0	100	1.5	3.0	
d. Concrete or concrete block walls, 8 to 12″ thick						
i. Unanchored	1.0	2.0	10	1.0	2.0	Displace and overturn rupture
ii. Anchored	1.5	3.0	80	2.0	4.0	
iii. Reinforced	2.0	4.0	80	3.0	5.0	
e. Brick wall						
i. Unanchored	0.75	1.5	10	1.0	2.0	Displace and overturn rupture
ii. Anchored	1.0	2.0	80	2.0	4.0	
iii. Reinforced	1.5	3.0	80	3.0	5.0	
f. Corrugated asbestos, steel or aluminum siding or paneling	0.5	1.0	100	3.0	5.0	Rupture
g. Conventional reinforced-concrete shear walls and slabs, heavy reinforced-concrete walls	3.0	5.0	100	5.0 10.0	10.0 20.0	Large cracks; no longer capable of carrying transferring load
h. Conventional reinforced-concrete and structural steel beams and columns[1]						
i. Nonmoment resist connections	2.0	4.0	200	3.0	5.0	Large cracks; no longer capable of carrying load
ii. Moment resist connections	3.0	5.0	200	5.0	10.0	
i. Furniture						Slides as overturns if aspect ratio greater than about 2.0
i. Not positively anchored	0.3	0.5	200	2.0	4.0	
ii. Positively anchored			100			
j. Mechanical and electrical cabinets, switch gears, MCC						Slides or overturns after anchors fail
i. Unanchored	0.5	1.0	100	2.0	4.0	
ii. Anchored	2.0	4.0	10	2.0	4.0	
k. Rugged mechanical components-pumps, valve, vessels, heat exchangers						
i. Anchored	10	25	50	3.0	5.0	
l. Mechanical and electrical distribution system						
i. piping (welded)	3.0	6.0	200	9.0	17.0	
ii. conduit	2.0	5.0	200	3.0	6.0	
iii. cable tray	1.5	3.0	150	3.0	5.0	
iv. duct	1.0	2.0	200	2.0	4.0	

1. Structural steel column ductility is limited to 2.0
2. Failure of significant quantities of the elements or components would occur at about 2.0 times the equivalent static threshold pressure.

TABLE 7.4. CRITERIA FOR DIRECT (PRIMARY) BLAST EFFECTS IN MAN FROM FAST-RISING, DURATION OF 100-MSEC. OR LONGER PRESSURE PULSES

Effect	Effective Peak or Equivalent Static Pressure (psi)
Lung damage:	
Threshold	12 (8–15)
Severe	25 (20–30)
Lethality:	
Threshold	40 (30–50)
50%	62 (50–75)
100%	92 (75–115)
Eardrum rupture:	
Threshold	5
50%	15–20 (more than 20 years old)
	30–35 (less than 20 years old)

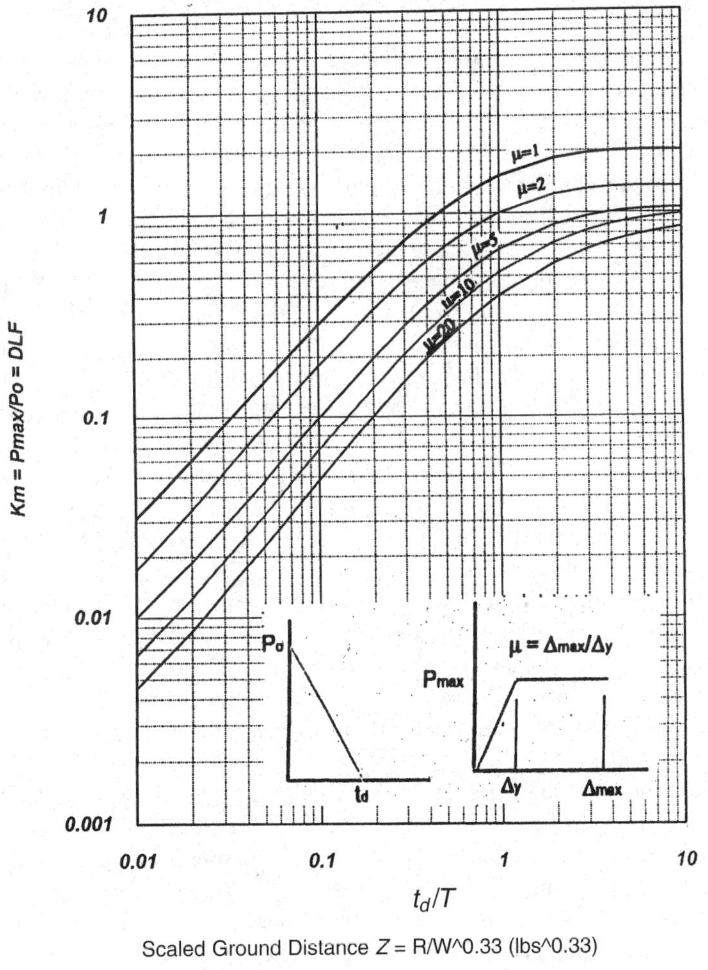

FIGURE 7.5. DYNAMIC LOAD FACTOR FOR SIMPLIFIED TRIANGULAR BLAST PULSE

significantly different durations of the impulse. The potential for damage or injury is a function of not only the pressure impulse magnitude, but also its duration where such impulsive loads are converted to an ESL. Thus, in this Chapter, where available, damage or injuries potential are given as a function of ESL not peak impulse pressure.

In order to compute the ESL, the peak dynamic pressure impulse is multiplied by the K_m coefficient derived from Figure 7.5 as a function of the load direction, ductility and natural period of the SSC structure, architectural feature or building contents being impacted.

7.3.1 Calculation of ESL

Figure 7.5 defines K_m (ESL) as a function of the duration of the impulse, t_d, divided by the dominant natural period of the structure, *T*, and ductility, μ. For a given t_d/T ratio, t_d is the duration of the pressure impulse, t_o or t_r from Figures 7.2 and 7.4 and the natural period of the structure, as listed in Table 7.3, determines coefficient K_{os} or K_{rs} equal to K_m. These values of K_m are then multiplied times P_{so} or P_r from Figures 7.1 or 7.3, respectively, to obtain the ESL.

7.4 DAMAGE AND FAILURE OF BUILDINGS' PRIMARY LOAD PATH, ARCHITECTURAL FEATURES, CONTENTS AND OCCUPANTS

As stated previously, the response of structures, architectural features, building contents, mechanical systems and components, and occupants to blast is a function of not only the dynamic characteristics of the load, but also the dynamic response characteristics of the structure, architectural features, building contents, mechanical systems, components and occupants. It should also be noted that the function, leak-tight and structural integrity of mechanical systems and components are often controlled by the building structure that houses and supports them. If, for example, the blast wave load is of very short duration relative to the natural vibrational period of the object excited by the blast, the blast effects will pass the object before the object has time to respond to the load. This is the major difference between chemical explosions of TNT equivalent of 5 tons or less and much larger (nuclear) explosions in the kiloton of TNT equivalent size range. Kiloton explosion blast waves are of much longer duration and result in both dynamic pressure pulses followed by dynamic wind loads of much greater intensity, and hence tend to cause much more damage for a given initial blast wave of the same peak magnitude from a 5 ton or less chemical explosion or detonation.

In the evaluation of blast damage to SSC it is necessary to distinguish between the local and global response of the SSC. Local response is also associated with response of SSC to local blast-driven missile penetration. Global structural elements are often engineered for dead, live and natural phenomena hazard loads in accordance with applicable structural and mechanical design codes and standards then evaluated for blast loading as applicable.

In Table 7.3 is a summary of the ESL that primary load path structural and architectural and mechanical elements and objects can withstand at the threshold of damage [high confidence low probability of failure (HCLPF)] and 50 percentile of significant damage and failure (median).

Table 7.4 contains the criteria for the effects of direct (primary) peak pressure on occupants subject to a blast wave pressure of 100-msec duration or longer. This pressure data being of 100-msec duration or longer can be considered as equal to the ESL.

Secondary blast effects on occupants that involve blast-related fragments impacting the occupant, or the occupant being projected into a structure, in most cases are more damaging than primary blast pressure effects. However, the computations of these velocities are quite complex and are beyond the scope of this chapter. Analytical methods for computing these velocities can be found in [5].

Damage levels in terms of equivalent static pressures on structures and associated structural elements and mechanical components are summarized in Table 7.3 along with applicable ductility factors and structural response periods. Ductility factors are a measure of how much a structural element can deform before it breaks or deforms to the point it loses its ability to function. It is determined numerically by dividing the deformation at yield (member begins to act plastic or inelastic) and the specified deformation. Allowable ductilities are often taken as one-quarter of the ductility at rupture to define median failure levels. Between one-half to one-third of these ductility values are typically used to define HCLPF levels (approximately 0.01 probability of failure), but these are not less than 1.0. Steel members tend to have high ductility in the range of 5 to 20 and brittle material such as glass or ceramics low ductility from 1.0 to 1.5.

7.5 EXAMPLE PROBLEMS

7.5.1 Example 1

Given 950 lbs of ammonium nitrate is exploded, what is the TNT equivalent?

a. From Table 7.1 ammonium nitrate has a 0.42 equivalency to TNT. Therefore, 950 lbs of ammonium nitrate equals: $950 \times 0.42 = 399$ lbs of TNT.
   What if the ammonium nitrate is mixed with fuel oil?
   From the footnote on Table 7.1, the TNT equivalency ammonium nitrate mixed with fuel oil is 0.7. Therefore, 950 lbs of an ammonium nitrate mixture with fuel oil is: $950 \times 0.7 = 665$ lbs of TNT equivalent.
b. What is the equivalent of 1,000 gallons of gasoline to TNT?
   Given that one gallon of gasoline weighs 7.5 lbs, then 1,000 gallons of gasoline weighs: $7.5 \times 1{,}000 = 7{,}500$ lbs.

However, from Table 7.2 the relative blast equivalence of gasoline is 1.2 times that of TNT: $1.2 \times 7500 = 9{,}000$ lbs.

But from Table 7.2, only 10% of the mass is assumed to detonate: $0.1 \times 9000 = 900$ lbs of TNT equivalency

7.5.2 Example 2

Given 350 lbs of TNT equivalent explodes 350 feet from the front wall of a building, what is the peak reflected pressure on the front face of the building wall and its duration?

a. For $W = 350$ lbs; $R = 350$ ft: $Z = R/W^{1/3} = 350/(350)^{1/3} = 350/7.05 = 49.6$
   From Figure 7.4, for $Z = 49.6$, $P_r = 1.8$ psi
b. From Figure 7.3 the duration of the reflected pressure for $Z = 49.6$: $t_r = 4$ msec • $W^{1/3}$; $t_r = 4 \times 7.05 = 28.2$ msec.

7.5.3 Example 3

Given the peak reflected pressure and duration from Example 7.2, determine the equivalent static pressure. Do these pressures exceed the threshold of damage for the following structural elements?

a. Reinforced-concrete shear wall
b Window glass
c Metal siding

Given: $P_r = 1.8$ psi $\quad t_{rd} = 28.2$ msec

a. From Table 7.3 for reinforced-concrete wall: period = 100 msec, ductility = 5.
   From Figure 7.5; period = 100 msec, ductility = 5. For $t_d = 28.2$ msec, $t_d/T = 28.2/100 = 0.282$, and $K_m = 0.27$
   Therefore, static equivalent load, $ESL = 1.8 \times 0.27 = 0.49$ psi ESL.
b. From Table 7.3 for window glass: period = 25 msec, ductility = 1.0

From Figure 7.5: for t_d = 28.2 msec, t_d/T = 28.2/25 = 1.128, μ = 1.0 and K_m = 1.5.
Therefore, equivalent static pressure = 1.8 × 1.5 = 2.7 psi ESL.

c. From Table 7.3 for metal siding: Period = 100 msec, ductility = 3.0.
From Figure 7.5: period = 100 msec, ductility = 3.0.
For t_d = 28.2 msec, t_d/T = 28.2/100 = 0.282 and K_m = 0.4.
Therefore, equivalent static load = 1.8 psi × 0.4 = 0.72 psi ESL.

While these values look rather small, it should be noted that uniform loads on buildings and building elements are usually given in pounds per square foot, or psf values. The psi values would have to be multiplied by 144 to get psf. For example, the design load on the floor of an automobile parking garage is 100 psf or 0.69 psi. A 100 mph wind would result in a free-field pressure load of q_c $0.00256V^2$, where 0.00256 = density of air, V = velocity of the wind mph; q_c = pressure in psf and $q_c = 0.00256 \times 100^2$ = 25.6 psf or 0.178 psi.

7.6 DESIGN ACTION OF SHIELD STRUCTURES

It has been observed that objects located in the lee or shadow of other structures typically sustain less damage than other unsheltered objects at the same distance from the source of the blast. This generally is the effect of the other shielding structures' interruption of the blast wave. It takes a finite time and distance for the blast wave to reform, and at a greater distance from the source of the blast. As a result the sheltered object is subject to side-on-only pressure until the pressure wave reforms. Since the side-on pressure peak pressure is always equal to or less than one-half the reflected pressure in general, the sheltered object will see significantly lower effective peak pressures. An approimate shelter or shielded volume can be defined as that volume developed by 45° angles projected from the top and sides of the sheltering structures.

REFERENCES

1. U.S. Department of the Army, 1990, "Structures to Resist the Effects of Accidental Explosions". *Army TM 5-1300, Navy NAVFAC P-397, AFR 88-22,* Departments of the Army, Navy and Air Force, Washington, D.C.
2. SAIC, 1994, *International Blast and Thermal Environment for Internal and External Explosions: A User's Guide for the BLASTX Code, Version 3.0,* (SAIC 405-94-2).
3. *DYNA3D User's Manual: Nonlinear Dynamic Analysis of Structures in Three Dimensions,* 1898, UCID-19592, Rev. 5, Lawrence Livermore National Laboratory, Livermore, California.
4. TRW, 2002, *Blast/FX Explosive Effects Analysis Software,* W.J. Hughes Technical Center, Atlantic City, NJ.
5. Baker, W.E., et.al., 1983, *Explosive Hazards and Evaluation,* Elsevier Publishing Co.

CHAPTER

8

HIGH-ENERGY SYSTEMS EXTREME LOAD DESIGN AND ANALYSIS

With the exception of postulated blast loading applied to petrochemical facilities [1, 2, 3], these types of analyses and design bases are normally applied only to hazardous nuclear facilities as part of the required design or evaluation procedure.

8.1 PIPE RUPTURE AND LEAKAGE EFFECTS

8.1.1 Introduction

In [4] is a detailed review of pipe failure that has occurred in U.S. Commercial Nuclear Reactor Plants between 1961 and 1997. In the 36-year period studied there were 2,247 recorded failure events, which included 118 pipe ruptures reported in the database. As a result of this experience, commercial nuclear power plant facilities are required to consider pipe rupture in high-energy piping and leakage in moderate energy piping as a design basis for the piping and the facility by the U.S. Nuclear Regulatory Commission (NRC).

The U.S. NRC has developed definitions for moderate- and high-energy piping [5].

High-energy fluid systems are fluid systems that, under normal plant conditions are either in operation or maintained pressurized under conditions where either or both of the following are met: (a) Maximum operating temperature exceeds 200°F; or (b) Maximum operating pressure exceeds 275 psig.

Moderate energy fluid systems are fluid systems that, under normal plant conditions, are either in operation or maintained pressurized (above atmospheric pressure) under conditions where both of the following are met: (a) Maximum operating temperature is 200°F or less; and (b) Maximum operating pressure is 275 psig or less.

unless they operate at these pressures or temperatures less than 2.0% of the time.

High- and moderate-energy fluid piping systems are required to consider piping failures and the mitigation of such failures in order to protect the safety function of other structures, systems or components (SSC) in its vicinity.

The U.S. NRC established these criteria in December 1972. Prior to that time, postulated nuclear reactor designers only considered pipe break as a design-basis event when design pressures typically exceeded about 1,000 psi and design temperatures exceeded about 400°F.

8.1.2 Leak Before Break

Leak before break evaluations have been used extensively to mitigate the assumption of a postulated pipe break [6,7]. The following factors as prescribed by the USNRC are necessary to utilize leak before break methodology as a substitute for a postulated design-basis pipe break.

1. The leak before break evaluation uses design-basis loading and is based on the as-built configuration as opposed to the design-basis configuration.
2. Degradation by erosion, erosion/corrosion and erosion/cavitation due to unfavorable flow conditions and water chemistry is considered.
3. Determination of leakage from a system under pressure involves uncertainties and, therefore, margins are needed. Unless a detailed justification can be presented that accounts for the effects of these sources of uncertainties, a margin of 10 on the leakage prediction is required for determining the leakage size flaw.
4. A systems evaluation of potential water hammer is made to ensure that pipe rupture due to this mechanism is unlikely. Water hammer is a generic term including various unanticipated high-frequency hydrodynamic events such as steam hammer and water slugging.
5. A review of creep and creep-fatigue is required. Operation below 700°F in ferritic steel piping and below 800°F in austenitic steel piping can satisfy concerns with creep.
6. The requirement that corrosion resistance of piping be demonstrated can rely on investigations of the frequency and degree of corrosion in the specific piping systems under review.
7. An assessment of potential indirect sources of pipe ruptures is required to demonstrate that indirect failure mechanisms defined in design are remote causes of pipe rupture.
8. The piping material is not susceptible to brittle cleavage-type failure over the full range of system operating temperatures.
8. The system(s) under evaluation do not have a history of fatigue cracking or failure. An evaluation is performed to ensure that the potential for pipe rupture due to thermal and mechanical induced fatigue is extremely low. Users of the methodology must demonstrate that there is adequate mixing of high- and low-temperature fluids so that there is no potential for significant cyclic thermal stresses. In addition, it must also be demonstrated that there is no significant potential for vibration-induced fatigue cracking or failure.

8.1.3 Pipe Rupture Effects

The pipe ruptures considered in high-energy pipe lines are of two types: circumferential and longitudinal. Circumferential breaks around the pipe cross section are assumed such that they cause complete severance of the pipe and result in the two ends of the severed pipe displacing at least one pipe diameter relative to each other such that a unrestricted fluid jet can form. This type of rupture is assumed at terminal ends of branch piping connecting to run pipe, or in run pipe connecting to component nozzle unless mitigated by a leak before break evaluation.

A longitudinal pipe break is parallel to pipe longitudinal axis with a length typically taken equal to twice the nominal diameter of the pipe and an area equal to the cross-sectional area opening in the pipe. This break is assumed in at least any one location in a pipe run (between anchor points) unless mitigated by a leak before break evaluation. Generally these types of breaks can be eliminated by use of leak before break criteria.

These postulated breaks result in two types of loads: (1) jet reaction loads acting on the broken pipng systems; and (2) jet impingement loads acting on other SSC in the vicinity of the break.

Jet Reaction

The jet reaction loads, as the name suggests, act on the ruptured system containing the fluid. In the simplest first order of analysis form, these reactions are represented as a dynamic force applied at the point of break. The dynamic nature of the load arises from the typical assumption of instantaneous break opening time. This results in a jet reaction load, which is applied with a very short rise time relative to the natural period of structural response of the piping system and, in many instances, remains on the system for a relatively long period of time.

Actually the method of analysis just outlined, while convenient to use, is a vast oversimplification of the jet reaction effect. This is true for the following reasons.

1. The dynamic load of the jet should contain an energy mass transfer term resulting from mass of the fluid being ejected from the opening at some velocity. This phenomenon can have as much as a doubling effect on the dynamic load, depending on such variables as flow rate, fluid phase and the shape of the break orifice.
2. The sudden break in the system results in pressure waves in the fluid being transmitted throughout the system, resulting in time-dependent fluid flow and differential pressure loads and momentum at changes of direction and areas in the system. This results in a whole series of time-dependent forcing functions acting on the system, rather than a single load acting at the point of break.

However, computer algorithms capable of defining such multiple time-history fluid loads have generally been available for only the past 15 years or so. Such programs are typically quite expensive, typically an order of magnitude more expensive than solid mechanics dynamic load structural analysis programs, and hence, have seen limited application to date. This is particularly true when it is recognized that the high-energy pipe lines, in a typical nuclear power plant or other type of high-energy process facility, exceed 80,000 feet of pipe.

Jet Impingement

Jet impingement loads result from the fluid that is ejected from the ruptured pressurized system impinging on an adjacent structural element. The effect is similar to that of a missile except that there is no finite amount of energy to be absorbed by the target, which is associated with the missile kinetic energy. The impinging jet acts as a suddenly applied sustained load.

Jet Load Design Procedures

Impulsive extreme load design in the form of steam and water jet reaction and impingement are typically associated with postulated rupture of a high-energy fluid system. As in the case of seismic loads, there are several levels of analytical rigor associated with postulated pipe break jet load analysis. These are presented as follows in order of usually increased complexity, effort and cost.

a. Static Analysis
   In this case, the pipe break load is defined at the point of break only and is applied statically to the system with a magnitude equal to the system operating pressure times the postulated area of the break. In this case, fluid characteristics and dynamic response of the system are not considered explicitly. Associated with this loading criteria, however, was, typically a conservative stress criteria developed before ASME code limits were set for the service level D condition, which limited membrane and bending stresses to yield or 90% of yield and assumed plastic hinge formation in the pipe at yield in the outermost fibers.
b. Quasi-Dynamic Analysis
   In this case, the pipe break load is defined only at the point of break and is developed as an equivalent static load based on the assumed one-degree-of-freedom (DOF) structural response of the piping system to a pulse load, as shown in Figure 8.1. The magnitude of the dynamic load is a function of fluid state and system pressure times the postulated break area and the response of the piping system as a one-DOF system. The equivalent static load thus determined is then applied statically at the point of break to the structural system to obtain resultant stresses. The stress criteria are typically those associated with the ASME Section III Appendix F for service level D condition. Implicit plastic response is considered by developing the equivalent static loads based on ductility limits greater than one. For example, the ductility assumed is much greater when the load results in membrane or axial tension than it would be if it resulted in compression or bending loading. The actual limits used depend on the material, structure and loading configuration under consideration. A variation of this method would be use of a dynamic forcing function defined at the point of break, which con-

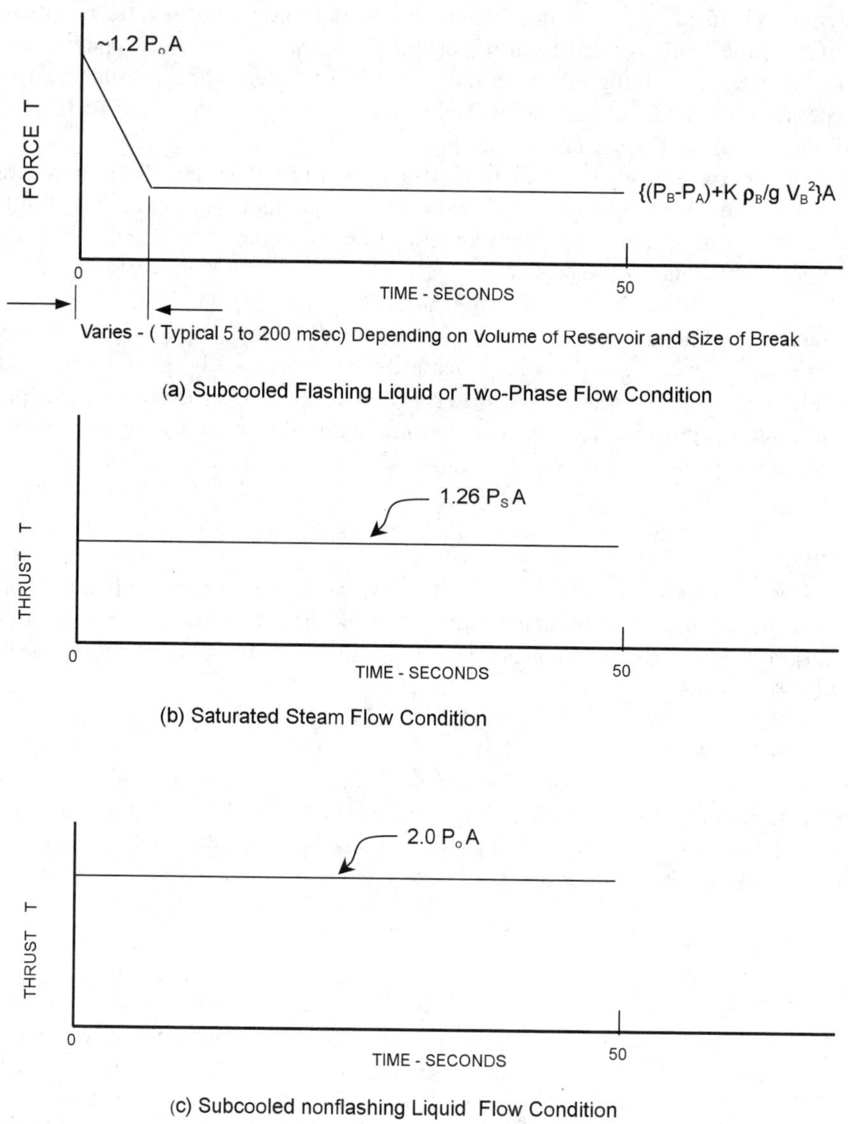

FIGURE 8.1. SIMPLIFIED DYNAMIC LOAD VS. TIME FOR VARIOUS FLUID CONDITIONS.

siders the geometry of the system being analyzed and, hence, alters the forcing function being used as a function of system geometry as well as fluid flow condition.

c. Simplified Dynamic Time History Analysis
This method is similar to method b, except that the structural system being analyzed is represented by a multi-degree-of-freedom (MDOF) model and includes elastic or inelastic response of the system to directly determine resultant stresses. Damping might also be considered here since the elements responding to the impulsive load in a characteristic mode may induce cyclic motion, which in turn could excite resonance in the structure.

d. Rigorous Dynamic time History Analysis
This method is similar to method c except that dynamic forcing functions are applied at changes in direction, area changes, etc., in the system as a function of fluid response in the system to the postulated break and not just at the assumed break location. This will typically result in a complex forcing function at each location, which includes rapid fluctuation

and changes in direction of forces with time as pressure and fluid momentum changes interact throughout the system. In such instances, the loading may become somewhat cyclic in nature and, therefore, structural damping should be included in the analysis in consideration of possible resonance as well as MODF response. It is only recently that computer programs have become generally available to perform this level of analysis.

For the most part, it is expected that method b will be the technique usually employed for practical design purposes in nuclear facilities, particularly for the bulk of *ASME B&PVC* Section III, class 2 and 3 or *ASME B31.1* or *B31.3* piping systems. However, the actual conservatism associated with such techniques may be subject to check verification by method d in a limited number of cases to ensure design adequacy.

Typical Fluid Forcing Functions

Peak dynamic forces or the magnitude of the forcing functions shown in Figure 8.1 are typically defined for the three fluid systems' conditions of interest in nuclear process facility design. They are detailed as follows for analytical levels of analysis b and c [8].

For subcooled flashing liquid or two-phase flow condition:

a.

1. Initial force $$T = 1.2\ P_o A \tag{8.1}$$
2. Sustained force

$$T = \left\{ (P_o - P_A) + K_1 \frac{p}{g} V_B^2 \right\} A \tag{8.2}$$

where A = cross-sectional area of pipe or break; T = jet dynamic force magnitude; K_1 = thrust coefficients from Moody [9] $K_1 = 1.0$; P_o = system pressure at point of break (operating pressure) (in psia); P_A = pressure of ambient outside system or back pressure (in psia); V_B = velocity of fluid at point of break; and P = fluid density at point of break.

b. For saturated steam condition: $T = 1.26\ P_S A$.
   Note: P_a is often conservatively assumed equal to zero.
c. For subcooled and nonflashing liquid:

$$T = K_1\ (K_2 P_o - P_a) A \tag{8.3}$$

where $K_1 = 2.0$; and $K_2 = 1.0$.

For discharge of flashing liquid or of steam/water mixtures, an estimate should be made of the steam content (quality) of the fluid in the vicinity of the break. If that quality, x_c, is greater than 2% Moody's critical flow model may be used for the evaluation of the fluid velocity and density at the break [9]. When the fluid quality is smaller than 2%, Fauske's model [10] is recommended as an upper bound for discharge flow (also for the subcooled and the saturated flashing liquid), in which the discharge coefficient is equal to 0.6 and the back pressure is a function of the length-to-diameter ratio.

When the fluid velocity, u_c, and its mass density, ρ_c, are determined at the break plane as a function of time, the thrust is found from:

$$T = A\ (P_c - P_a + \rho_c u_c^2) \tag{8.4}$$

When $x_c < 0.02$, i.e., Fauske's model is utilized, set $P_c = P_b$, where P_b is the back pressure, which is a function of the length-to-diameter ratio. In the past, this load has often been approximated as:

$$T = 1.2\ P_o\ A \tag{8.5}$$

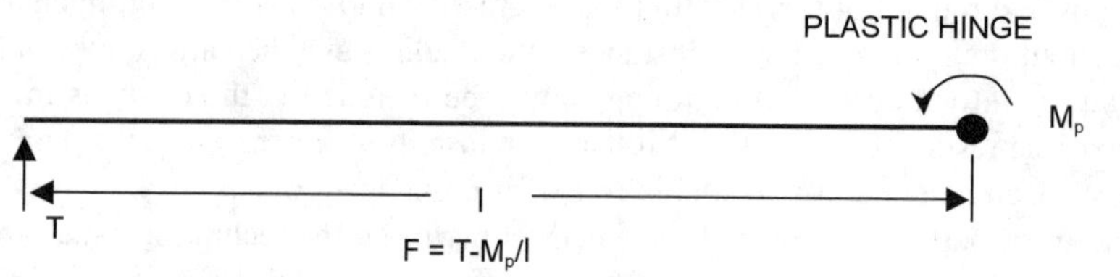

FIGURE 8.2. MODEL FOR DETERMINING PIPE WHIP DESIGN PARAMETERS.

It should be noted that in a typical subcooled or flashing system, the value of P_o decays rapidly in time, which gives rise to the typical ramp shape of the forcing function. The loads thus determined are applied as loads on the system in order to determine the structural response of the fluid system and its supports.

Pipe Whip

Another phenomenon associated with a pipe rupture is pipe whip. The geometry of the piping systems, both upstream as well as downstream of the postulated break plus the magnitude of the break loads, may be such that the pipe run cannot resist the resultant pipe reactor loads without plastic hinges forming. If a plastic hinge mechanism forms then rigid body motions can occur and the pipe is said to whip, as shown in Figures 8.2 and 8.3.

Effects of Pipe Whip

The procedure described here is for particular pipe geometry of a cantilever pipe loaded transversely at its tip. For other configurations, the rigid body dynamics associated with that particular geometry must be used. Consider a simple cantilever as shown in Figure 8.2. The reaction load that will cause pipe whip will be that load in excess of the load necessary to cause a plastic hinge and, in this case, to form a plastic hinge mechanism.

$$F' = \frac{M_P}{1} \tag{8.6}$$

$$F\ell = (T - F')\ell \tag{8.7}$$

where F' = force required to form a plastic hinge mechanism; F = unbalanced force used to determine pipe whip acceleration and impact velocity; T = jet force as determined from Eqs. (8.1) through (8.4); ℓ = length of pipe between hinge and load; and M_p = plastic hinge moment.

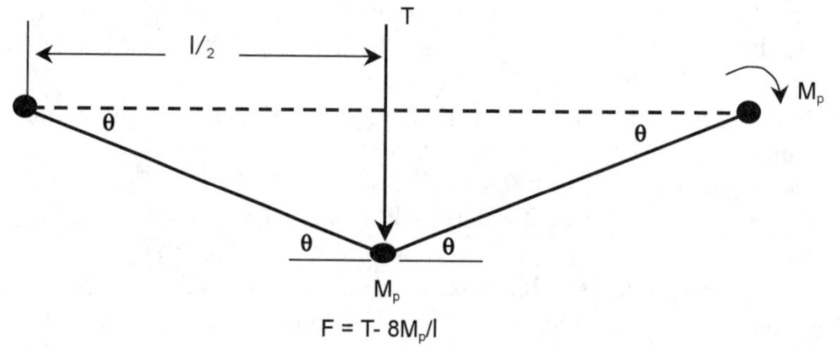

FIGURE 8.3. MODEL FOR DETERMINING MEMBRANE EQUILIBRIUM LOAD CAPACITY.

A simple rigid body rotation is assumed. Conservation of angular momentum requires that:

$$F\ell = J_d \frac{d\theta}{dt^2} \tag{8.8}$$

where J = mass moment of inertia of the pipe about the hinge; and $(d^2\theta/dt^2)$ = angular acceleration of pipe about the rigid support.

Integrating Eq. (8.8) twice for initial conditions $\theta_o = 0$, $(d\theta/dt) = 0$

$$\frac{d\theta}{dt} = \frac{F1}{J}t \tag{8.9}$$

$$\theta = \frac{F1t^2}{2J} \tag{8.10}$$

$$t = 2J\theta/F\ell \tag{8.11}$$

Solving Eq. (8.11) in time and substituting this result into Eq. (8.9).

$$\frac{d\theta}{dt} = \frac{2FL\theta}{J} \tag{8.12}$$

Since

$$J = \frac{1}{3}ml^3 \tag{8.13}$$

where m = mass of pipe per unit length,

then

$$\frac{d\theta}{dt} = \frac{1}{\ell}\frac{6F\theta}{m} \tag{8.14}$$

Since $(d\theta/dt) = \omega$, the angular velocity (in radians/sec), the velocity of the pipe tip is determined

$$V = \omega\ell \tag{8.15}$$

$$V = 6F\theta/m \tag{8.16}$$

where θ = the angle (in radians) between the original position of the pipe and the pipe when it strikes the target.

In this way, the velocity of the pipe whip for the cantilever geometry acting as a missile can be determined. The effective mass of the pipe is determined based on the desire to have the kinetic energy of the distributed mass of the pipe, whose linear velocity varies as the length to the hinge point equal to energy of the system, where the mass is concentrated at the end of the pipe impacting at tip velocity.

It should be understood that the rigid steel piping systems usually encountered do not behave in the same manner as a whipping fire hose, which has little or no bending stiffness. Often there is a tendency to equate the behavior of thc two.

Early in the development of pipe break criteria, it was convenient to assume the full displacement of a circumferentially broken pipe to permit the free flow of jets formed and this simplifying conservative assumption became the basis of much fluid system response evaluation, which overstated the effects of the pipe break transient on the system.

In fact, in most cases the fluid forces acting on the assumed broken ends of pipe are not sufficient to cause the pipe to displace significantly and permit unrestrained jet flow. This is particularly true if pipe whip restraints are added, which usually limit displacements to a few inches. In an environment of ever-more rigorous analysis, it is strongly recommended that this simplifying overly conservative nonmechanistic assumption be replaced by evaluations that do consider the actual deformation characteristics of the broken pipe in the inelastic as well as elastic range.

Membrane Stability of Plastic Hinged Pipes

In some instances, it is possible for a piping system in which a plastic hinge mechanic has formed to reach a stable membrane equilibrium state and preclude the development of a pipe whip. An example of such a case would be the formation of a three-hinge truss as shown in Figure 8.3. The formation of a stable truss is dependent on the ability of the end-point restraints to resist very large longitudinal load and the pipe to undergo relatively large plastic strains before rupture in order to develop the truss geometry. Such strains are normally limited to 50% of the ultimate strain of the pipe material in the longitudinal direction.

8.1.4 Current Analytical Procedures for Generation of Pipe Rupture Loads and Jet Forces

Currently simplified analytical procedures with examples for generating

- Fluid thrust forces
- Fluid jet geometry
- Fluid jet impingement effects

can be found in [8].

8.1.5 Jet Impingement Loads

In addition to the jet reaction loads acting on the piping system, a fluid jet as the result of either a postulated cross-section pipe break or longitudinal crack containing both a pressure and a velocity profile is formed. If this jet strikes a barrier, the momentum of the fluid traveling at impact velocity and the differential pressure in the jet acting on the barrier as an impulsive force are developed as shown in Eq. (8.4).

The velocity, density and pressure terms are those defined at the barrier. The problem then is to define these parameters at the barrier. Saturated steam jets have been observed to disperse at a total angle of approximately 20°. In this process, the steam jet is mixing with air and thereby changing its velocity, density and pressure profile. Until recently, jet loads were typically taken for design purposes as:

$$T = 1.2\, P_o A \tag{8.17}$$

Dispersing at a total angle of 20°. Recently, a free turbulent jet theory was introduced [8], where more than 10 break diameters, velocity and pressure profiles have been developed.

These profiles may be used to further define barrier loads. In addition, they provide the necessary data to analyze the loading effect on a small obstruction in the jet stream. In this case, the load or the object is in the form of:

$$F = C_D \rho V^2 \tag{8.18}$$

where F = dynamic force on the object; C_D = drag coefficient for the object; ρ = jet density; and V = jet velocity.

The pressure drop across the object can usually be neglected.

8.1.6 Pipe Whip Restraint Design

Pipe whip and break loads are dynamic, impulsive and impactive in nature, and are usually several orders of magnitude larger than either dead or seismic loads. There are typically three kinds of pipe break restraints used, defined as:

a. Steel frames
b. Steel cables
c. Steel bands

Steel Frames

Moment and shear loadings in frames do not take advantage of the inherently largest load-carrying capacity of steel, that of membrane or axial tension. In addition, bending and shear structural response may be subject to instability-type failure long before the full plastic capacity of the steel structure is mobilized.

Because of these potential instability modes of failure the ductility permitted in bending and shear members must be limited. This has the effect of increasing dynamically applied load, particularly the load resulting from impact. Typical ductility limits are shown in Table 8.1 from [11].

Because of these two disadvantages, rigid-frame-type restraints as shown in Figure 8.4 are not recommended for use as pipe whip restraints, except in the case where restraint displacement is severely limited.

Steel Cables

The flexible cable restraint overcomes the first deficiency of the (a) type frame restraint since it carries load in membrane tension and because its flexibility makes it quite easy to apply. However, most commercial ferritic carbon steel and to a lesser extent stainless-steel cable has been significantly cold-worked during fabrication, which increase its strength characteristics but reduces its ductility. The end result is ductility limits that may be severely limited, which tends to increase the effect of the dynamically applied loads.

TABLE 8.1. ALLOWABLE LOCAL DUCTILITY,[1] μ, TO BE USED IN DESIGN OF CONCRETE AND STEEL STRUCTURAL ELEMENTS FOR IMPACT-TYPE LOADS

1. Structural-Steel Tension Members

$$\mu'' \; 0.25\varepsilon\mu/\varepsilon y \le 0.1/\varepsilon y$$

where $\varepsilon\mu$ = strain corresponding to elongation at failure (rupture); and εy = strain corresponding to yield stress.

2. Structural-Steel Flexural Members
   A. Open sections (*W, S, WT,* etc.)
      μ'' 12.5
   B. Closed sections (pipe, box, etc.)
      μ'' 20.0
   C. Members where shear governs design
      μ'' 5.0

3. Structural-Steel Compression Members

$$\mu = 0.255/\lambda^2 \le 10 \le \varepsilon_{st}/\varepsilon_y$$

$$\lambda = (Jk/\pi r)\sqrt{\frac{F_y}{E}}$$

where ε_{st} = strain corresponding to the onset of strain hardening; *Kl* = effective length of compression member in the plane producing the largest Kl/r ratio; and *r* = radius of gyration corresponding to plane of bending used to define *kl*

4. Reinforced Concrete
   See Appendix C of ACI Standard 349-99

[1] Allowable local ductility factor is defined as the ratio of permitted strain or deformation to the strain or deformation at yield in the member.

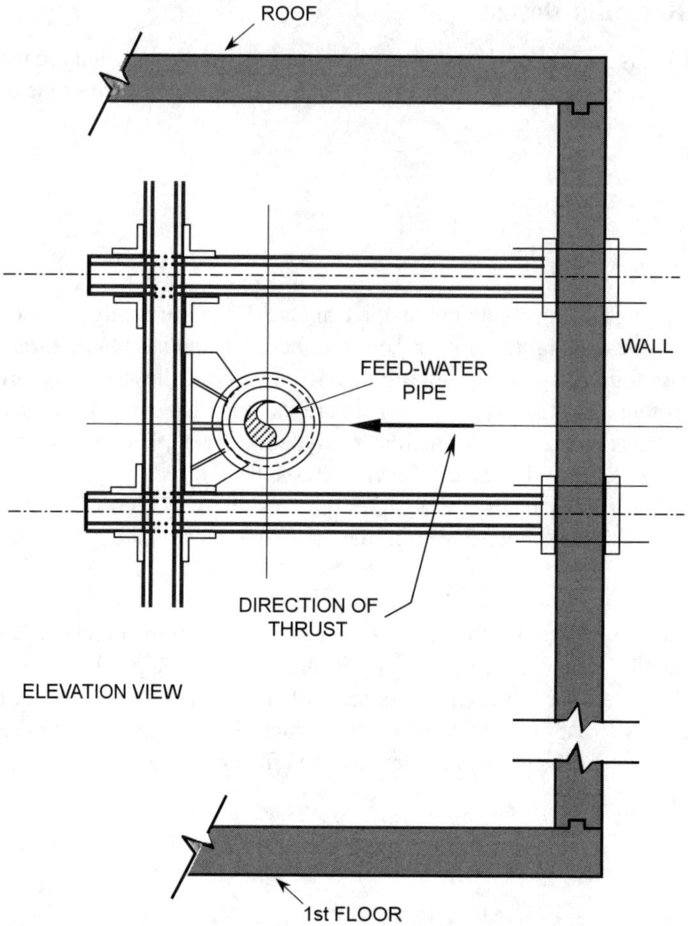

FIGURE 8.4. TYPICAL BEAM AND COLUMN-TYPE RUPTURE RESTRAINT TIED TO REINFORCED-CONCRETE WALL

Steel Bands

Finally, ductile, low-carbon or stainless-steel bands have been used. These bands are quite flexible, hence load is transferred by membrane tension and extremely ductile with uniform ultimate strain limits typically greater than 20% for carbon steel and 35% for stainless steel. For this reason, such bands are the preferred pipe break restraints. It should be noted, however, that rather large local deflections of the bend could occur before the load capacity is realized. A typical band-type restraint is shown in Figure 8.5.

To aid in the dynamic load-carrying capacity of the pipe break restraint structure, crushable material can also be used to absorb impact energy and limit the impactive load on the restraint, which in turn can be transmitted through the crushable material into the support structure. Such crushable material may take the form of steel or other metallic honeycomb collars, which are fitted to the pipe at points of potential contact with the restraint structure or as lining of the impact surfaces of the restraint structure.

8.1.7 Simple Sample Problem Illustrating the Effect of Ductility in Restraint Design

The first terms of Eq. (8.18) represent the kinetic energy of the pipe moving through the gap between the pipe and the restraint, which must be absorbed by impact on the restraint. Such gaps are generally necessary to permit unrestricted movement of the pipe due to temperature effects. The second term in Eq. (8.19) rep-

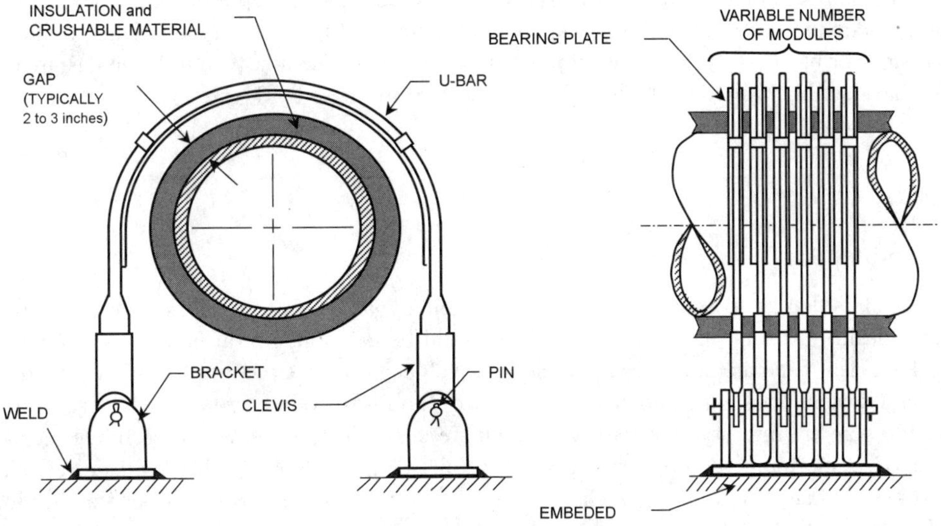

FIGURE 8.5. TYPICAL MEMBRANE TENSION-TYPE PIPE WHIP RESTRAINT.

resents the sudden application of the pulse load on the restraint. Another and more convenient way to express the first term of Eq. (8.18) may be in terms of work done on the support by the load, R_n:

$$R_m = \frac{T \cdot GAP}{\Delta L} + T\left(\frac{1}{1 - 1/2\mu}\right) \tag{8.19}$$

where T_1 = fore applied at the point of break from Eq. (8.1); GAP = gap between pipe and its pipe whip restraints; ΔL = change in length of the support due to impact (0.001 μL); μ = ductility factor; and L = initial length of a steel support.

If μ = 1.0, and T is defined in Eqs. (8.1) to (8.4) as a function of fluid condition and L = 60 inches and GAP = 3.0 inches:

$$R_m = \frac{T \cdot 3}{0.001x1.0x60in} + T\left(\frac{1}{1 - 1/2}\right) \tag{8.20a}$$

$$R_m = 50T + 2T = 52T \tag{8.20b}$$

or if μ = 10 or a total strain in the support of less than 1.5%:

$$R_m = \frac{T \cdot 3}{0.001x10x60in.} + T\left(\frac{1}{.95}\right) \tag{8.21a}$$

$$R_m = 5T + 1.05T = 6.05T \tag{8.21b}$$

If μ = 100, $R_m = .5T + 1.0T = 1.5T$.

where μ = ductility permitted in the restraint. A μ of 100 would typically represent about 15% strain in a type of restraint loaded in axial tension of a type shown in Figure 8.5. Such ductility levels are possible for stainless-steel bands or bands with energy-absorbing impact surfaces. The box-type restraint design as shown in Figure 8.4 has ductility limited to about 12.5 where design is limited by flexure in the structural-steel beam.

8.2 MISSILES

8.2.1 Introduction

Missiles are classified as to source in determining their dynamic characteristics. There are two primary sources of missiles: natural phenomena hazard as would be developed in a tornado or other severe wind storm; and accident-generated missiles as would be developed during structural failure or rupture of a high-pressure fluid system, rotating equipment, or impact of aircraft or other types of vehicles. Excluded from consideration here are military-type missiles, which are designed to penetrate or perforate metal and concrete targets. Missile source description is essential in defining the dynamic characteristics of missile as to velocity, dimension and mass. A fourth characteristic of missiles is the crushability or fragility level of the missile. It is known that a significant amount of the kinetic energy of a missile may be utilized in its crushing and plastic deformation. Unfortunately, quantitative evaluation of this energy absorption by crushing of the missile is not available in the technical literature, except for a few cases where experiments have been performed to define specific crushing strength and energy absorption capability, or to define the impactive forcing function developed by that missile upon impact.

8.2.2 Missile Source Definition

Listings of tornado-type missiles, which have been considered in nuclear power plant design, are presented in [12].

Pressurized Equipment Missiles

Small equipment items, such as valves, reactor bolts, instrument thermowells, valve stems, and large items, such as valve bonnets, control rods and their drive mechanisms, comprise the most likely list of equipment rupture missiles associated with high-energy nuclear reactor coolant and heat transport systems. Since many of these systems are located inside containment, they constitute a particular threat to the containment liner where leak-tight as well as structural integrity must be maintained. Other systems that must be maintained are the safety features associated with the containment system, the reactor coolant system and systems necessary to maintain the reactor in a safe shutdown condition.

Tornado and Extreme Wind Generated Missiles

At least two tornado missiles are specified for a given site. The first is meant to represent essentially a penetrating effect on structures and, as such, it tends to be of higher velocity and have a smaller area of impact taken at a velocity equal to 0.6 times maximum wind velocity, such as a length of steel pipe. The second missile is used to represent large impactive load effects on structures and, as such, it tends to be defined as having relatively lower velocity and high mass typically taken at 0.2 times the maximum tornado wind speeds, such as in an automobile [12]. In general, there is no attempt to rationally define tornado missiles, which may be defined as being generated at the site. Wooden poles or planks have also been defined as design-basis missiles. However, experience has shown such missiles tend to disintegrate when impacting concrete walls typical of industrial facility structures.

Aircraft Accident Crash Missiles

With the ever-increasing air traffic and the occasional proximity of hazardous facilities to airports, there has been specific as well as general concern for the possibility of accidental aircraft crashes into hazardous facilities. Generally, only commercial jet aircraft are considered in these examinations and have been applied only to a relatively few number of facilities dictated by their close proximity or in the approach pattern to commercial airports. Historically, small, less than 11,000 lbs. (5,000 kg), aircraft accidental crash effects traveling at speeds no greater than 150 mph (71m/sec) have been enveloped by tornado load effects.

In general the aircraft fuselage and wings represent an impactive-type missile, which is limited by the crushing strength of the airframe. The engines represent a more typical penetrating-type missile.

Recent concerns over the potential for deliberate or malevolent aircraft impacts on hazardous facilities are expected to become a more common design basis in the future.

Rotating Machinery Missiles

The turbines, rotors, fans and flywheels of production-type hazardous industrial facilities, which are the largest pieces of rotating equipment in the system, are considered to be the possible sources of high-energy missiles in nuclear power plants and to a lesser extent in other nuclear facilities. Most manufacturers have concentrated on redundant control systems to preclude overspeed as a potential source of missile generation. In addition, they have improved quality control and inspection procedures for the material making up the rotating equipment to the point where missiles from such sources are in many cases no longer considered credible.

In many instances, rotating machinery and potential pressurized equipment missiles have been orientated in such a way that potential line-of-sight missiles are precluded from interacting with any safety-related SSC. This procedure results in the elimination of missiles for design purposes based on their reduced probability of impact on safety-related SSC.

Design-Basis Missiles

As recommended in the *ANS 58.3* [13], all potential missiles that could affect a hazardous facility should be evaluated to demonstrate that either the probability of occurrence of the missile and subsequent damage to critical targets is extremely low, or that the potential damage to critical targets is acceptable.

The joint probability shall be considered extremely small or not to exceed the upper probability limit for design for nuclear power plants if:

a. It is less than or equal to 10^{-7} per year for the total of all plants at a site for all missiles from any single major source with associated damage that could be in excess of limiting extreme events.
b. It is less than or equal to 10^{-6} per year for missiles whose generation is considered a prescribed limiting event.
c. It is less than or equal to 10^{-5} per year for missiles whose generation is considered a lesser event and has effects that could upset limiting events.

8.2.3 Determination of Missile Design Velocity

Generation of missiles requires energy in one form or another to propel objects. The formulas for calculating the velocities of these missiles, as presented by Gwaltney [14], are summarized as follows:

Missiles Resulting from Plant Operation

Piston-Type Missile

$$v = \left(\frac{2PAL}{m}\right)^{1/2} \tag{8.22}$$

where v = the initial velocity at the end of the piston stroke; P = pressure of the fluid; A = cross-sectional area of the piston; L = length of the stroke; and m = mass of the missile.

Jet-Propelled Missiles

For a liquid jet:

$$V_T = \left[V^2 + 2g\left(\frac{P}{w} - H_L\right)\right]^{1/2} \tag{8.23}$$

where V_T = velocity of the liquid in the jet at the nozzle; V = velocity of the liquid in the container or pipe; g = gravitational constant; P = pressure in the container or pipe; w = specific weight of the liquid; and H_L = energy lost by the expansion of the liquid inside the nozzle.

For a gas jet:

$$V_T = \left\{V^2 + 2g\left\langle\left(\frac{K}{K-1}\right)\frac{P}{w}\left\{1 - \left(\frac{P_T}{P}\right)^{(k-1)/K}\right\} - H_L\right\rangle\right\}^{1/2} \tag{8.24}$$

where P = pressure in the container; P_T = pressure in the throat of the jet; W = specific weight of the gas; and K = ratio of specific heat at constant pressure to specific heat at constant volume.

At the point of minimum cross-sectional area, the velocity of the gas is limited to the acoustic velocity at the temperature and pressure of the gas at that point.
Then

$$\frac{P_T}{P_O} = \left(\frac{2}{K+1}\right)^{K/(K-1)} \tag{8.25}$$

where P_O = pressure in the container.

The acoustic velocity is calculated with the formula

$$V_C = \left(\frac{KP_T g}{w_T}\right)^{1/2} = (gKRT)^{1/2} \tag{8.26}$$

where V_C = acoustic velocity; w_T = specific weight of the gas at the throat; R = universal gas constant; and T = absolute temperature at throat.

For a two-phase fluid jet:

Gwaltney [14] describes a simplified approach for determining the velocity of a two-phase fluid jet, which involves the experimental determination of V/V_O vs. L/D, where V = actual velocity, V_O = average velocity; L = tube length; and D the tube diameter. Having established this experimental relationship and assuming the critical exit pressure, the average velocity can then be calculated with the following expression:

$$V_O = C_d\left[\frac{2g}{\rho_i}(P_i - P_e)\right]^{1/2} \tag{8.27}$$

where V_O = average two-phase fluid velocity; C_d = discharge coefficient or Euler number i = initial density of the fluid in the vessel; P_i = initial pressure of the fluid in the vessel; and P_e = exist pressure.

The velocity of a jet-propelled projectile can also be obtained by computing the force on the projectile by multiplying the stagnation pressure of the fluid relative to the missile by the drag coefficient. The velocity of the missile, V_m, is given by:

$$V_m\left(\frac{C_D A V^{2\rho X}}{m}\right)^{1/2} \tag{8.28}$$

where C_D = drag coefficient; A = cross-sectional area of the missile; V = relative velocity of the missile with respect to the fluid; ρ = density of the fluid; m = mass of the missile; and A = distance traveled by the missile.

Self-Propelled Missiles

These types of missiles can be generated from pipes, valves or other hollow structures that contain fluid at the time of rupture and are driven by the expansion of the fluid they contain.

The usual technique employed in determining missile velocities is to apply a dynamic force equal to the saturation pressure of the contained fluid times the assumed break opening area times some coefficient (typically 1.26 for steam and 2.0 for nonflashing water), which represents the mass transfer effect of the fluid ejected from the pipe. The resultant force, which is in excess of that required to form a plastic hinge mechanism, is assumed to act on the component causing it to accelerate as a function of its mass and as a function of time-resultant velocities as a function of distances traveled or gap, can be determined.

Missiles from Rotating Machinery

Under certain postulated conditions, ruptures of turbines, pumps, motors, fans and other rotating machinery have the possibility of generating missiles. For instance, overspeeding could lead to overstressing and failure of the rotating parts, or overspeeding may cause the rotating equipment to reach the critical resonance speed resulting in large vibrations and possible generation of missiles. The problem is to determine what part might become a missile and then determine if such a potential missile has sufficient energy to escape the equipment housing, and, if so, at what residual velocity.

Missiles Originating from Stored Strain Energy

Pressure vessel head bolt nuts, other nuts restraining tendons for prestressed concrete should be thought of as potential missiles. The large amounts of strain energy stored in these structural components could lead to high velocities in case of failure.

By assuming that all the strain energy is converted to kinetic energy, the velocity can be calculated

a. In terms of strain

$$V = \left(\frac{gE}{W}\right)^{1/2} \varepsilon \tag{8.29}$$

or

b. In terms of stress

$$V = \left(\frac{g}{EW}\right)^{1/2} \sigma \tag{8.30}$$

where V = velocity of projectile; E = modulus of elasticity; W = specific weight of projectile; ε = strain in the bolt or tendon before the accident; and σ = stress in the bolt or tendon before the accident.

Missiles from Ruptured Vessels

High-velocity missiles can be generated directly by the rupture of a vessel operating at high pressure and indirectly by the shock waves resulting from the rupture. The missiles thus generated can be random fragments or certain parts of the vessel. The velocity of such missiles is proposed to be:

$$V = 1.092\,(E_f R)^{1/2} \tag{8.31}$$

where V = velocity of projectile (in ft/sec); and E_f = available energy released by isentropic expansion of pressurized fluid to one atmosphere for one mass unit (in ft^2-lb/lb-sec^2).

$$R = \frac{\frac{C}{M}}{1 + \frac{C}{2M}} \quad \text{for cylinders,} \tag{8.32}$$

$$R = \frac{\frac{C}{M}}{1 + \frac{3C}{5M}} \quad \text{for spheres} \tag{8.33}$$

in which C = mass of the fluid (in lb-sec^2/ft); and M = mass of vessel (lb-sec^2/ft).

8.2.4 Missile Effects

Missile effects on structures once their source characteristics of mass, velocity, dimension and crushability have been defined are divided into two categories: penetration and impact.

Missile Penetration

In this discussion the term "penetration" is reserved for the entry of a projectile into a target without passing completely through it. The targets considered are typically structural elements such as walls, floors, roofs, etc., but may also consist of protective housing surrounding equipment. Penetration and perforation are typically referred to as local missile effects.

Missile penetration of structures is typically determined by empirical formulas. These empirical formulas are based on extensive experimental study of terminal ballistic missile effects on concrete and steel structures performed in the 1940s. However, it should be understood that the high-speed computer algorithm capable of solving real plastic-dynamic design problems are just now becoming available to structural designers. These new computer programs should have the capability of solving the penetration problem analytically, where in the past no such practical tools of solution existed.

For penetration of steel plate, the following empirical relationships are typically used:

$$\frac{E}{DF^2} = \frac{S}{46{,}500}\left(16{,}000\frac{T^2}{F^2} + 375\frac{WT}{F^2}\right) \tag{8.34}$$

where E = critical kinetic energy required for penetration (in ft-lb); D = missile diameter (in in.); S = ultimate tensile strength of target plate (in psi); T = target plate thickness (in in.); W = length of side of square window in the target frame (in in.); and F = scale factor = W/W_s; W_s = 4 in.

Within the following range of parameters:

$$\begin{array}{lllll}
0.1 & < & T/D & < & 0.8 \\
0.002 & < & T/L & < & 0.05 \\
10 & < & L/D & < & 50 \\
5 & < & W/D & < & 8 \\
8 & < & W/T & < & 100 \\
0.2 & < & W/L & < & 1.0 \\
70 & < & V_c & < & 400
\end{array}$$

where V_c = critical penetration velocity (in ft/sec); and L = projectile length. These ratios are not all independent but must be satisfied simultaneously.

Missile Impact

The effective missile impact load is typically referred to as a global effect. There are three basic simplified techniques generally available to determine the effect of impactive missile loads on a target SSC. They are solution by: (1) energy method; (2) momentum method, and (3) empirical methods. In its most simplistic form, the energy method takes the kinetic missile energy at the instant of impact and divides it by the maximum permitted deformation of the target to determine the equivalent impulsive force applied to the target. To determine the duration of the loading, twice the depth of penetration plus any plastic deformation of the missile is divided by the velocity at impact to determine duration of loading. In this way, a forcing function is defined and the elastoplastic response of the target to the loading function is determined. Iterative design is performed until the total deformation assumption is consistent with calculated behavior. The momentum method is used to determine the energy imparted to the missile as a function of either elastic or plastic impact and as a ratio of missile-to-target mass. The potential energy capacity of the target is determined by computing the area of the resisting force-displacement curve.

The energy method tends to be used more realistically when the missile considered is of the high-velocity, small-mass, large-penetration type. It tends to break down for low-velocity, large-mass missiles with little or no penetration. For the latter type of missile, the momentum method is considered to give more realistic answers. It should be understood that in the immediate vicinity of the missile impact (typically taken within 10 missile diameters), empirical penetration formulas rather than stress criteria should govern design. Missile design procedures are discussed in more detail in [15] and [16].

8.2.5 Missile Impact Simplified Design Procedures

Solution by Energy Methods

This method particularly lends itself for solution by relatively high-energy, low-mass, penetration-type missiles.

In the solution of design for missile impact by the energy method, it is necessary to first define the energy input to the target SSC. This energy is defined as the initial kinetic energy of the missile:

$$E_m = \frac{1}{2} m_1 V_1^2 \tag{8.35}$$

where m_1 = mass of the missile; V_1 = impact velocity of the missile; and E_m = kinetic energy of the missile at impact.

Next, it is necessary to determine the duration of the impact load:

$$t_d = \frac{2D}{V_1} \tag{8.36}$$

where t_d = duration of impact; and D = depth of missile penetration in slab plus depth of crushing of the missile.

Thirdly, determine the effective dynamic force:

$$F_1 = E_m/D \tag{8.37}$$

The fourth step is to calculate the fundamental period, T, of the target structure in the vibration: mode excited by the impactive load. For a selected ductility factor:

$$\mu = \frac{X}{X_e} \tag{8.38}$$

where X = total deflection; and X_e = limit of elastic deformation.

Knowing the peak dynamic force F_1, the equivalent static force, R_m, can be determined from an assumed triangular or square forcing function force, F_1, acting for a duration, t_d, taken from [20].

Solution by Momentum Method

This method of solution should be used for missiles that tend to be relatively low-velocity, high-mass relative to the structure and little penetration.

First, it is necessary to determine the amount of energy absorbed by the structure. The technique used assumes conservation of momentum and plastic impact [17]:

$$E_s = \tfrac{1}{2} (m_1+m_2) V_2^2 \tag{8.39}$$

where E_s = energy to be absorbed by the target SSC; m_1 = mass of the missile; m_2 = effective mass of the target SSC (i.e., a volume typically taken as 10 times the missile diameter times the surface thickness of the target); and V_2 = the final velocity of the combined target and missile.

$$V_2 = V_1 \frac{m_1}{m_1 + m_2} \tag{8.40}$$

In Figure 8.6 from [17] is shown a plot of the ratio of the kinetic energy absorbed by the SSC as a function of the weight ratio W_1 of the missile to W_2 of the target due to elastic impact.

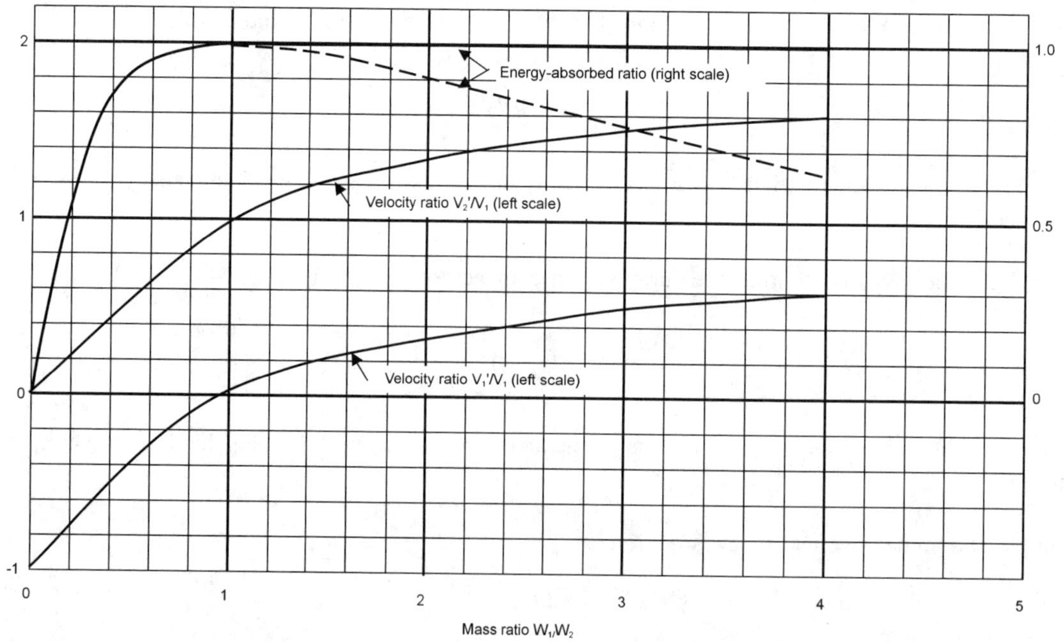

FIGURE 8.6. ENERGY-ABSORBED RATIO AND VELOCITY RATIOS VS. MASS RATIO, ELASTIC IMPACT; V_2=0.

Once the energy to be absorbed by the target structure has been determined, the maximum allowable ductility factor, μ, can be selected from Table 8.1.

Having selected μ, then

$$R_m = \frac{E_s}{(\mu - 1/2)X_e} \quad (8.41)$$

where R_m = equivalent static load resistance function or load capacity that the target structure must develop using limit analysis methods for structure steel.

For both the energy and momentum methods, it is necessary to define that region of the target SSC that is effectively loaded by the missile so that the effective mass and the period or frequency characteristics of the target structure can be determined. This region has been suggested as a volume contained within a cylinder whose diameter is 10 times that of the minimum dimension of the impacting missile at the conclusion of the impact and whose height is equal to the target slab or shell thickness. This diameter was selected based on the observed mean radius of damage or visible cracking associated with impact of numerous missiles on concrete walls or slabs. A second criterion is also used to determine the region affected by the missile as $5\sqrt{t}$, where t is the thickness of the target SSC contact surface. The smaller of the two regions defined should be used for design purposes. Within this diameter it is also assumed that the empirical penetration formulas govern design. Outside this region, calculated stresses, forces, moments, etc., whether determined by momentum or energy method, govern design. Obviously, this method may become an iterative design process since the depth of the target member is changed as might be the case in design.

Another approach to effective mass is to determine the volume of the target structure traversed by a shear wave in the time a compression wave has traversed the missile and returned.

8.2.6 Ductility Factor

The ductility factor is a measure of a material's ability to absorb energy in the plastic region as a function of its fabricated shape and type of loading. Structural-grade low-carbon steel, for example, typically has uniform ultimate strain capacity when loaded in unaxial tension in the range of 20 to 25%. This means that within the gauge-measuring distance (typically 2 inches), the material has the capability of 20 to 25% elongation before it begins to neck down and lose load-carrying capacity.

The material would then have a ductility factor determined as follows: X = total deformation; and X_e = elastic deformation.

$$\frac{X}{X_e} = \frac{0.2 \text{ in/in} \ \ (\text{uniformultimate strain})}{0.001 \text{ in/in} \ \ (\text{strain at yield})} \quad (8.42)$$

Structural steel is also used in bending and compression as well as in membrane tension. In such cases, depending on the shape of the cross section and the amount of strain required to form a plastic hinge, there exists a potential for local instability in the compression zone (i.e., compression flange buckling for wide flange shapes). Table 8.1 presents some suggested values of μ for various steel shapes. The values shown contain, when known, a nominal safety factor of 2.

8.2.7 Allowable Behavior Criteria

Design for missile impact effects typically assumes implicit plastic response of the structure. When simplified methods are used, this plastic response is usually defined as an acceptable ductility ratio or ductility factor. When more rigorous analytical methods typically associated with finite elements are used, equivalent plastic strain limits are usually defined. Currently such limits are being defined in the draft *ASME B&PVC*

Section VIII, Division 3. In the interim it is recommended that these strains be limited to 1% membrane strains, 2% membrane plus bending and 5% localized strains. These strain limits primarily are used to define equivalent static loads to be applied to the target.

8.2.8 Design Approaches

The general design approach for the various missile and target characteristics are as follows:

A. Soft missile and hard target (i.e., airplane or auto crash into relatively rigid SSC):
In such instances a dynamic time history forcing function is developed that is a function of the crushing strength of the missile and its velocity at impact as it continues to crush.
B. Hard missile and hard target (i.e., turbine missile impact on relatively rigid SSC):
In these instances the analysis is more likely to consider conservation of momentum transfer between missile and target. In such a case, the effective impact mass of the target is determined by the volume a shear wave in the target structure has traversed in the time it takes a compression wave to travel the length of the missile and return.
C. Soft missile and soft target (i.e., airplane or auto into steel tank):
In this analysis a forcing function is defined that is based on the time it takes to penetrate the target, which is also used to define the effective mass of the target by defining the volume the shear wave has traversed in penetrating the target and the local buckling mode, plus crushing or deforming of the missile.
D. Hard missile and soft target (i.e., turbine wheel into steel tank):
Generally the same approach will be taken as that defined for C, except that deformation of the missile can be neglected.

8.3 STEAM AND WATER HAMMER

Water and steam hammer typically occurs in a fluid system when there is a sudden stoppage of fluid flow in the system due to valve closure or pump stoppage such that the velocity of the fluid is converted to a stagnation pressure in the fluid. Another particularly damaging form of hammer is where a slug of water or other liquid is accelerated by the application of steam at relatively high pressure to the slug, causing rapid acceleration and high velocity of the slug, which is then suddenly stopped at a closed valve or redirected by an elbow or other fitting in the system. This situation is typically found when a bypass or other unused line may collect water at a low spot and steam is suddenly introduced in to the line at startup or shutdown.

According to Antaki in [18], if the change of flow velocity occurs over a short period of time, t_d, of the order of magnitude or shorter than the propagation time, t_p, equal to 2 l/a, where l is the distance between the reservoir and a value of the speed of sound in the fluid, then the flow transient is a water (fluid) hammer. In this case, the pressure rise due to the sudden, instantaneous valve closure ($t_D < t_p$) is given by the formula:

$$dP = \frac{\rho a(\Delta v)}{g} \tag{8.43}$$

dP = pressure change due to instantaneous valve closing; ρ = fluid density; Δv = change of fluid velocity at disturbance; g = gravity; and a = speed of sound in the fluid.

for water flowing at ambient temperature at 10 ft/sec (ρ = 62.4 lb/ft^3, a = 4,860 ft/sec, Δ = 10 ft/sec, g = 32.2 ft/sec^2), the sudden closure of a valve will cause a pressure rise dP = 654 psi, a very significant pressure spike. Note that the pressure rise does not depend on the initial pressure, only on the initial velocity. Response of the pipe to the spike is also a function of the duration of the spike and the period of the pipe excited by the spike.

Conversely, for an instantaneously opening valve, the pressure drop is:

$$H_D = [n_F (2H + n_F]^{0.5} - n_F \quad (8.44)$$

where

$$n_f = H_{R,max}^2/(2H) \quad (8.45)$$

HD = pressure drop; $H_{R,}$max = maximum pressure rise from a water hammer defined by Eq. (8.50); and H = system static head

The large pressure rise dP that accompanies a water hammer has two damaging effects: first, some pipe fittings or components, such as cast-iron valve bodies, instruments or expansion joints, or even the pipe itself, may not be able to withstand the increased pressure $P+dP$ effects, and may fail by bursting or significant distortion and leakage at joints under the transient pressure. Second, on change in direction or cross section, the pressure rise will cause unbalanced axial loads $A(dP)$, where A is the cross-section area of the obstruction or the pipe area at an elbow. Under the effect of this unbalanced force, the pipe may move and bend, potentially causing the failure of pipe supports.

In Reference [18] can be found a more detailed discussion of the water or steam hammer phenomena and is an excellent reference for piping engineering.

8.4 RELIEVE VALVE DESIGN

8.4.1 Introduction

The load computations applied to relieve valves consider: evaluation of momentum effects and evaluation of pressure effects. The response computation of the piping system includes consideration of transient, dynamic loading effects of the sudden opening and closing valve action. The load computation is combined with other specified loads and translated to a force and moment calculation in the discharge pipe, the valve and the valves attachment to the line being relieved. Computations are performed that convert the relief valve load operation and other applicable loads to stress in the pipe and their comparison to the code-allowable stresses.

The basic principles of analysis of a pressure relief valve station are generally applicable to both open- and closed-discharge systems. However, the application of these basic principles may be quite different.

8.4.2 Open System – Discharge Thrust Load Effects

The steady-state load due to fluid reaction force from the opening and subsequent venting of the relief valve includes consideration of both momentum and pressure effects and may be computed by the formula:

$$F = (W/g)\, V_e + (P_e)A \quad (8.46)$$

where F = reaction force; W = mass flow rate [relieving capacity of the valve (in lbs/sec)]; g = gravitational constant; V_e = exit velocity at point e; P_e = static gage pressure at point e; and A = exit flow area at point e.

The reaction force F is a design mechanical load that requires structural equilibrium for system stability and is applied as shown in Figure 8.7 [19]. It should be noted that discharge pressure vector is also dictated toward the branch connection in order to reduce the bending moment on the branch connection.

To ensure consideration of the effects of the suddenly applied load, F, a dynamic load factor (DLF) can be determined based on the relief valve opening time and system dynamic characteristics used to determine the applicable forces and moments due to the suddenly applied reactor force, F. In the limit the DLF may be taken equal to 2.0.

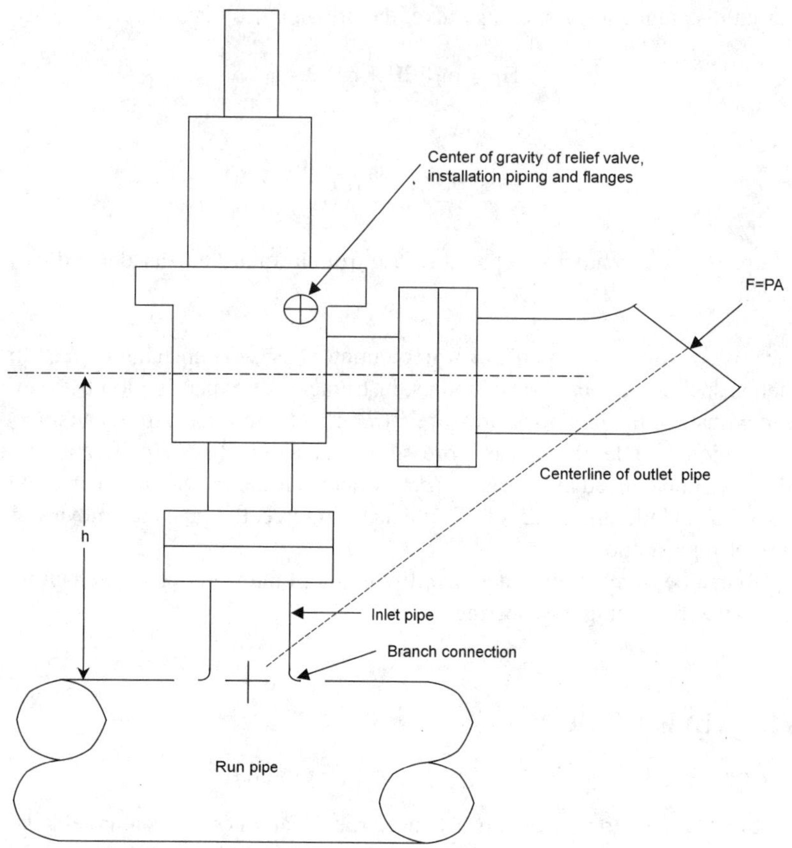

FIGURE 8.7. SAFETY VALVE INSTALLATION (OPEN DISCHARGE SYSTEM)

8.4.3 Other Loads

Other mechanical loads to be considered concurrent with the relief valve discharge loads are:

a. Operating condition loads in the pipe being relieved.
b. Interaction loads on the run pipe being relieved when more than one valve releases.
c. The transient impacting of the valve mechanism opening and closing, if applicable.
d. Seismic load if valve is actuated by seismic effects.

8.4.4 Closed-System or Long-Discharge Pipe

When the valve is suddenly opened, there is a period of transient flow until the steady-state discharge condition is reached. During this transient period, the pressure and flow will not be uniform. When the relief valve is initially opened, the discharge pipe may be filled with air. If the relief valve is on a steam system, the steam discharge from the valve must purge the air from the pipe before steady-state steam flow is established and, as the pressure builds up at the valve outlet flange and pressure waves start to travel down the discharge pipe, the pressure wave initially emanating from the valve will steepen as it propagates, and it may steepen into a shock wave before it reaches the exit.

Relief valves discharging into an enclosed piping system or long-discharge pipe create momentary unbalanced forces, which act on the piping system during the first few milliseconds following relief valve opening. The pressure waves traveling through the piping system following the rapid opening of the safety valve will cause bending moments in the safety valve discharge piping and throughout the remainder of the piping system. In such a case, the designer must compute the magnitude of the loads as they vary in time and location, and perform appropriate evaluation of their effects as described in Section 8.1.3.

8.5 HEAVY LOAD DROP

In hazardous material storage and processing facilities that often transfer heavy loads by cranes or other overhead devices, postulated load drop is often considered of high enough probability that that it becomes a design-basis accident. Loading characteristics of the resultant dropped load as to weight, drop height or impact velocity need to be defined. The dropped load is then treated as a missile impacting an SSC surface.

The structural and leak-tight integrity of the dropped load may or may not be required. In general, other than the experiment or test, this effect is determined by a finite-element analysis into the inelastic range. The limits for acceptability are typically defined as equivalent strain with a factor of safety of four against equivalent strain at rupture of the material.

ASME has developed a crane code, NOG-1 which defines 3 classes of cranes. A class 1 crane is intended to be a failure proof crane. Hence, a load drop would not be postulated for such a crane.

REFERENCES

1. Center for Chemical Process Safety, 1993, "Guidelines for Engineering Design for Process Safety," American institute of Chemical Engineers,
2. Center for Chemical Process Safety, 1994, "Guidelines for Evaluating the Characteristics of Vapor Cloud Explosions, Fires and BLEVE's," American Institute of Chemical Engineers,
3. Center for Chemical Process Safety, 1996, "Guidelines for Evaluating Process Buildings for External Explosion and Fires," American Institute of Chemical Engineers,
4. Bush, S. H., 1999, "A Review of Nuclear Piping Failures and Their Use in Establishing the Reliability of Piping Systems," *PVP Vol. 392 Service Experience in Fossil and Nuclear Power Plants,* American Society of Mechanical Engineers, New York, NY.
5. *SRP 3.6.1,* 1990, "Plant Design for Protection Against Postulated Pipe in Fluid Systems Outside Containment," Rev. 2, NUREG-0800, U.S. Nuclear Regulatory Commission,
6. *Draft SRP 3.6.3,* 1997, "Leak Before Break Evaluation Procedure," Federal Register, August,
7. Cloud, R. L. Assoc., 1987, "Applying Leak-Before-Break to High Energy Piping," *NSAC 114,* Electric Power Research Institute,
8. *ANS 58.2,* 1988, "Design Basis for Protection of Light Water Nuclear Power Plants Against the Effects of Postulated Pipe Rupture," American Nuclear Society,
9. Moody, F. J., 1965, "Maximumt Two-Phase Vessel Blowdown from Pipes," *APED 4827,* General Electric Co.,
10. Fauske, H. K., 1965, "The Discharge of Saturated Water Through Tubes," *Chem. Engring. Proj. Symp. Ser.* American Institute of Chemical Engineers, Vol. 61, No. 59.
11. *AISC N690,* 1994, "Specifications for the Design, Fabrication and Erection of Steel Safety-Related Structures for Nuclear Facilities," American Institute of Steel Construction, Chicago, IL.
12. *SRP Plan 3.5.1,* 1981, "Missiles Considered in Design of Nuclear Power Plants," Rev. 2, NUREG-0800, U.S. Nuclear Regulatory Commission, Washington, DC.
13. *ANS 58.3,* Appendix B, 1992, "Physical Protection for Nuclear Safety-Related Systems and Components – Plant Design Against Missiles," American Nuclear Society, La Grange Park, IL.
14. Gwaltney, R. C., 1968, Missile Generation and Protection in Light Water Cooled Power Reactor Plants," *ORNL-NSIC-22-Sept.,* Oak Ridge National Laboratory, Oak Ridge, TN.
15. *ANS 58.3,* Appendix B, 1992, "Plant Design Against Missiles," American Nuclear Society, La Grange Park, IL.

16. "Report of the ASCE Committee on Impactive and Impulsive Loads," 1980, Vol. V, *Proc., Civ. Engrg. and Nuclear Power Conf., Vol. 5,* ASLE, Reston, VA.
17. Norris, C. H., et. al., 1959, *Structural Design for Dynamic Loads,* McGraw Hill Book Co., New York, NY.
18. Antaki, G. A., 2003, *Piping and Pipeline Engineering – Design Construction Maintenance Integrity and Repair,* Marcel Dekker, New York, NY.
19. Non-Mandatory Appendix II, 2002, "Rules for the Design of Safety Valve Installation," *ASME B31.1-2001,* American Society of Mechanical Engineers, New York, NY.
20. U.S. Army Corps. of Engineers, 1957, "Design of Structures to Resist the Effects of Atomic Weapons," *Manual EM 1110-345-415,* Washington, DC.

CHAPTER

9

VIBRATION RESPONSE OF FLUID DISTRIBUTION SYSTEMS

9.1 INTRODUCTION

The analysis of fluid distribution systems normally does not consider vibration as a design-basis requirement. However, once the system is put into service vibrations may be observed due to active equipment (pumps or valves) operation or fluid flow, and analysis may be required to determine whether the observed vibration may cause fatigue types of failure of the system involved [1]. Each type of response may correspond to different failure modes; hence, they should be addressed in a specific way corresponding to the type of vibration being observed or considered.

The most typical case is represented by the beam-type response as a result of the excitation of the structural mode shapes. Piping system structural vibrations typically correspond to the low-frequency range (less than 20Hz), while ductwork or other thin-walled distribution systems may respond with higher-frequency plate or shell modes of vibration.

Experience shows that the most efficient way to reduce the low-frequency vibrations can be achieved by proper design of support type, spacing and locations, while higher-frequency plate and shell modes of vibration can be controlled by the use of thicker plate or modified design to increase stiffness.

Another type of stress, which can effect this type of failure, not normally considered in design is large residual stresses caused by fit-up of piping systems during fabrication or erection. Residual stresses may affect the fatigue evaluation since they affect the nominal state of stress in the SSC. Support relocation and devices such as dampers used for reduction of piping structural vibrations caused by cycling loads produced by operating and transient vibrations may be required to reduce stresses within code-defined margins to failure during the specified service life of the system or component associated with normal, severe or extreme types of loads.

9.1.1 High-Frequency Vibration Mechanics

Turbulent Flow

The high-frequency vibrations are typically induced by turbulent flow and connected rotating equipment that could affect large zones of the distribution systems. The high-frequency vibrations have very small amplitudes. For this reason the use of rigid supports may not be an effective way for reduction of the high-frequency vibrations. In Real installations turbulent flow, if present, may produce an excitation force (pressure impulses) with high frequency. Using rigid supports will typically reduce low-frequency structural vibrations and tend to increase the distribution systems frequency. If structural frequencies increase, they will tend to be closer to the high-frequency flow-induced vibration frequency, and resonance effects could amplify them.

The most effective way to reduce high-frequency flow-induced vibration is to eliminate the root cause by reducing flow turbulence, usually by reducing flow velocity, eliminating cavitation effects and propagation of the pressure pulses produced by pumps, compressors, etc. Pressure pulsations could be amplified by resonance effects (acoustic resonance) and resonance with the pipe shell frequencies. The resonance effects could be avoided by changing the pipe diameter and/or wall thickness. At the design stage these factors could be considered by controlling the flow, pipe layout (size of pumps and pipes), using dampers and silencers, low noise control valves, etc. After-construction vibration is usually controlled by modifying support spacing or installing energy-absorbing pipe support or restraint devices such as pipe dampers.

Rotating and Reciprocal Motion of Attached Active Components

The unbalanced rotation of active components (pumps, fans, etc.) can cause either low (below 20 Hz or 1,200 rpm) frequency vibrations or high-frequency vibrations above these values. Alternating current motors tends to rotate at 1,500 to 1,800 or 3,000 to 3,600 rpm so vibration from these sources tends to be high frequencies. Geared motor drives typically result in low-frequency vibration.

Cavitation

Cavitation is often the source of pipe vibrations and also produces noise, pressure variations, erosion damage and loss of flow capacity.

Vaporization cavitations form when the pressure of the liquid drops below the level at which liquid changes to a vapor state. This can appear at pressure reducing orifices and control sluices. Cavitations appear when a vapor cavity collapses under a pressure higher then the vaporizing pressure. This can occur when a vapor cavity travels downstream from the orifice/control sluice. Vapor cavity break produces pulsations that can generate pipe vibration, surface erosion and accelerated corrosion. The sounds produced by the cavitations differ depending on its severity and can vary from snaps to sounds that resemble rubble being transported through the pipe. At higher levels, it can be harmful to hearing.

Erosion or corrosion can occur when vapor cavities form near the pipe or component surface. Cavitations damage the rusty protection surface, which allows corrosion to advance at an accelerated rate.

Components of piping systems that contribute to lowering the pressure required to produce cavitations are valves, orifices, nozzles, pumps and bends. Damages can be reduced by maintaining a low cavitation level; eliminating the boundary surface in the cavitations area; treating the boundary surface to make it damage-resistant; dissipating flow energy in steps; or ejecting the fluid in separate fluxes. The safest treatment for cavitations is reducing or eliminating the source.

9.2 EXCITATION MECHANISMS AND DISTRIBUTION SYSTEM RESPONSE

Pipes and other types of distribution systems can respond to vibrations as beams or through vibrations of system walls in plate or shell models. Each of these responses affects piping in its own way; consequently, each type of vibration requires specific design or post-construction corrective actions.

9.3 VIBRATION ANALYSIS ACCORDING TO ASME NUCLEAR STANDARD [1]

This section establishes the requirements for the checks that need to be made in order to evaluate the vibrations of different piping systems used in process installations. This chapter may be used as a guide to determine the vibration levels of operational piping systems. Its scope includes those piping systems that need to perform safety functions or whose failure could affect other systems that perform safety functions. This section also discusses checking methods, parameters that need to be measured and evaluated, acceptance criteria, data collection and corrective actions.

9.3.1 General Requirements

Vibrations are of two categories: steady-state vibrations and transient vibrations. According to the importance of the system and the type of vibrations, piping systems can be included into three vibration monitoring groups, according to *ASME OM-S/G-2000* Part 3 [1]. A system can be included into a vibration monitoring group for steady-state vibrations and also in another vibration monitoring group for transient vibrations. The testing requirements, acceptance criteria and recommendations for corrective actions associated with these categories are presented below. The measurement and assessment of the vibration level can be performed before plant operation (preoperational, initial start-up testing) or during plant operation, according to the testing requirements.

For preoperational, initial start-up and operational testing, test specifications need to be drafted; these will include the following minimum requirements:

a. Test objectives
b. The systems that will be tested
c. Conditions or requirements for the pretest phase
d. Applicable documents and designs (drawings)
e. Cautions
f. Quality assurance and control (including the necessary documents and authorized signatures)
g. Acceptance criteria
h. Testing conditions and checkpoints
i. Measurements that will be performed and acceptable limits (including visual inspection)
j. The equipment used (including technical specifications)
k. Evaluation and recording of data
l. Bringing the system back to its initial state

The test specifications shall be drafted in a manner that shall ensure the achievement of test objectives and the precise or conservative assessment of the results. An inspection of components and supports will be performed before testing, in order to check for proper installation, according to plans, specifications, instructions and applicable codes.

When test results are correlated with specific analysis, the test conditions and measurements will be carefully specified to ensure the integrity of the assumptions used in the analysis. Correlation between test and analysis must confirm the validity of the analysis and must indicate that the analytical results are conservative. If the test results indicate an inadequate analysis, or when the measured data show that the real potential function is not conservatively enveloped by the potential function used for the analysis, the analysis must be repeated.

Two options are available if the test results indicate exceeding the values defined in the testing specification: performing the test again using more exact methods, or applying corrective actions. Competent personnel shall participate in developing requirements for testing specifications, selecting the equipment, establishing acceptance criteria, review, evaluation and approval of test results.

Selecting the locations of testing equipment and the type of measurements that will be performed shall be based on the stress analysis, including fatigue of the piping system, the response of similar systems, or experience gained from other similar tests and analyses, and shall reflect all peculiarities of operating characteristics for the system that is to be tested. The evaluation of test data shall also take into consideration the equipment use.

9.3.2 Classification of Vibratory Regimes

The vibrations of piping systems are divided into two categories: steady-state vibrations and transient vibrations. Within each category, piping systems will be classified into three vibration monitoring groups, according to the criteria described in Section 9.3.3 (according to *ASME OM-S/G 2000 Part 3*).

Adding to the requirements presented in Section 9.3.3, the safety classification or the mission importance of the system, or both, must also be taken into consideration when dividing the system into vibration monitoring groups.

Steady-State Vibrations

Vibration Monitoring Group 1 (VMG 1)

The monitoring program for systems included in this group generally requires sophisticated monitoring devices and extensive data collection, in order to determine with as much accuracy as possible the vibration-induced pipe stresses or other limitations of the specified component.

The determination of fundamental modes, the amplitude of modal response and the total system response can be used within this analysis method. The methods and devices required by this vibration monitoring group shall be used when highly accurate measurements of the system response characteristics are necessary.

All the areas of the pipe system that experience steady-state vibrations and meet one or more of the following requirements will be included in VMG 1:

a. Pipe systems exhibiting a response that cannot by characterized by simple fundamental modes of the pipes (like radial vibrations of the pipe wall or strong high-frequency vibrations).
b. Pipe systems for which VMG 2 and VMG 3 methods are not applicable.

Vibration Monitoring Group 2 (VMG 2)

The methods and devices used in evaluating VMG 2 give point-wise measurement and evaluation means.

All areas of pipe systems that meet one or more of the following requirements will be included in VMG 2:

a. All piping systems that show a significantly increased vibration response compared to prior experience on similar systems or similar operating conditions of the system.
b. Piping systems for which VMG 3 methods are not applicable and for which unacceptable vibrations are noticeable.

Vibration Monitoring Group 3 (VMG 3)

The visual method for evaluating VMG 3 is elementary in nature and provides the simplest means to determine any significant vibrations in the system. Evaluating the level of vibrations with this method is based on experience and judgment and provides an acceptable basis for the assessment. Should precise quantitative evaluations be necessary, methods VMG 1 and VMG 2 will be used.

All the areas of the piping systems that meet one or more of the following requirements will be included in VMG 3:

a. Systems that are not included in VMG 1 and VMG 2 for which measurements or testing data are available on prototype or duplicate systems and for which minimum unacceptable vibrations are noticeable.
b. Parts of ASME B&PVC Section III class 1, 2, 3, and *ASME B31* pipe systems, which are not expected to show a significant vibration response deduced based on prior experience on similar systems or based on similar operating conditions of the system.

Transient Vibrations

Vibration Monitoring Group 1 (VMG 1)

This group includes systems for which it is anticipated, based on prior experience accumulated in existing plant operations, that a significant nonsteady dynamic regime will occur due to the component design, the functioning of the system or to different design elements, for which a transient analysis was not conducted.

Vibration Monitoring Group 2 (VMG 2)

Areas of piping systems that experience nonsteady vibrations, which include systems designed and analyzed for known dynamic loads and for which loading (mechanical or hydraulic) is based on a simplified methodology that conservatively estimates the transient potential functions and the associated structural response.

Vibration Monitoring Group 3 (VMG 3)

Includes systems subjected to nonsteady vibrations during the entire operating period (systems subjected to start-up transients, opening and closing of valves) and for which a significant vibration regime is not expected to occur based on prior experience on similar systems or operating conditions.

9.3.3 Monitoring Requirements and Acceptance Criteria

Acceptance criteria are based on *ASME OM-S/G.* More severe criteria can be applied if the following assumptions are considered inadequate for the system under consideration:

a. The vibrations induce elastic stresses; as a result, no corrections for cyclic incursions in plastic domain (i.e., ratcheting) are introduced.
b. Thermal transient effects, if they occur simultaneous to the vibration regime, were taken into consideration when evaluating the pipe system.
c. Membrane stresses produced by pressure fluctuations are insignificant when compared to stresses due to vibration-induced moments.
d. The usage factor due to the vibrations does not significantly affect the cumulative usage factor computed for other predefined transient regimes.
e. Fatigue curves described in Section III of *ASME BPVC* code represent the fatigue characteristics for the considered material and loading.

Vibration Monitoring Group 1

The vibration response of systems in group 1 will be estimated using detailed evaluation methods.

For steady-state vibrations, the computed maximum intensity of the alternate stress Salt S_{alt} will be limited as defined below:

Steady-State Vibrations

For class 1 ASME B&PVC Section III pipe systems:

$$S_{alt} = \frac{C_2 \cdot K_2}{Z} M \le \frac{S_{el}}{\alpha} \tag{9.1}$$

where C_2 = secondary stress index as defined by *ASME BPVC* code, Section III α = allowable stress reduction factor: 1.3 for materials included in *ASME BPVC* code, according to Section III, Figure I.9.1; or 1.0 for materials included in *ASME BPVC* code, according to Section III, Figure I.9.2-2; K_2 = local stress index as defined by *ASME BPVC* code, Section III; M = maximum zero to peak dynamic moment loading, due to vibrations only, or in combination with other loads as required in the system design specifications; S_{el} = $0.8Sa$, where Sa is the alternating stress at 10^6 cycles from ASME BPVC Code, Section III, Fig. I.9.1; or Sa at 10^{11} cycles from *ASME BPVC* code, Section III, Figure I.9.2-2 (The user shall consider the influence of temperature on the modulus of elasticity); and Z = section modulus of the pipe.

For class 2 and 3 ASME Section III Pipes, and *ASME B31:*

$$S_{alt} = \frac{C_2 \cdot K_2}{Z} M \le \frac{S_{el}}{\alpha} \tag{9.2}$$

where $C_2K_2 = 2i$; and i = stress intensification factor described in older editions of the *ASME BPVC* code, Section III Subsections NC and ND or *ASME B31.* Figures I.9.1, I.9.2-1 and I9.2 of *ASME B&PVC* Section III Appendix I are reproduced here as Figures 9.1, 9.2 and 9.3

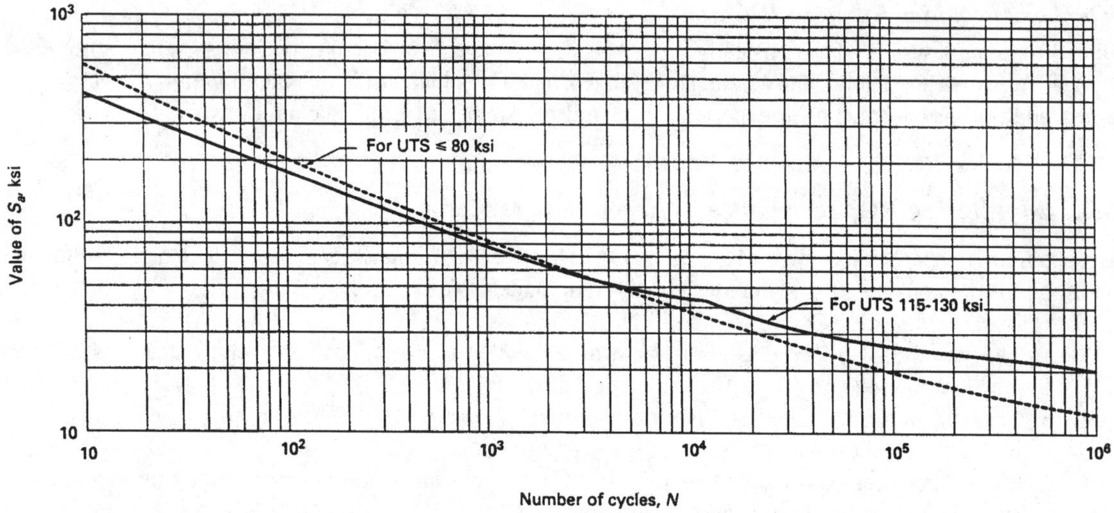

GENERAL NOTES:
(a) $E = 30 \times 10^6$ psi
(b) Interpolate for UTS 80.0-115.0 ksi
(c) Table I-9.1 contains tabulated values and a formula for an accurate interpolation of these curves.

FIGURE 9.1. DESIGN FATIGUE CURVES FOR CARBON, LOW ALLOY AND HIGH-TENSILE STEELS FOR METAL TEMPERATURES NOT EXCEEDING 700°F

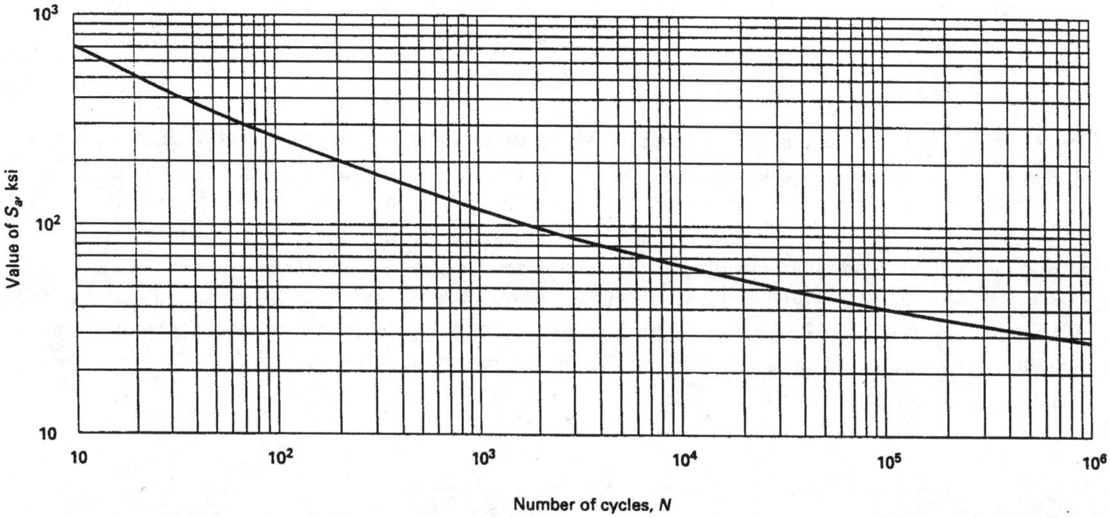

GENERAL NOTES:
(a) $E = 28.3 \times 10^6$ psi
(b) Table I-9.1 contains tabulated values and a formula for an accurate interpolation of this curve.

FIGURE 9.2. DESIGN FATIGUE CURVES FOR AUSTENITIC STEELS, NICKEL-CHROMIUM-IRON ALLOY, NICKEL-NONCHROMIUM ALLOY AND NICKEL-COOPER ALLOY FOR S_A < 28.2 KSI FOR TEMPERATURES NOT EXCEEDING 800°F

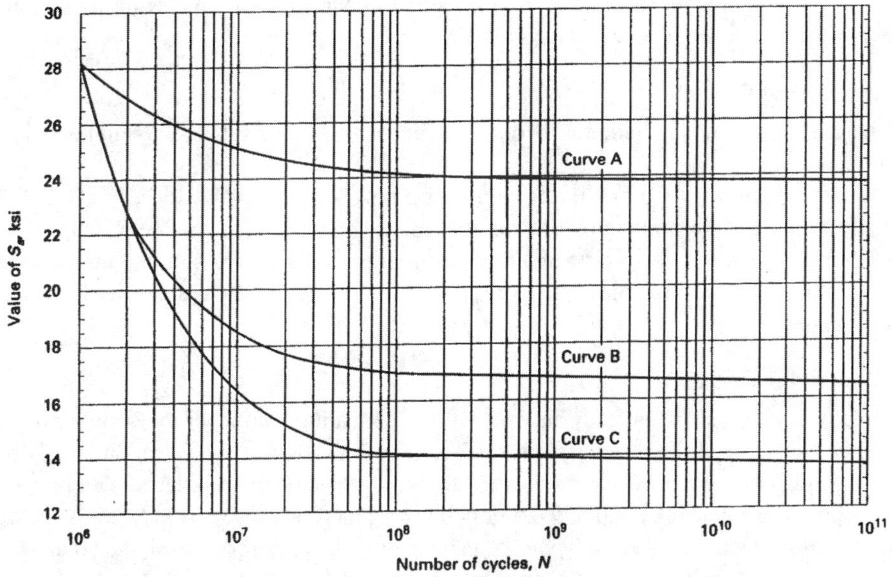

FIGURE 9.3. DESIGN FATIGUE CURVES FOR AUSTENITIC STEELS, NICKEL-CHROMIUM-IRON ALLOY, NICKEL-IRON-CHROMIUM ALLOY AND NICKEL-COPPER ALLOY FOR S_A " 28.2 KSI FOR TEMPERATURES NOT EXCEEDING 800°F

Transient Vibrations

For transient vibrations, the maximum alternating stress intensity must be limited to the value defined in this section. Before determining the allowable value of the maximum alternating stress intensity, the maximum number of loading cycles *"n"* shall be determined.

ASME B&PVC Section III Class 1 Piping Systems

The maximum alternating stress intensity shall be limited to a value that does not affect system design requirements. If the transient event has not been taken into consideration in the system design requirements, it shall be evaluated. The reserve in the reduction factor, U_v, will be determined as follows:

$$U_v = 1 - U$$

where U = cumulative usage factor from ASME class 1 analysis, which excluded vibratory load. The maximum allowable equivalent vibratory load cycles, N_v, shall be calculated from:

$$N_v = \frac{n}{U_v} \tag{9.3}$$

Using N_v, the maximum alternating stress intensity S_{alt} shall be limited to S_a where: S_a = allowable alternating peak stress value from *ASME BPVC* code, Section III as shown in Figures 9.1, 9.2 or 9.3.

For transient vibrations that have not been previously analyzed and for which separate evaluation of loads is not adequate, a new fatigue analysis may be required according to *ASME BPVC* code, Section III.

ASME Class 2 and 3 and ASME B31 Piping Systems

Stresses may be evaluated according to the ref 1 displacement limits requirements. Alternately, the appropriate section of the *ASME B&PVC* Subsection NC and ND and B31 piping may be used to determine stresses in transient vibrations as a functioning of applied displacements. Cycles in nonpressure-retaining components may be evaluated in accordance with AISC [2].

Vibration Monitoring Group 2

Vibration response for systems included in Group 2 are measured using relatively simple devices for monitoring vibration amplitudes.

For steady-state vibrations, the vibration response of piping systems included in VMG 2 shall be evaluated according to the allowable deflection or velocity limits.

For transient vibrations, simple devices for measuring vibration velocity can be employed as screening instruments, but these can lead to conservative results. If these limits are passed, criteria corresponding to VMG 1 shall be used.

Vibration Monitoring Group 3

If an acceptable level of transient or steady-state vibrations is identified, no other measurements or evaluations are needed. The observer shall be responsible for the correct evaluation of the vibration level. The principles and requirements for determining the acceptable vibration level must be in accordance with limits specified by the allowable velocity presented in Section 9.4.5 or deflections method presented in Section 9.4.4.

When the vibration level is too low to be noticed and if there is a minimum evaluated damage possibility, the system is acceptable. Judgment on the system acceptability can only be performed by evaluating all requirements described below:

a. Location, magnitude and rate of vibrations
b. The existence of sensitive equipment in the vicinity of the system
c. Branch behavior
d. Component support capability

Any operating characteristic specific to the system shall be considered during the evaluation.

If an acceptable evaluation of the observed deflections cannot be made, the vibrations shall be accepted based on the data recorded by measurements.

9.4 SIMPLIFIED METHODS

9.4.1 Visual Inspection

The capability of piping systems to withstand steady-state and transient vibrations can be determined and accepted by visual inspection. This section describes various methods and some of the simple devices that can be used in this evaluation applied to as-built systems.

Evaluation Methods

Maximum deflection points can be established visually. The size of deflection can be estimated using simple measurement instruments (scribing displacement instrument). If this is not possible (high frequencies, relatively low amplitudes), then electronic measurement devices should be used. Analogy with beam-type elements can be used to support the developing judgment of the acceptability of observed displacement.

Local vents and drainage systems generally have one or two isolation valves that behave as concentrated masses. Special attention shall be given to vibrations in these areas.

Minor main run pipe vibrations can lead to significant vibrations of connecting branch pipes. These lines must be evaluated together with the system that is to be qualified.

Vibrations that affect functionality, operability and structural capability of flow-sensitive equipment (like pumps, valves, heat exchangers) shall be carefully evaluated.

Special attention shall be given to those areas in the vicinity of the elements additionally welded on the route of piping systems subjected to vibrations. If these elements modify the system configuration in such a manner that they induce additional local moments due to vibrations, local stress effects shall be evaluated.

9.4.2 Displacement Method

This simplified method requires that displacements due to vibrations be determined in representative system locations. The piping system is divided into a sufficient number of subsystems or vibratory characteristic spans containing adequate or conservative boundary conditions.

A handheld or temporarily mounted transducer that is suited for multiple measurements of displacement should be used. For example, an accelerometer can be used, with velocity and speed obtained by single and double integration of the recorded acceleration signal.

Measurements are taken at points along the piping to determine maximum deflection locations and to establish node points where deflections are minimum. The node points define the characteristic span lengths. Node points (zero deflection points) are found at restraint points, but can also be found at points between constraints on long runs of piping.

Branch piping is attached to process piping and has a smaller diameter then the process piping. Three of the potential problems are presented below:

1. Branch piping can be excited near or even at its resonance frequency, due to the movement of the main run piping, fluid pulsation or other sources. This situation is characterized by high amplitude vibrations with clearly defined frequency and mode shape. The amplitude measured on the branch piping is generally higher than that measured on the main piping. Due to the phasing, the relative movement of the branch piping is approximated accurately enough by adding the displacement measured on the main piping to the motion of the branch pipe.
2. The attachment point of the branch piping with the main run piping displaces relative to a branch line support.
3. The main run piping may induce a high level of accelerations in the branch piping. This situation is often associated with a cantilevered mass. The maximum acceleration at the center of gravity of the branch piping has to be measured to establish the inertial load that acts at the center of gravity of this pipe. The mass and center of gravity of the branch piping need to be conservatively estimated, and the resultant stress computed and compared to acceptance criteria.

Deflection Limits

The vibratory deflection limits of a piping system depend on a large number of geometric and material characteristics and of various combinations between them. A method that can deal with this complexity is subdivision of piping systems into a number of characteristic spans that can be defined and modeled. The measurement of a deflection can be conservatively checked against an allowable deflection calculated for the span. The deflection limit is dependant on the configuration coefficient, K, the outer diameter D_0, the length of the span, L, according to *ASME OM-S/G,* paragraph 5.1.1.5 [1].

9.4.3 Velocity Method

The method requires consecutive measurements of the velocity at several points on the piping system to identify the point of maximum velocity. Once this point is located, a final measurement of maximum velocity (V_{max}) at that point is made and compared to the maximum allowable velocity (V_{allow}).

The acceptance criteria is:

$$V_{max} < V_{allow} \tag{9.4}$$

Instrumentation

The instruments used should be portable and capable of making a number of consecutive measurements of the velocity at different locations on the piping system. The instrument shall be capable of tracking the real signal in velocity from which the maximum velocity can be determined.

Initial measurements shall be made at points on the piping system that are anticipated to be subjected to large displacement. These generally correspond to high speed points. In each of these points, measurements can be taken around the circumference of the pipe to obtain the value of maximum velocity.

Maximum velocity should be obtained exclusively from the real time-velocity signal. Recording of the signal should be of sufficient duration to ensure a high probability of correct evaluation of the maximum velocity direction and location.

Maximum Allowable Velocity

The maximum allowable velocity is as defined in Section 5.1.2.4 of [1].

Precautions

The fundamental relationship between allowable velocity and stress is developed from the assumption that the vibratory mode shape matches the mode shape at the first natural frequency. The user is cautioned against indiscriminate use of the velocity criterion without taking into account the frequency of velocity amplitudes and the mode shape of the vibration. For example, if the pipe span natural frequency was 20 Hz and the piping was vibrating at 10 Hz, the stresses predicted from a velocity measurement could be un-conservative by a factor equal to 2, without a C_5 correction factor as defined in Section 5.1.2.4 of C1).

For multiple span piping systems, usually encountered in process plant piping, special attention must be given to evaluating stresses induced by resonant excitation from adjacent spans. An estimate of the natural frequencies of characteristic spans must be made before deciding to use the velocity criterion. If the ratio of the first natural frequency of the span to the measured frequency is equal to or less than two, then the velocity method can be used. Values of this ratio higher than 2.0 are not covered by [1].

Velocity Criterion used for Screening

If conservative levels of the correction factors defined in [1] are combined, a criterion is obtained that indicates safe levels of vibrations for any kind of piping configuration. Using this criterion, different piping systems can be checked, and those that indicate vibratory levels with velocities lower than the screening values need no further analysis; for those that exceed the screening values, detailed analyses are required to establish their acceptability.

A screening value of 0.5 in./sec (12 mm/sec) has been established for velocities to be used in the screening process. Piping systems with maximum velocities lower than 0.5 in./sec (12 mm/sec) are considered safe with regard to vibratory stresses and do not require additional analysis to determine their acceptability.

The first step that must be made when speeds exceed 0.5 in./sec (12 mm/sec) is to determine more accurate values for the correction factors C_1, C_3, C_4, C_5 of [1] and the stress indexes C_2K_2 of the *ASME BPVC* Section III Subsection NB of Reference [3], in order to establish precisely the velocity criteria applicable to the piping system.

9.5 INACCESSIBLE PIPES

For inaccessible piping systems that require monitoring, the search procedure for locating maximum response is not necessary. Locations of the maximum expected response shall be defined during construction in order for the measurement devices to be mounted. Special attention will be given to the compatibility of the assumptions used to select the locations of the maximum expected response with the as-installed system response.

9.6 EVALUATION OF THE FATIGUE EFFECT DUE TO OPERATING VIBRATIONS BY ANALYSIS

ASME BPVC Section III Subsection NB [3] identifies the fatigue effect as a possible failure mode. The cyclic load effect that might lead to failure can be described as follows:

Due to the combined effect of the secondary and primary stresses, local stress concentrations and residual stresses, the stress values may locally exceed the yield limit.

If the yield limit is exceeded due to cyclic loads, then the plastic strain is accumulated for each load cycle leading to additional residual strain.

This process of plastic strain accumulation may continue during a number of cycles as ratcheting occurs until the ultimate strain is reached and a crack is initiated.

After the crack is initiated it will start to develop (crack growth) while the plastic strain accumulation is occuring in the vicinity of the crack (zone of high concentration of stresses or strains).

The initial cracks in the early stage after initiation typically are very small and very localized and are very difficult to detect using standard non destructive examination techniques.

The process is accelerated after the crack is initiated due to stress concentration in the vicinity of the crack.

The velocity of the crack growth depends on the vibration regime, material properties and loading conditions corrosion, erosion, residual mean stresses, etc.

Under actual conditions, particularly in a small pipe less than 150 mm (6 inches) in diameter, a critical crack length can develop and rupture occur before the crack is detected.

ASME Section III NB-3222.4 provides a method for estimating by analysis the number of cycles or life-duration function of the usage-factor-calculated function of number of load cycles over allowable number of load cycles. When the usage factor is equal to 1.0 it means that the component has spent 100% of its fatigue design life.

The usage factors could be evaluated for different type of load cycles and they can be combined as a cumulative usage factor.

9.6.1 Evaluation of the Usage Factor for Cycling Loads

According to ASME NB-3222.4 [3] the cumulative usage factor is:

$$U = \sum_{i=1}^{m} \frac{n_i}{N_i} \tag{9.5}$$

where U = cumulative usage factor; n_i = number of cyclic loads of type *"j"*; and N_j = allowable number of load cycles of type *"j"* corresponding to alternate stress, S_{1j}.

The peak stress S_p for pipe is defined by Eq. (11.7) of ASME Section III NB3652 as follows:

$$S_p = K_1 C_1 \frac{P_0 D_0}{2t} + K_2 C_2 \frac{D_0}{2I} M_i + \frac{1}{2(1-\nu)} K_3 E\alpha|\Delta T_1| + K_3 C_3 E_{ab}|\alpha_a T_a - \alpha_b T_b| + \frac{1}{1-\nu} E\alpha|\Delta T_2| \tag{9.6}$$

where C_1,C_2,C_3 = secondary stress indices for the specified component under investigation; K_1,K_2,K_3 = local indices for the specified component under investigation; D_o, t,I = outer diameter; t_n = nominal wall thickness; I = moment of inertia of the pipe cross section; E_{ab} = average modulus of elasticity of the two sides of a gross structural discontinuity or material discontinuity at room temperature; M_i = resultant range of moment, which occurs when the system goes from one service loads set to another; P_o = range of service pressure; T_a(T_b) = range of average temperature on side *a(b)* of gross structural discontinuity or material discontinuity; ΔT_2 = absolute value of the range for the portion of the nonlinear thermal gradient through the wall thickness not included in ΔT_1. ΔT_2 = absolute value of the range of the temperature difference of the outside surface T_0 and the temperature of the inside surface T_i of the piping product assuming equivalent linear temperature distribution; $\alpha_a(\alpha_b)$ = thermal expansion coefficient on side *a(b)* of a gross structural discontinuity or material discontinuity at room temperature; and v = poison ratio for pipe material.

The alternate stress for load cycle of type *"i"* is:

$$S_{alt} = S_p/2 \tag{9.7}$$

Using S_{alt} and applicable design fatigue curve (ASME Section III NB, Figure 1.9.1), find the corresponding number of allowable cycles, *N*. The usage factor, U_i, for load set *"i"* is given by:

$$u_i = n_i/N_i \tag{9.8}$$

The cumulative effect of all load sets is given by a summation of all U_i. If the operating vibrations govern the fatigue failure mode, a simplified evaluation can be performed considering only:

$$S_p = K_2C_2\ M_i/Z \tag{9.9}$$

where M_i = is the bending moment produced by the operating vibrations and Z in the elastic section modules.

9.7 CORRECTIVE ACTIONS

Possible corrective actions for reduction of stress-induced vibrations to within allowable limits may include:

1. Identification and reduction or elimination of the vibration source or excitation mechanism
2. Structural modification by changing frequency to avoid resonance effects
3. Change operating procedure to eliminate or reduce the transient's effect
4. Installation or modification of support to increase clamping to reduce vibrations.

If structural changes are proposed for reduction of operating vibrations, stress analysis should be performed again to check whether the acceptance criteria are met by the modified piping system.

9.7.1 Stress Analyses and Supplementary Measurements

Identification of the root cause and solutions for reduction of vibrations usually requires detailed testing and analysis. Also, detailed analyses could reduce the degree of conservatism associated with simplified evaluations. For example, the assessment of the vibration level using simplified methods could be refined to check the acceptance criteria using detailed analyses. The simplified methods are useful and efficient for a quick assessment and screening process.

Detailed evaluation involves linear/nonlinear finite-element dynamic analysis developed using analytical models calibrated based on test data and resultant. Detailed tests may include direct measurements of vibration-induced accelerations, displacements strains and stresses. The measured accelerations, displacements, or resultant strains and stresses could be used for calibration of the analytical models in order to reduce the conservatism of the evaluation. Calibrated analytical models are used to study the optimum solution for reduction of the vibration level to within the allowable limits. In some special cases it could be necessary to use a continual monitoring system during plant operation and to prescribe (based on detailed analyses) warning and alarm limits and associated operating instructions.

9.8 USE OF PIPEWORK VISCOUS DAMPERS TO REDUCE VIBRATION OF PIPING

Pipework viscous dampers have seen successful application to the vibration restraint of piping in process facilities and power plants primarily in Europe for the past 40 years. They also have the capability similar to snubbers of restraining pipe due to the sudden application of load or deformation resulting from earthquakes,

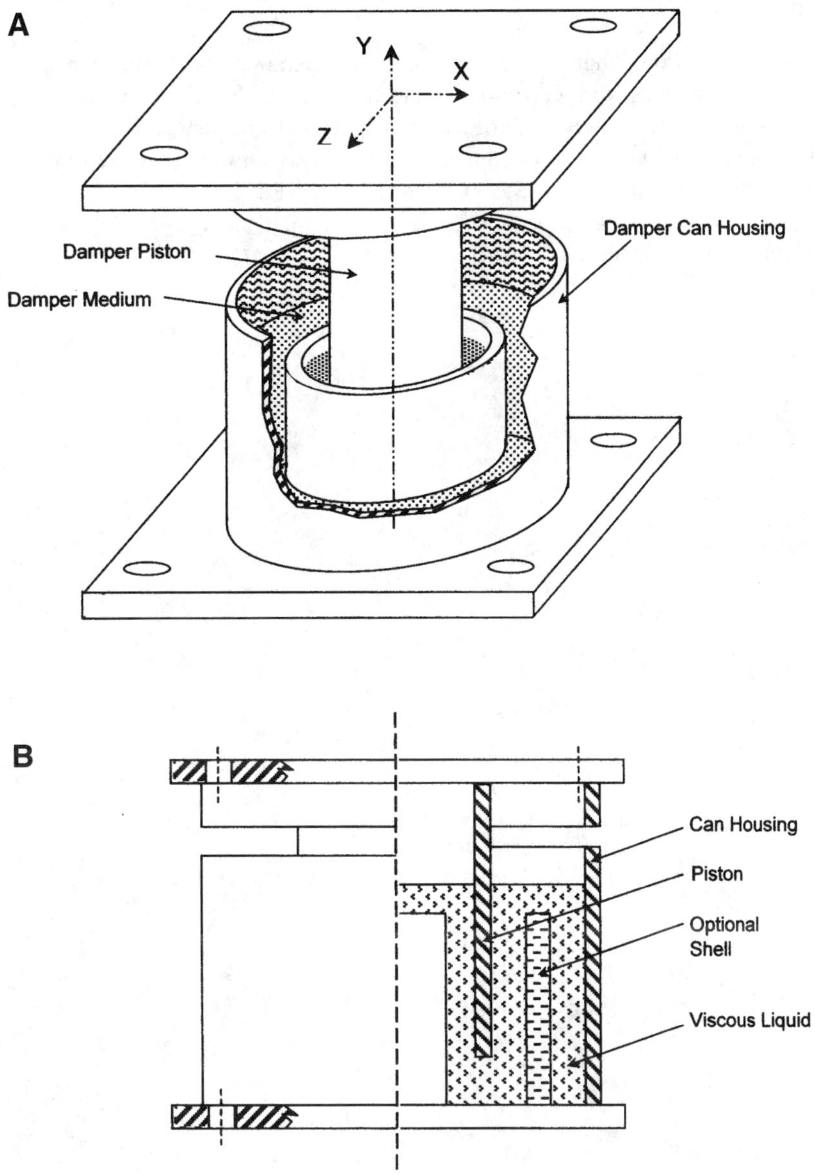

FIGURE 9.4. VISCOUS DAMPER DYNAMIC RESTRAINT

pipe rupture, water and steam hammer, while not restraining the pipe due to application of slowly applied thermal deformation. They have the added advantage of being effective in resisting or reducing vibratory effects on piping where some snubber designs because of relatively large dead bands have not been effective against vibration.

In Figure 9.4 is a sketch of a typical damper-type restraint. The device comprises two parts: The part attached to the pipe is an attachment plate plus a piston. The part attached to the supporting structure is an open pot containing a viscous fluid, and in some models, thin shells attached to the base of the pot that are used to increase the cavitation load and stiffness of the device in its resistance to suddenly applied load or deformation. Commercial models of the device are capable of restraining sudden motion in all six degrees-of-freedom. In such devices restraint of motion varies from ± 0.75 to 7.5 inches with sudden application of the load, which ranges from 0.1 to 65 kips. The device is also capable of providing a significant increase in piping system damping.

REFERENCES

1. *ASME OM-S/G-2000,* 2000, "Standards and Guides for Operation and Maintenance of Nuclear Power Plants – Part 3 Requirements for Preoperational and Initial Start-Up Vibration Testing of Nuclear Power Plant Piping Systems," American Society of Mechanical Engineers, New York, NY.
2. AISC, 1989, *Manual of Steel Construction-Allowable Stress Design, Appendix K,* "Strength Design Considerations," American Institute of Steel Construction, *9th* Ed. 1989.
3. *ASME Boiler and Pressure Vessel Code,* 2004, "Nuclear Components Class I, Subsection NB," American Society of Mechanical Engineers, New York, NY.

CHAPTER

10

SEISMIC EVALUATION OF BURIED PIPELINES, TUNNELS AND VAULTS

Buried pipeline systems are commonly used to transport water necessary for heat removal and fire protection safety function for hazardous facilities. In some instances they may also conduct radioactive or other toxic and pathogenic fluids.

Safety-related cables such as emergency power supply and instrumentation and control lines between buildings are often routed in buried reinforced-concrete tunnels or vaults. All these types of structures are referred to in this chapter as buried systems.

These buried systems are often required to remain functional in case of an earthquake not exceeding a design-basis earthquake. This means that failure modes are considered that may induce large deformation in the pipes, tunnels or vaults; or that may produce pipe, tunnel or vault support failure; or that may produce tunnel or vault wall or roof collapse over the pipes or cables.

An important characteristic of buried systems' seismic loading, which distinguishes them from above-ground structures, is that the principal seismic loads are transmitted through soil deformation resulting from seismic wave passage and the inertia forces due to the weight of the buried systems, and its content, discussed in Chapter 5, are relatively unimportant.

It should be understood the seismic deformations and strains induced in buried systems in the absence of differential soil motion or faulting are a function of deformation of the soil as the result of wave passage, and hence, are deformation-limited and could be defined as producing secondary stresses. However, it is common practice to require such stresses to meet primary stress limits of the construction code applicable to the buried system. While ASME codes used to design piping recognize the difference between primary and secondary stress limits, and the potential for failure the ASCE and ACI codes generally used to design buried concrete tunnels and vaults in general do not.

The procedures and methodology contained in this chapter are based on a book authored by O'Rourke and Liu [1]. This book should be consulted to develop a more thorough understanding of the material presented in this chapter.

10.1 WAVE PROPAGATION

There are two types of seismic waves: body waves and surface waves. The body waves propagate through earth while the surface waves travel along or near the ground surface. Body waves include compression waves (*P*-waves), with particle motion in the direction of wave propagation, and shear waves (*S*-waves), with particle motion perpendicular to the direction of wave propagation typical of ocean wave motion, while the surface waves include Love waves (*L*-waves) and Rayleigh (R-waves) with the same respective particle motion.

When a seismic wave travels along the surface ground, any two points located along the propagation path will experience differential displacement motions. These motions may induce axial and bending strains in a buried system due to buried system-soil interaction.

The main goal, in order to assess the buried system stresses induced by wave motion, is to determine the soil buried system interaction parameters as well as loading, which is the ground deformation expressed as axial strain and/or curvature.

10.1.1 Effective Propagation Velocity—Body Waves

Body wave motions considered are usually limited to *S*-waves since *S*-waves transmit most of the energy and tend to generate larger ground motion than *P*-waves. The apparent propagation velocity for *S*-waves is:

$$C = \frac{C_s}{\sin \gamma_s} \qquad (10.1)$$

where γ_s = incidence angle of *S*-waves with respect to the vertical; and C_s = shear-wave velocity of the soil media.

The apparent propagation velocity for *S*-waves in hard rock typically range from 2,100 to 5,300 m/sec, with an average of about 3,400 m/sec. as shown in Figure 10.1. In soil deposits, depending on the stiffness of the soil, the *S*-wave propagation is much slower. Table 10.1 presents shear-wave velocities for various types of foundation media, which are taken from Table 1615.1 of [1].

TABLE 10.1. SITE FORMATION CLASS DEFINITIONS AND SHEAR-WAVE VELOCITIES FROM [10]

		Average Properties in Top 100 feet, as per Section 1615.1.5		
Site Class	**Soil Profile Name**	**Soil shear wave Velocity, $\bar{v}_s$, (ft/s)**	**Standard penetration resistance, $\bar{N}$**	**Soil undrained shear strength, $\bar{s}_u$, (psf)**
A	Hard rock	$\bar{v}_s > 5,000$	N/A	N/A
B	Rock	2,500 < $\bar{v}_s$ ″ 2,500	N/A	N/A
C	Very dense soil and soft rock	1,200 < $\bar{v}_s$ ″ 2,500	$N > 50$	$\bar{S}_u \geq 2,000$
D	Stiff soil profile	600 ″ $\bar{v}_s$ ″ 1,200	15 ″ N ″ 50	1,000 ″ $\bar{S}_u$ ″ 2,000
E	Soft soil profile	$v_s < 600$	$N < 15$	$\bar{S}_u < 1,000$
E	—	Any profile with more than 10 feet of soil having the following characteristics: 1. Plasticity index (PI) > 20 2. Moisture content $w \geq 40\%$ 3. Undrained shear strength $\bar{S}_u < 500$ psf		
F	—	Any profile containing soils having one or more of the following characteristics: 1. Soils vulnerable to potential failure or collapse under seismic loading such as liquefiable soils, quick and highly sensitive clays, collapsible weakly cemented soils. 2. Peats and/or highly organic clays (H >10 feet of peat and/or highly organic clay where H = thickness of soil) 3 Very high plasticity clays [H >25 feet with plasticity index (PI)>75] 4. Very thick soft/medium stiff clays (H>120 feet)		

Note: For SI: 1 foot = 304.8 mm, 1 square foot = 0.0929m^2, 1 pound per square foot = 00.0479 kPa. N/A = not applicable.

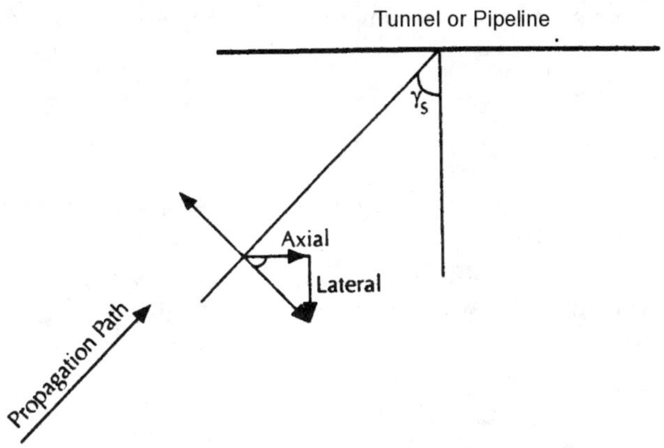

FIGURE 10.1. *S*-WAVE PROPAGATION.

10.1.2 Effective Propagation Velocity—Surface Waves

For surface waves, usually only *R*-waves are considered since *L*-waves normally generate bending strains in buried systems that are significantly less than axial strains induced by *R*-waves. Since *R*-waves always travel parallel to the ground surface, the phase velocity of *R*-waves, C_{ph}, is the apparent propagation velocity. The phase velocity is defined as the velocity at which a transient vertical disturbance at a given frequency, originating at the ground surface, propagates across the surface of the medium. The phase velocity is a function of the variation of the shear-wave velocity with depth and also is a function of frequency.

The variation with frequency of phase velocity (typically quantified by a dispersion curve) for layered soil profiles in which the shear-wave velocity increases with depth is:

$$C_{ph} = \begin{cases} 0.875C_H & \dfrac{H_s f}{C_L} \le 0.25 \\ 0.875C_H - \dfrac{0.875C_H - C_L}{0.25}\left(\dfrac{H_s f}{C_L} - 0.25\right) & 0.25 < \dfrac{H_s f}{C_L} \le 0.50 \\ C_L & \dfrac{H_s f}{C_L} > 0.50 \end{cases} \tag{10.2}$$

where f = frequency (in Hz); for long wavelengths, C_L = mean shear-wave velocity of the surface layers; C_L = for short wavelengths, C_L = shear-wave velocity of the first layer; C_H = for long wavelengths, C_H = shear-wave velocity of the half space; for short wavelengths, C_H = shear-wave velocity of the second layer; and H_s = thickness of the surface layers.

10.2 GROUND STRAIN AND CURVATURE DUE TO WAVE PROPAGATION

Newmark [3] developed a simplified procedure to estimate the ground strain. He considered a simple traveling wave with a constant wave shape. Newmark's approach is based on three assumptions:

1. The ground motion (acceleration, velocity and displacement time histories) at two points along the propagation path are assumed to differ only by a time lag.
2. System inertia terms are small and may be neglected.
3. There is no relative movement at the soil-system interface, and hence, the buried system strain equals the ground strain.

The maximum ground strain ε_g (tension and compression) in the direction of wave propagation is given by:

$$\varepsilon_g = \frac{V_m}{C} \tag{10.3}$$

where ε_g = maximum ground strain (tension and compression); V_m = maximum horizontal ground velocity; and C = propagation velocity of the seismic wave.

The maximum ground curvature, Φ_g, which is the second derivative of the transverse displacement with respect to distance, is given by:

$$\Phi_g = \frac{A_m}{C^2} \tag{10.4}$$

where Φ_g = maximum ground curvature; A_m = maximum horizontal ground acceleration perpendicular to the direction of wave propagation; and C = propagation velocity of the seismic wave.

10.3 SIMPLIFIED APPROACHES TO ASSESS THE STRAIN AND CURVATURE OF A STRAIGHT CONTINUOUS BURIED SYSTEM DUE TO WAVE PROPAGATION

10.3.1 Newmark Approach

A simplified procedure for assessing the response due to wave propagation can be developed based on the Newmark approach. Taking into account the previous assumptions, the ground strain parallel to the system axis is:

$$\varepsilon_g^{C_s} = \frac{V_m}{C} = \frac{V_m}{C_s}\sin\gamma_s;\ \varepsilon_g^{C_{ph}} = \frac{V_m}{C_{ph}} \tag{10.5}$$

where $\varepsilon_g^{C_s}$ = maximum ground strain induced by shear wave; V_m = peak ground velocity; C_s = shear-wave velocity; C = apparent propagation velocity with respect to the ground surface and the system axis; $\varepsilon_g^{C_{ph}}$ = maximum ground strain induced by *R*-wave; and C_{ph} = phase velocity.

It should be noted that the above equations overestimate system strain; for most cases slippage occurs at the system-soil interface and the member strain is less than the ground strain.

Since the maximum effects of *R*-wave and *S*-wave cannot occur in the same moment and place, only the maximum ground strain between the two is considered:

$$\varepsilon_{system} = \varepsilon_g = \max\left(\varepsilon_g^{C_s}, \varepsilon_g^{C_{ph}}\right) \tag{10.6}$$

Similarly, for the evaluation of ground curvature:

$$\Phi_{system} = \Phi_g = \max\left(k_g^{C_s} = \frac{A_m}{C^2}, k_g^{C_{ph}} = \frac{A_m}{C_{ph}^2}\right) \tag{10.7}$$

10.3.2 M. O'Rourke and EL Hmadi Approach (Applicable to Pipe)

A different approach was developed by M. O'Rourke and El Hmadi [4] (If the ground strain between two points separated by a distance L_s is modeled by a sinusoidal wave with wavelength $\lambda = 4L_s$.)

As long as the pipe strain is below its yield strain and the maximum force in the soil spring is less than the frictional resistance at the pipe-soil interface, the strain is:

$$\varepsilon_p = \frac{\pi}{2}\varepsilon_g^{ave}\frac{\beta^2}{\beta^2+\left(\frac{\pi}{2L_s}\right)^2}\cos\frac{\pi x}{2L_s} \tag{10.8}$$

where $\beta^2 = (K_g/AE)$; $K_g = [t_u/(x_u/2)]$; L_s = separation distance ($\lambda/4$); K_g = ground spring stiffness for axial movement of the pipe; A = cross-sectional area of the pipe; E = elasticity modulus of the pipe; ε_g^{ave} = average ground strain over a separation distance L_s; t_u = maximum resistance in horizontal axial direction; and x_u = maximum elastic deformation in horizontal axial direction.

10.4 FRICTION STRAIN MODEL GROUND STRAINS AND DISPLACEMENTS DEVELOPED ALONG THE BURIED SYSTEM

Considering the case where slippage occurs over the whole system's length, the uniform frictional force per unit length is t_u; therefore, the maximum buried system strain due to friction is given by:

$$\varepsilon_p = \frac{t_u L_s}{AE} \tag{10.9}$$

10.4.1 Straight Continuous Buried Systems—Cross-Section Resultant Forces and Moments

Once the maximum axial ground strain, ε_g, is calculated, the axial force N induced by the ground strain shown in Figure 10.2 is:

$$\sigma_{\text{pipeline}} = \varepsilon_g \cdot E \tag{10.10}$$

$$N_{\text{pipeline}} = \varepsilon_g \cdot EA \tag{10.11}$$

If the buried system deflects under uniform curvature, the inverse of the radius of curvature at any point of the elastic curve is obtained as:

$$\frac{1}{R} = \frac{M}{EI} \Rightarrow M = \frac{EI}{R} = \phi \cdot EI \tag{10.12}$$

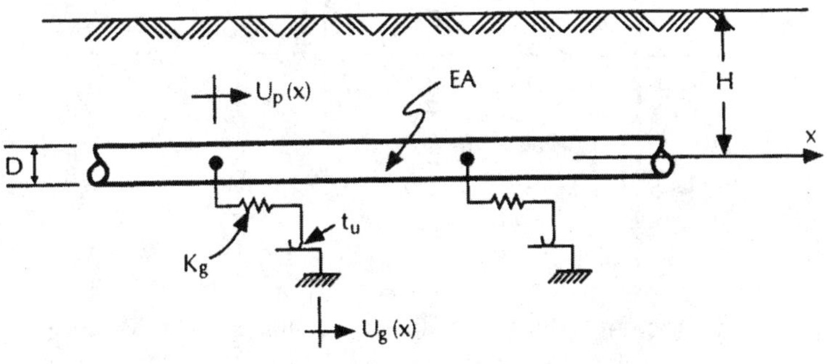

FIGURE 10.2. CONTINUOUS PIPELINE MODEL.

in which ϕ = curvature (defined as the angle change per unit length at any given location along the axis of a buried system subjected to bending loads); M = bending moment at the point; and EI = flexural rigidity of the cross section. Based on previous assumptions (no relative movement between soil and buried system etc.), the buried system curvature can be equated to the ground curvature; hence:

$$M_{pipeline} = \Phi_g \cdot I \tag{10.13}$$

10.4.2 Buried System Bends and Tees—Cross-Section Resultant Forces and Moments

Assuming a buried system moving with the soil at the location with zero ground movement (point B in Figure 10.3), Shinozuka and Koike [4] developed simple equations to estimate buried system strain at bends. The total ground deformation within the quarter wavelength is:

$$\Delta g = \frac{\lambda}{2\pi} \cdot \varepsilon_g \tag{10.14}$$

The axial force in element 1 of Figure 10.3 (shear force in element 2), S, can be expressed by:

$$N = \frac{3\lambda}{8\pi} \cdot \frac{K_g}{\zeta} \cdot \frac{1}{1+Q}(1-\beta_c)\varepsilon_g \tag{10.15}$$

where $\zeta = \sqrt[4]{K_g / 4EI}$; and $Q = (3/16)(K_g \lambda / AE\zeta)$.

Critical shear strain is:

$$\gamma_{cr} = \frac{t_u}{\pi D G_s} \tag{10.16}$$

Maximum shear strain at pipe soil interface:

$$\gamma_0 = \frac{2\pi}{\lambda} \frac{Et}{G_s} \varepsilon_g \beta_0 \tag{10.17}$$

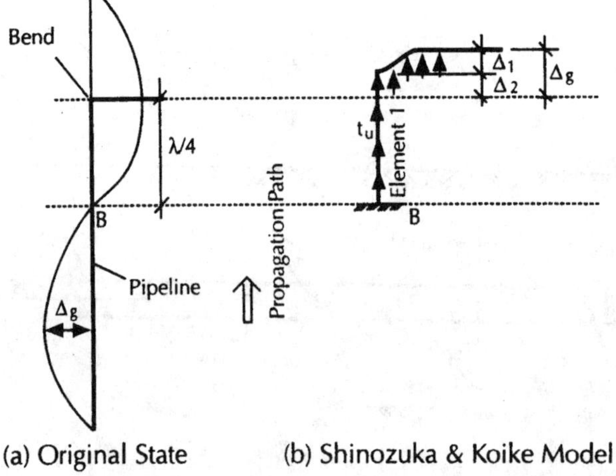

FIGURE 10.3. DISPLACEMENT AND FORCES IN PIPE AT ELBOW.

The ground-to-pipe conversion factor is:

$$\beta_0 = \frac{1}{1+\left(\frac{2\pi}{\lambda}\right)^2 \cdot \frac{AE}{K_g}} \tag{10.18}$$

If $\gamma_{cr} > \gamma_0$, then no slippage occurs at the buried system soil interface; therefore, the conversion factor β_c is equal to β_0. If the shear strain at the buried system soil interface, γ_0, is larger than the critical shear strain, γ_{cr}, the ground-to-pipe conversion factor is

$$\beta_c = q \cdot \frac{\gamma_{cr}}{\gamma_0} \cdot \beta_0 \tag{10.19}$$

where q = a factor that ranges from 1 to $\pi/2$ and quantifies the degree of slippage at the buried system soil interface. Conservatively, $q = 1$ has been assumed.

The axial strain for a straight buried system is

$$\varepsilon_p = \beta_c \cdot \varepsilon_g \tag{10.20}$$

The bending moment at bend, M, is:

$$M = \frac{N}{3\zeta} \tag{10.21}$$

The flexural displacement, Δ_1, at bend is:

$$\Delta_1 = \frac{4\zeta N}{3K_g} \tag{10.22}$$

10.5 FINITE-ELEMENT APPROACH TO EVALUATE SECTIONAL FORCES ACTING ON BURIED SYSTEMS

In order to evaluate the section forces acting on buried systems subject to wave propagation, a series of finite-element models can be used. Axial and lateral soil springs are used to model the interaction at the buried system soil interface as shown in Figure 10.4. The quasi-static seismic excitation is modeled by displacing the bases of the soil springs with respect to the propagation path. The movement of the soil spring bases parallel to the propagation path varied along the buried system matching a sinusoidal pattern characterized by wavelength, λ, and maximum soil deformation, Δg.

10.5.1 Load-Deformation Relations at Buried System Soil Interface

Soil interaction forces for a buried system surrounded by competent, nonliquefied soil are well established. The ASCE Technical Council on Lifeline Earthquake Engineering (TCLEE) Committee on Gas and Liquid Fuel Lifelines [6] has suggested idealized elastoplatic models for the purpose of analysis. (Note that an elastoplastic model is fully characterized by three parameters: the maximum resistance, t_u, p_u, q_u, in the horizontal axial, horizontal transverse and vertical transverse directions, respectively, having units of force per unit length and the maximum elastic deformation, x_u, y_u, z_u, having units of distance.)

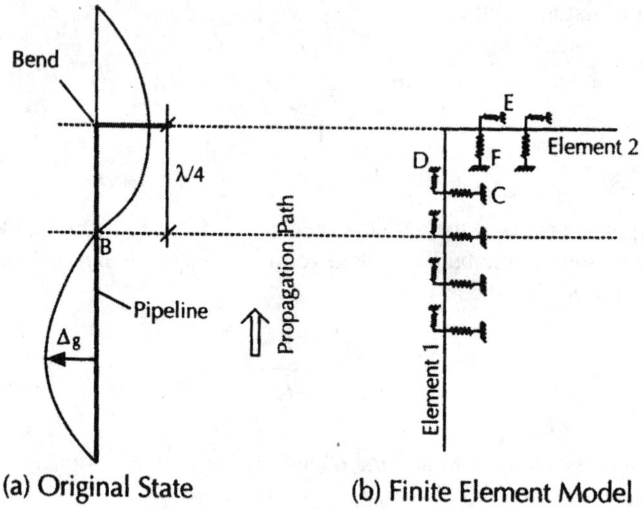

FIGURE 10.4. FINITE-ELEMENT MODEL FOR ELBOW SUBJECT TO WAVE PROPAGATION.

10.5.2 Longitudinal Movement

Relative movement parallel to the buried system axis results in longitudinal (horizontal axial) forces at buried system soil interface. The elastoplastic spring parameters are given as follows:
For sand:

$$t_u = \frac{\pi}{2} D\gamma H \cdot (1 + k_0) \cdot \tan(k\phi) \tag{10.23}$$

$$x_u = (0.1\sim0.2) \text{ in} = 2.54\sim5.08 \times 10^{-3} \text{ m} \tag{10.24}$$

For clay:

$$t_u = \pi D \alpha S_u \tag{10.25}$$

$$x_u = (0.2\sim0.4) \text{ in} = 5.08\sim10.16 \times 10^{-3} \text{ m} \tag{10.26}$$

where D = pipe diameter; γ = effective unit of the soil; H = depth to centerline of the buried system; ϕ = angle of shear resistance of the sand; k_0 = coefficient of lateral soil pressure at rest (for normally consolidated cohesionless soil, k_0 = 0.35–0.47; as a conservative estimate k_0 = 1.0) (reduction factor depending on the outer-surface characteristics and hardness of the buried system); k = – concrete pipe, steel or cast iron pipe with cement coating: k = 1.0; – cast iron or rough steel: k = 0.7~1.0; smooth steel, pipe with smooth relatively hard coating: k = 0.5 ~ 0.7; α = empirical adhesion coefficient varying with S_u, taken from Figure 10.5. S_u = undrained shear strength of the surrounding soil; and m = meters.

10.5.3 Horizontal Transverse Movement

Relative movement perpendicular to the buried system axis in the horizontal plane results in horizontal transverse forces at the buried system soil interface. The elastoplastic spring parameters are given as follows:

For sand:

$$p_u = \gamma H N_{qh} D$$

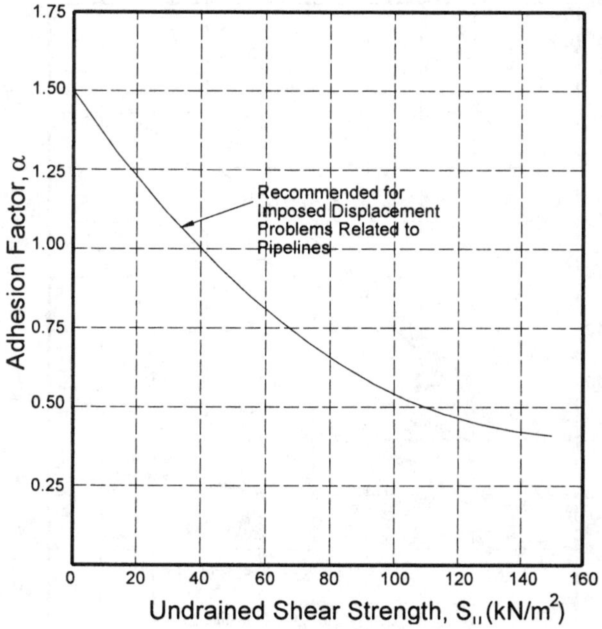

FIGURE 10.5. RECOMMENDED FOR IMPOSED DISPLACEMENT PROBLEMS RELATED TO PIPELINES.

$$y_u = \begin{cases} (0.07 \sim 0.10)(H + D/2) & \text{for loose sand} \quad (10.27) \\ (0.03 \sim 0.05)(H + D/2) & \text{for medium sand} \quad (10.28) \\ (0.02 \sim 0.03)(H + D/2) & \text{for dense sand} \quad (10.29) \end{cases}$$

For clay:

$$p_u = S_u N_{ch} D \tag{10.30}$$

$$y_u = (0.03 \sim 0.05)(H + D/2) \tag{10.31}$$

where N_{qh} = horizontal bearing capacity factor for sand, taken from the figure; N_{ch} = horizontal bearing capacity factor for clay; N_{ch} is equivalent to N_{qh} with ϕ=30°; S_u = undrained shear strength of the surrounding soil; H = depth to centerline of the buried system; and D = buried system diameter, as shown in Figure 10.6.

10.5.4 Vertical Transverse Movement, Upward Direction

Relative upward movement perpendicular to the buried system axis results in lateral forces at the buried system soil interface. The elastoplastic spring parameters are given as follows:

For sand:

$$q_u = \gamma H N_{qv} D \tag{10.32}$$

$$z_u = (0.01 \sim 0.015)H \tag{10.33}$$

For clay:

$$q_u = S_u N_{cv} D \tag{10.34}$$

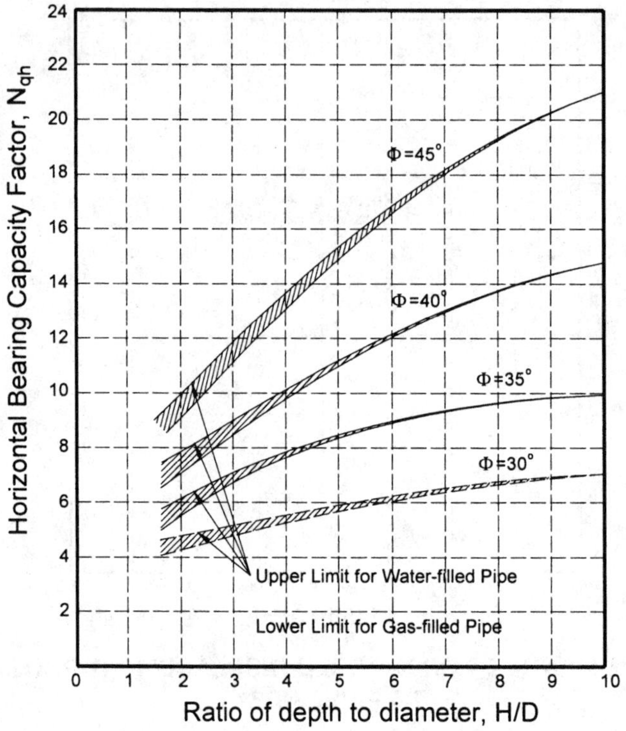

FIGURE 10.6. HORIZONTAL BEARING CAPACITY AS A FUNCTION OF ANGLE AND DEPTH-TO-DIAMETER RATIOS.

$$z_u = (0.1 \sim 0.2)H \tag{10.35}$$

where N_{qv} = vertical uplift factor for sand, presented as a function of depth over diameter ratio; N_{cv} = vertical uplift factor for clay, presented as a function of depth over diameter ratio; S_u = undrained shear strength of the surrounding soil; H = depth to centerline of the buried system; and D = buried system diameter, as shown in Figure 10.7.

10.5.5 Vertical Transverse Movement, Downward Direction

Relative downward movement perpendicular to the buried system axis in the vertical plane results in lateral forces at the buried system soil interface. The elastoplastic spring parameters are given as follows:
For sand:

$$q_u = \gamma H N_q \mathrm{D} + \frac{1}{2}\gamma D^2 N_y \tag{10.36}$$

$$z_u = (0.10 \sim 0.15)D \tag{10.37}$$

For clay:

$$q_u = S_u \mathrm{N_c D} \tag{10.38}$$

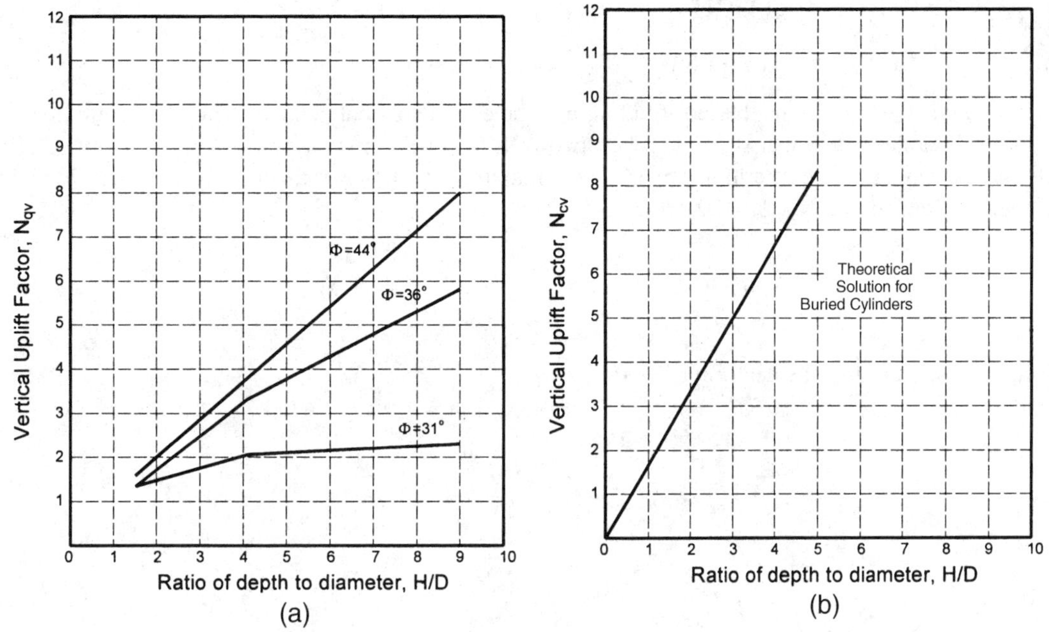

FIGURE 10.7. VERTICAL UPLIFT AS A FUNCTION OF DEPTH TO DIAMETER

$$z_u = (0.10 \sim 0.15)H \tag{10.39}$$

N_q = bearing capacity factors for horizontal strip footings on sand loaded in vertically N_y downward direction, as shown in Figure 10.8; N_c = bearing capacity factor for horizontal strip footings on clay; S_u = undrained shear strength of the surrounding soil; H = depth to centerline of the buried system; and D = buried system diameter.

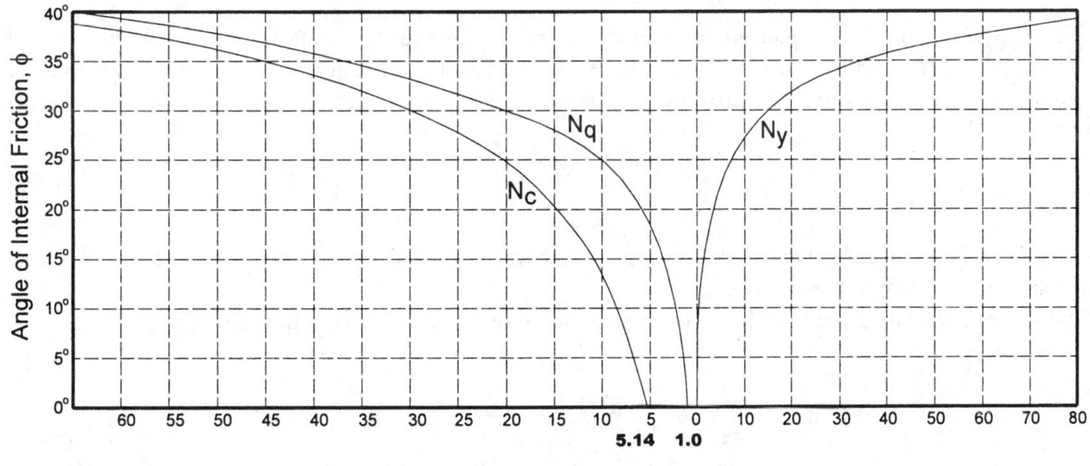

FIGURE 10.8. BEARING CAPACITIES AS A FUNCTION OF ANGLE OF INTERNAL FRICTION.

10.6 ACCEPTANCE CRITERIA

10.6.1 Straight Continuous Pipeline

The stress state at any point in a buried structure may be completely defined by giving the magnitudes and directions of the three principal stresses. When two or three of these stresses are different from zero, the proximity of yielding can be determined by means of maximum shear stress theory (Tresca criterion) including a capacity reduction factor $\phi = 0.9$, which is expressed:

$$\left. |\sigma_1 - \sigma_2| \, , \, |\sigma_2 - \sigma_3| \, , \, |\sigma_3 - \sigma_1| \, \right|_{\text{MAX}} = 0.9 \cdot Sy \tag{10.40}$$

The principal stresses may be determined as follows:

$$\sigma_{\text{axial}} = \left. \left(\frac{N}{A} \, , \frac{M}{Z} \right) \right|_{\text{MAX}} \tag{10.41}$$

$$\sigma_{\text{hoop}} = \frac{P_{\max} D_o}{2t_n} \tag{10.42}$$

$$\sigma_p = -P_{\max} \tag{10.43}$$

where N = maximum axial force due to ground strain; M = maximum bending moment due to ground curvature (note that N and M are not acting simultaneously), A = pipe cross-section area; D_o = nominal outside diameter of pipe; $P_{\max}$ = peak pressure; and t_n = wall thickness.

10.6.2 ACCEPTANCE CRITERIA—PIPELINE BENDS AND TEES

The effects of pressure, weight, other sustained loads and occasional loads, including earthquake loads, is given by:

$$S_{OL} = B_1 \frac{P_{\max} \cdot D_o}{2t_n} + B_2 \left(\frac{M_A + M_B}{Z} \right) \tag{10.44}$$

where D_o = nominal outside diameter of pipe; $P_{\max}$ = peak pressure; t_n = wall thickness; Z = pipe section modulus; B_1 and B_2 = primary stress indexes; M_B = resultant moment loading on cross section due to occasional loads; and M_A = resultant moment loading on cross section due to weight and other sustained loads.

$$M_A, M_B = \sqrt{M_1^2 + M_2^2 + M_3^2} \tag{10.45}$$

M_1, M_2, M_3 = orthogonal moment loading components at a giving position in the piping system with respect to the applied resultant moment loading.

The stress ratio (SR), calculated based on consideration of level D service limits, $S_{OL} < \min(3S_h; 2S_y)$, as given below:

$$SR = \frac{S_{OL}}{\min(3S_h; 2S_y)} \tag{10.46}$$

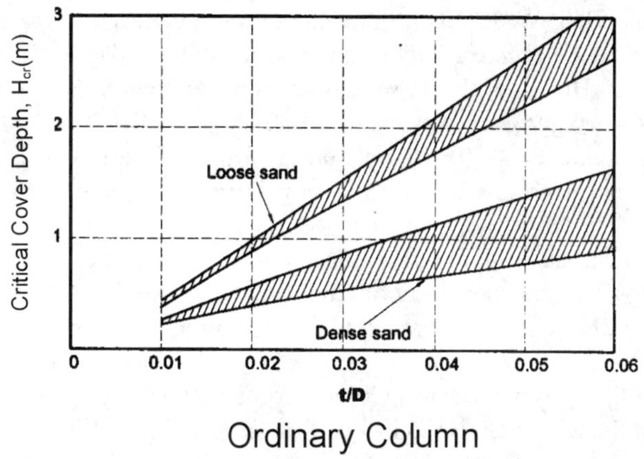

FIGURE 10.9. CRITICAL COVER DEPTH TO AVOID BEAM BUCKLING.

10.6.3 Buckling Failure Criteria of Pipelines

Local Buckling

Local buckling (wrinkling) involves local instability of the pipe wall. After initiation of local shell wrinkling, all further geometric distortion caused by ground deformation or wave propagation tends to concentrate at the wrinkle. In terms of failure criterion, the onset of wrinkling occurs at strains in the range:

$$0.15 \cdot \frac{t}{R} \leq \varepsilon_{cr} \leq 0.20 \cdot \frac{t}{R} \tag{10.47}$$

Beam Buckling

Beam buckling of a pipeline is similar to Euler buckling of a slender column in which the pipe undergoes a transverse upward displacement. The relative movement is distributed over a large distance and hence the compressive pipe strains are not large.

Figure 10.9 shows the critical cover depth for grade B and X-60 steel pipes. Any pipe buried with less cover than the critical depth would experience beam buckling before local buckling.

10.7 ACCEPTANCE CRITERIA—CONCRETE PIPE TUNNELS AND VAULTS

Acceptance criteria for concrete buried structures can be found in ACI Standard [7].

10.8 ALTERNATE DESIGN PROCEDURE

REFERENCES

1. O'Rourke, M. J. and Liu, X., 1999, "Response of Buried Pipelines Subject to Earthquake Effects," *Monograph Series 3,* Multidisciplinary Center for Earthquake Engineering Research, University at Buffalo, New York.
2. *ASCE-7-02,* 2002, "Minimum Design Loads for Buildings and Other Structures," American Society of Civil Engineers, Reston, VA.

3. Newmark, N. M. and Hall, W. J., 1975, "Pipeline Design to Resist Large Fault Displacement," *Proc., 1975 U.S. Nat. Conf. on Earthquake Engerg.,* Ann Arbor, MI, pp. 416–425.
4. O'Rourke, M. J. and El Hmadi, K. E., 1988, "Analysis of Continuous Buried Pipelines for Seismic Wave Effects," *J. Earthquake Engerg. and Struc. Dyn.,* Vol. 16, pp. 917–929.
5. Shinozuka, M. and Koike, T., 1979, "Estimation of Structural Strains in Underground Lifeline Pipes," *Lifeline Earthquake Engineering – Buried Pipelines, Seismic Risk, and Instrumentation, PVP-34,* ASME, pp. 31–48.
6. ASCE Committee on Gas and Liquid Fuel Lifelines, 1984, "Guidelines for the Seismic Design of Oil and Gas Pipeline Systems," American Society of Civil Engineers, Reston, VA.
7. *ACI-318,* 2002, "Building Code Requirements for Reinforced Concrete," American Concrete Institute, Detroit, MI.
8. ASCE Standard 43-05, Seismic Design Criteria for Structure, Systems and Components in Nuclear Facilities, American Society of Civil Engineers, Reston, VA.

CHAPTER

11

Quality Assurance and Control in Construction and Procurement of Safety-Related Structures, Systems and Components in Hazardous Facilities

11.1 INTRODUCTION

Hazardous facilities are defined as those facilities that process, store, handle or transport detonable, toxic or pathogenic materials and waste in a form and quantity such that a life-threatening or serious injury hazard potentially exists for collocated workers, the public, or significant damage to the environment. These facilities are divided into three hazard categories as described in Chapter 1. For safety-related structures, systems and components (SSC) in the more hazardous facilities (hazard category 1 and SSC performance categories 4 and 3) characteristic of high-hazard nuclear facilities, a formal written QA program plan should be required. Hazard category 2 and performance category 2 SSC characteristic of petrochemical and biomedical and moderate hazard nuclear facilities also should have a written, somewhat less formal, QA plan. For the lower and PC-1 and PC-0 SSC, a formal written QA plan is not usually required. This is typically referred to as the graded approach to QA. It should be understood the QA usually is the procedures and documentation that ensure that the QC examination and tests contained in applicable industry standards or owner specifications are properly implemented and evaluated.

Section 11.2 discusses Safety Classification and a graded and defense-in-depth approach to QA. Section 11.3 discusses quality in the constructed project, Section 11.4 provides observations, conclusions and recommendations, and Section 11.7 provides References. Appendix 11.A provides a detailed comparison of *ISO-9001-94* [1] with *NQA-1-2000-10CFR50* [2] Appendix B and *10CFR830.122* [3] which define the basic nuclear QA requirements in the U.S. Appendix 11.B of this chapter provides details concerning the existing Nuclear Procurement Issues Committee (NUPIC) and the role this organization plays in auditing nuclear safety-related product, material and services supplied to nuclear power plants in the U.S. Appendix 11.C is a comparison of *ISO-9001-94* requirements and *ASME NQA-1* requirement prepared by ASME. Appendix

11.D contains a summary of IAEA QA requirements in comparison to British nuclear QA requirements. Appendix 11.E provides a discussion of QA excesses that have provided the basis for less emphasis on document QA and more emphasis on project and facility performance responsibility for QA. Finally, Appendix 11.F is the Forward to a typical QA program that identifies the different levels of QA that are part of a graded approach to QA. It also distinguishes between the format and content references for a QA program manual applicable to a professional services organization which typically follow the 10CFR832.122 [3] or IAEA QA [4] and that which is meant to guide a vendor of nuclear safety related products and services [2] which generally follows the QA format of the USNRC[4].

By far the most formal application of QA has been its application in the nuclear industry.

11.2 SAFETY-RELATED CLASSIFICATIONS

Historically safety-related SSC in high and moderate-hazard facilities have required a level of quality including QA and QC above that required for ordinary industrial applications.

11.2.1 Nuclear Hazard Facilities

Nuclear facilities are typically defined into three hazard categories; 1, 2 and 3 which correspond to high, moderate and low hazard.

High hazard are typically large or high energy density core power nuclear reactors requiring, forced-cooling of the case during operation and in the core in the shutdown condition and spent fuel and high-level radioactive waste or isotope processing or storage facilities that require forced cooling.

Moderate-hazard nuclear facilities are typically defined as small or relatively low-energy density reactor spent fuel storage facilities and nuclear waste and isotope processing or storage facilities not requiring forced cooling whose unmitigated release of radioactivity is within facility site boundary regulatory limits.

Low-hazard facilities are typically radioactive material storage or process facilities whose unmitigated release would approach yearly normal radiological worker dose limits at the facility boundary.

11.2.2 Petrochemical Hazard Facilities

Principal among petrochemical high-hazard facilities are large liquefied natural gas storage or process facilities whose unmitigated release of LNG in gaseous form and its subsequent deflegrations or detonation could cause mass casualties on the site and beyond the site boundary. If anticipated casualties are limited to site, the facility should be considered a moderate-hazard facility. Low-hazard facilities are those whose casualties of a postulated release are limited to a small area of the site. Other petrochemical facilities are typically considered high, moderate or low- hazard as a function of the size of the site and beyond where casualties are expected to occur in the event of a postulated unmitigated release.

11.2.3 Biomedical Hazard Facilities

These facilities that process or store material whose unmitigated release could cause casualties at the facility boundary, should be at least identified as moderate-hazard facilities. Otherwise, they should be identified as low-hazard facilities.

11.2.4 Graded Approach and Defense-in-Depth

It should be understood that in hazardous construction and procurement projects typically much less than half of their SSC are classified as safety-related. Hence, in such projects most items being constructed or procured are normal commercial grade that historically have not required a formal QA program. Typically, desired quality for ordinary commercial items is assured by adherence to a series of standard operating procedures which are enforced by project or facility management and which must comply with the QC requirement of applicable industrial standards.

The level of QC and QA required is directly related to the magnitude and consequences of hazards and incorporates considerations of the nature and complexity of the facility or activity. A higher-hazard facility with complex systems generally requires a more formalized QA program.

The concept of defense-in-depth is usually applied to multiple SSC designed to mitigate or prevent the same design basis event. Thereby, as a result of redundancy and some level of independence, it significantly reduces the probability of unacceptable safety-related consequences from such events. The use of robustness in the design of single SSC for that purpose is intended to accomplish the same end. When multiple SSC are used, it is a design objective to make each one as functionally independent as practical, so that a common mode failure does not occur. In general, multiple SSC performing the same safety function permit the use of more relaxed acceptance criteria and safety-related classifications associated with design-basis extreme events than for a single SSC without redundancy.

Robustness in defense-in-depth design usually employs a single SSC. Such SSC are conservatively designed to normal or anticipated event acceptance criteria even though the loads are typically associated with an extreme or abnormal event. Extreme or abnormal event design bases generally permit the use of lower margins to failure in recognition of the reduced probability of failure as the result of the convolving of a low probability of an extreme or abnormal initiating event, with failure probability given the event as a function of the capacity of the SSC.

11.3 QUALITY IN THE CONSTRUCTED PROJECT

QA as it applies to construction and procurement of safety-related SSC in hazardous facilities is divided into two activities: quality administration and QC. [7]

11.3.1 QA

In general currently there are three QA systems in existence in the U.S., according to which safety-related SSC in hazardous facilities are constructed and procured.

ASME NQA-1

The NQA-1 quality system developed by the American Society of Mechanical Engineers is in compliance with and organized in a manner similar to the U.S. Nuclear Regulatory Commission's (NRC's) *10CFR50* Appendix B requirements for nuclear facilities. *10CFR50* Appendix B was first published in 1969, and nuclear plants receiving construction permits after that date were required to have formal written QA programs. NQA-1 was first published in 1979 three years after the last NRC construction permit was issued as a nuclear industry standard to meet the *10CFR50* Appendix B requirements. As a result, very few nuclear power plants are formally committed to the NQA-1 program. Most existing U.S. nuclear power plants that received construction permits between 1969 and 1979 are currently committed to *10CFR50* Appendix B and/or the *ANSI N45.2* [8] industry program that preceded NQA-1. Compliance with NQA-1 is an in-house process requiring no external certification but does recognize the applicability of external audits. There is an NUPIC voluntary organization, as described in attachment B, funded and staffed by nuclear power plant utilities that perform audits in accordance with EPRI NP6030 [9] The current *NQA-1-2000* standard was published in May 2001 and its implementation was expected 6 months after that date for any new construction and associated procurement. However, it should be understood that portions of the NQA-1 standard may not be acceptable to either the *ASME BPVC* Section III or the NRC. It is recommended that NQA-1's format for QA be applied to suppliers of Safety Class and Performance Category 4 and 3 non professional services and products. This standard is intended to be applicable to organizations providing nuclear safety-related products and services to nuclear facilities.

10CFR830.120 and 122 and doe std. 411.1

The 10 CFR 830.120 and .122 regulation and DOE Implementation Guides 411.1-1 and 411.1-2 developed by the U.S. Department of Energy employs 10-QA requirements or criteria. The 10 requirement format has seen wide application outside the U.S. for nuclear facilities since this QA system was also adopted in 1996 by the International Atomic Energy Agency in its QA codes and safety guides for application to nuclear power plants in developing countries. The N411.1 standard is meant primarily to be applicable to professional services organizations providing those services to safety-related nuclear facilities.

ISO-9001-94

The *ISO-9001-94* [1] QA standard is implemented by the International Standards Organization through a number of registrar organizations that are authorized to provide *ISO-9001-94* registration to an organization that wishes to be *ISO-9001*-certified and contains 20 identified requirements. An *ISO-9001*-2000 standard is a major revision of the *ISO-9001-94* standard, which effectively eliminates the 20 requirements or criteria and places much more emphasis on less-prescriptive QA goals and processes that makes it a much less definitive document than *ISO-9001-94*. It should be understood that the *ISO-9001* QA system is not acceptable to the U.S. NRC, NUPIC or *ASME NQA-1*, because requirements are not considered detailed enough to be used as the basis of an effective QA program. However, the *ISO-9001* QA system has seen wide application to a large number of suppliers of safety-related non-nuclear items and services both inside and outside the U.S. and for nuclear facilities outside the U.S.

11.3.1.4 Implementation of QA Programs

The NQA-1 and *ISO-9001* QA systems are specifically applicable to construction and procurement of nuclear safety-related SSC. The NQA-1 system is organized and based directly on the *10CFR50* Appendix B requirements first defined by the U.S. NRC in 1969. The 10 CFR 830.122 DOE Std. 411system was first specified as DOE Std. 5700.6C in 1989 and has been specified primarily for QA associated with the supply of professional (engineering) services to nuclear facilities.

An *ISO-9001* QA system certification requires review by an outside authorized registration organization hired by the organization to be certified to assure *ISO-9001* compliance. Such outside review and registration seal and certification are not required of *10CFR50* Appendix B, *ASME NQA-1* or 10 CFR 830.122 compliant QA systems. Together with deficiencies that have been observed by ASME in most ISO-9901 QA requirements as summarized in Appendix 11C the U.S. NRC and NUPIC to date have not permitted *ISO-9001* to be used as the basis in U.S. NRC-regulated facilities and projects.

It should be understood that in all cases QA is meant to provide a framework, organization and procedures to ensure that sufficient QCs are in place, and that those which have been committed relative to quality associated with the supply of products, materials and services are in fact being provided. A QA program by itself does not ensure quality. This is provided by the detailed *QC* standards specified for construction or procurement, and the enforcement that those controls are being applied. For example, an effective QA program would require that necessary construction standards be specified. It does not ensure that implementation of the QC requirements of these standards would result in a high-quality product, material or service, particularly when a graded approach to construction or procurement is permitted.

On January 1, 1993, the European Community issued a set of directions that required *ISO-9001* certification by suppliers of regulated products. The ninth annual survey of *ISO-9001* registered companies in the U.S. as of December 1999 listed approximately 33,800 organizations that were *ISO-9001* registered, with a total of 340,000 *ISO-9001* registered organizations worldwide. This compares with approximately 975 U.S. organizations listed in the Nuclear News 2001 Buyers Guide that are presumably NQA-1-qualified.

A listing of the 18 QA criteria of NQA-1 2000 – *10CFR50* Appendix B and the 20 QA criteria of *ISO-9001-94* are given in Table 11.1 along with the ten 10 CFR 830.120 criteria for comparison purpose.

It should be understood that while there is a requirement for documentation of all QA activities, the type, quantity and cost of documentation can vary significantly. For example, one type of documentation is to require a certified material test report on all steel material supplied for safety-related applications on a project. This type of documentation is generated when the metal heat is poured and must follow the steel products made from that heat through all processes to their end use. A simpler and far less expensive form of documentation is a certificate of compliance, which is simply a certification by the material or product supplier that there is a procedure in place that ensures the material by ASTM or ASME-designated number and grade is being supplied. An even simpler and cheaper documentation, which is less susceptible to counterfeiting, is testing and documentation of the test results of a statistically significant sample of the material at the point of end use by the purchasing organization for minimum required chemical and mechanical properties as defined by the procurement specification. This end use testing form of material certification is often called performance rather than document-based QA.

11.4 ROLE OF A PROJECT QA GROUP

Project management often employ a project quality assurance group, (PQAG) when the project line organization has relatively little or limited experience in working under a formal QA program. In such a case, initially the project assurance group assists the project line organization by development of a *Project QA Program and Manual* which is compatible with the project management or project implementation plan as well as a referenced QA system as described in Section 11.3. The PQAG also assists in development, review and comment concerning Project Procedures that have an impact on quality. Most important, the PQAG acts as a mentor to the project line organization in understanding and implementing the QA Program.

Initial duties of a P QAG typically are as follows:

- Prepare a QA Program manual compatible with existing applicable and referenced Q.A. requirements and project implementation or management plan.
- Assist in the preparation, overview and comment of project procedures that impact quality.
- Mentor the project line organization to understand and implement the project QA program.
- Provide training in QA requirements.
- Provide QA input to design, construction and procurement reviews.
- Act as an agent of project management in performing management QA assessments.
- Provide audits of lower-tier suppliers of safety-related products, materials and services to the project in the absence of a corporate procurement organizations supplied audits.
- Provide independent QA audits of other projects as requested.

As the project line group matures and routinely adapts and implements required QA procedures, the monitoring role of the PQAG should begin to decrease along with a reduction of the PQAG personnel. Mature project organizations with a good QA performance record often no longer have a PQAG. Independent assessment and supplier audits in such cases are typically supplied by matrixed QA personnel from outside the project organization.

It is important to recognize that project QA personnel are expected to assist the project line organization in meeting its prime responsibility for project quality and QA and not take on this responsibility themselves.

11.5 QC

QC is typically defined as all those activities associated with measuring, examining, and testing or verifying of construction or other materials, processes and activities as prescribed by the specified industrial construction or procurement standard for SSC. Consistent with the concept of the graded approach, different QC requirements may be applied as a function of the standard specified in meeting associated functional and performance classifications or categorizations.

QC requirements are normally addressed in detail in industry procurement or construction standards, their referenced daughter standards or as specified by the owner's agent in procurement specifications. Daughter standards are typically those that are referenced from another standard. For example, *ASME Boiler and Pressure Vessel Code* Section III, Subsection NF for hot-rolled linear-steel supports specifies as a daughter standard the American Welding Society (AWS) [11] standards for detailed structural welding requirements and American Society of Testing Materials (ASTM) [12] for member material properties and requirements.

Differences in standards specified may also include different design conditions or load combinations. As an example, *ASME BPVC* Section III PC-5 and PC-4, and in some cases PC-3 SSC, has more extreme design loads as compared to *ASME B31.1* [13] or *ASME BPVC* Section VIII, Division 1 [14] which is specified for PC-2, 1 and 0 pressure retaining mechanical systems and components. The *ASME BPVC* Section III is generally considered more rigorous and more restrictive than *ASME BPVC* Section VIII and B31.1 or B31.3 [15] requirements.

It should be understood that there are many items and services for which there are no applicable standards associated with their construction or procurement. In such cases, their procurement specifications would be required to contain specific QC requirements, usually developed by the design organizations which would be necessary to ensure the desired level of quality.

11.6 RECOMMENDATIONS AND CONCLUSIONS

There are three independent QA systems in general use in the U.S. for hazardous facilities that could be used to develop QA programs for the construction projects and their suppliers of safety-related items. The *10CFR50* Appendix B-NQA-1 2000 and *ISO-9001* standards are applicable to vendors of nuclear safety-related products and services and *10CFR830.*, 122 and *DOE 0 414* are particularly applicable to supply of professional services.

There is much evidence that direct and excessive involvement of project QA groups in project line organizations responsible for QA can significantly increase cost and negatively impact performance with no increase in quality; hence, it should be carefully controlled by project Management.

Project QA personnel should be used to mentor project line managers and personnel until they demonstrate their capability to implement the QA program and produce a quality product. After this their major role should be management assessment.

In the nuclear field the weight of current and future commercial considerations, not withstanding current NRC and NUPIC opposition, suggests that the *ISO-9001* QA system supplemented by an external audit requirement may become the dominates QA system applicable to vendors of safety-related products, materials and services.

Perhaps a more pressing need, rather than the comparison of criteria between the three reference QA systems associated with construction and procurement of safety class SSC products and services in hazardous facilities, is to define those QA differences that should exist between graded high-moderate, and low-hazard "safety-related" and "safety significant" SSC QA program requirements. This issue is briefly addressed in Appendix 11.F.

REFERENCES

1. *ANSI/ASQC Q 9001*, 1994, "Quality Systems Model for Quality Assurance in Design, Development, Production, Installation and Servicing," American Society for QC, Milwaukee, WI.
2. *ASME NQA-1-2000*, "Quality Assurance Requirements for Nuclear Facility Applications," American Society of Mechanical Engineers, New York, NY.
3. *10CFR 830.122*, "Quality Assurance Requirements," Department of Energy, Washington, DC.
4. *IAEA Nuclear Safety Series Code 50-C-Q*, "Quality Assurance for Safety in Nuclear Power Plants and Other Nuclear Installations," International Atomic Energy Agency, Vienna, Austria, 1996.
5. *10CFR50* Appendix B, "Quality Assurance," Nuclear Regulatory Commission, Washington, DC.
6. Paragraph NA-4131 Quality Assurance, *ASME Boiler and Pressure Vessel Code* Section II, Subsection NA, New York, NY, 1974.
7. *ANSI N45.2.11*, "Quality Assurance Requirements for the Design of Nuclear Power Plants," American National Standards Institute, New York, NY, 1974.

8. *DOE STD-1027-92*, 1992, "Hazard Categorization and Accident Analysis Techniques for Compliance with DOE Order 5480.23, Nuclear Safety Analysis Reports", U.S. Department of Energy, Washington, D.C.
9. *EPRI NP 6030*, "Guidelines for Performance Based Supplier Audits," Electric Power Research Institute, Palo Alto, CA.
10. *AWS D1.1*, 2001, American Welding Society Structural Welding Code – Steel, 2005.
11. *ASTM Standards Section 1*, Iron and Steel Products.
12. *ASME B31.1*, "Power Piping," American Society of Mechanical Engineers, New York, NY
13. *ASME BPVC* Sect6ion VIII Div. 1, "Unfired Pressure Vessels," American Society of Mechanical Engineers, New York, NY, 2004.
14. *ASME B31.3*, "Process Piping, American Society of Mechanical Engineers, New York, NY, 2004.
15. *DOE O 414*, "Quality Assurance," U.S. Department of Energy Order, Rev. 1, Washington, DC, 2000.

APPENDIX 11A: A COMPARISON OF QUALITY ASSURANCE SYSTEMS BASIC REQUIREMENTS BETWEEN *ISO-9001-94* AND *ASME NQA-1-2000 – 10CFR50* APPENDIX B

The following is a comparison between *ISO-9001-94* and *ASME NQA-1-2000 – 10CFR50,* Appendix B, Basic Requirements.

ISO-9001-94	*10CFR50t* Appendix B, *ASME NQA-1-2000* and *DOE/RW-0333P*	*10CFR830.120, 830.122* and *DOE 0.0414*
I. Management Responsibility: Management must have a written policy statement of their commitment to quality. This policy must be communicated to and understood by all employees. Management must clearly define quality-related organization responsibilities and interrelationships. A management representative must be assigned to oversee the implementation and continuous improvement of the quality system. Senior management must routinely review the quality system, assess compliance and identify opportunities for improvement.	**Organization:** The supplier/client shall be responsible for the establishment and execution of the quality assurance (QA) program. The supplier/client may delegate to others, such as contractors, agents or consultants, the work of establishing and executing the QA program, or any part thereof, but shall therefore retain responsibility. The authority and duties of persons and organizations performing activities affecting the safety-related functions of structures, systems and components (SSC) shall be clearly established and delineated in writing. These activities include both the performing functions of attaining quality objectives and the QA functions. The QA functions are those of: Ensuring that an appropriate QA program is established and effectively executed; and verifying, such as by checking, auditing, and inspecting, that activities affecting the safety-related functions have been correctly performed. The persons and organizations performing QA functions shall have sufficient authority and organizational freedom: to identify quality problems; to initiate, recommend or provide solutions; and to verify implementation of solutions.	**Quality Assurance Program and Personnel Training and Qualification:** Management shall develop, implement and maintain a QA program. This program shall include details of how work is to be managed, performed and assessed, consistent with the basic requirements in this code. The QA program shall include the organizational structure, function, responsibilities, levels of authority and interfaces for those managing, performing and assessing the adequacy of work. The QA program shall address management measures, including planning, scheduling and resource considerations. Management in the entire and constituent areas of work shall provide and demonstrate support for the effective implementation of the QA program consistent with specified time schedules for accomplishing project activities. The responsible organization shall also be responsible for the establishment and implementation of the overall QA program. If the responsible organization delegates to other organizations the work of establishing and implementing all or part of the overall program, it shall retain responsibility for the effectiveness of the program under all circumstances. Nuclear safety shall be the fundamental consideration in the identification of the items, services and processes to which the QA program applies. A graded approach based on the relative importance to the nuclear safety of each item, service or process shall be used. The graded approach shall reflect a planned and recognized difference in the applications of specific QA requirements. Personnel shall be trained and qualified so that they are competent to perform their assigned work and understand the safety consequences of their activities. Items, services and processes that do not meet specified requirements shall be identified, and the safety impact of the nonconformances assessed and reported to the appropriate level of management. Depending on the results of the assessment, the items shall be either accepted, rejected, repaired or reworked, and services and processes accepted or rejected.

ISO-9001-94	*10CFR50t* Appendix B, *ASME NQA-1-2000* and *DOE/RW-0333P*	*10CFR830.120, 830.122* and *DOE 0.0414*
II. Quality System and Training: The quality system must be documented. A quality manual should be developed to document the policy and the quality system requirements. Quality plans should be developed describing the quality control provisions for products and services. Documentation of the quality system should include an overall description of the quality system and details about the organization of the supporting documentation.	**Quality Assurance Program and Personnel Indoctrination and Training:** The supplier/client shall establish at the earliest practicle time, consistent with the schedule for accomplishing the activities, a QA program that complies with the requirements of this project. The program shall be: Documented by written policies, procedures or instructions and shall be carried out throughout the project and if applicable, facility life in accordance with those policies, procedures or instructions. The supplier/client shall identify: The SSC to be covered by the QA program and the major organizations participating in the program, together with the designated functions of these organizations.	
A system must be developed to identify necessary skills, establish required training programs, plan and conduct required training, and maintain records of such training.	The QA program shall provide control over activities affecting the quality of the identified SSC to an extent consistent with their importance to safety. Activities affecting quality shall be accomplished under suitably controlled conditions, which include: The use of appropriate equipment; suitable environmental conditions for accomplishing the activity, such as adequate cleanness; and assurance that all prerequisites for the given activity have been satisfied. The program shall take into account the need for: special controls; processes; test equipment; tools, and skills to attain the required quality and the need for verification of quality by inspection and test. The program shall provide for indoctrination and training of personnel performing activities affecting quality as necessary to ensure that suitable proficiency is achieved and maintained. The applicant shall regularly review the status and adequacy of the QA program. Management of other organizations participating in the QA program shall regularly review the status and adequacy of that part of the QA program that they are executing.	

ISO-9001-94	*10CFR50t* Appendix B, *ASME NQA-1-2000* and *DOE/RW-0333P*	*10CFR830.120, 830.122* and *DOE 0.0414*
III. Design Control: To ensure that client requirements are met, procedures must be established and maintained for controlling and verifying the design of new and modified products. This includes any client-provided design and material requirements	**Design Control:** 1. Measures shall be established to ensure that applicable regulatory requirements and the design-basis application for those SSC to which this standard applies are correctly translated into specifications, drawings, procedures and instructions. 2. These measures shall include provisions to ensure that appropriate quality standards are specified and included in design documents and that deviations from such standards are controlled. 3. Measures shall also be established for the selection and review of suitability of application of materials, parts, equipment and processes that are essential to the safety-related functions of the SSC. 4. Measures shall be established for the identification and control of design interfaces and for coordination among participating design organizations. 5. These measures shall include the establishment of procedures among participating design organizations for the review, approval, release, distribution and revision of documents involving design interfaces. 6. The design control measures shall provide for verifying or checking the adequacy of design, such as by the performance of design reviews, by the use of alternate or simplified calculation methods, or by the performance of a suitable testing program. 7. The verifying or checking process shall be performed by individuals or groups other than those who performed the original design, but who may be from the same organization. 8. Where a test program is used to verify the adequacy of a specific design feature in lieu of other verifying or checking processes, it shall include a suitable qualification testing of a prototype unit under the most adverse design conditions. 9. Design control measures shall be applied to items such as the following: reactor physics, stress, thermal, hydraulic and accident analyses; compatibility of materials; accessibility for in-service inspection, maintenance, and repair; and delineation of acceptance criteria for inspections and tests. 10. Design changes, including field, shall be: a. Subject to design control measures commensurate with those applied to the original design. b. Approved by the organization that performed the original design unless the applicant designates another responsible organization.	**Design:** Design, including subsequent changes, shall be carried out in accordance with established engineering codes and standards, and shall incorporate applicable requirements and design bases. Design interface shall be identified and controlled. The adequacy of design, including tools and inputs and outputs, shall be verified or validated by individuals or groups other than those who originally performed the work. Verification, validation and approval shall be completed before implementation of the design.

ISO-9001-94	*10CFR50t* Appendix B, *ASME NQA-1-2000* and *DOE/RW-0333P*	*10CFR830.120, 830.122* and *DOE 0.0414*
IV. Document Control: Procedure must be established and maintained for controlling all documents and data that relate to the requirements defined in the quality manual. Before they are issued, all documents related to the quality manual requirements must be reviewed and approved by authorized personnel. The authorized personnel also must review and approve any changes to those documents.	**Document Control:** 1. Measures shall be established to control the issuance of documents, such as instructions, procedures and drawings, including changes thereto, which prescribe all activities affecting quality. 2. These measures shall ensure that documents, including changes: a. Are reviewed for adequacy and approved for release by authorized personnel. b. Distributed to and used at the location where the prescribed activity is performed. 3. a. Changes to documents shall be reviewed and approved by the same organizations that performed the original review and approval. b. Unless the applicant designates another responsible organization.	**Document Control and Records:** Documents such as procedures, instructions, specifications and drawings, or other media that describe processes, specify requirements or establish design, shall be prepared, reviewed, approved, issued, distributed, authorized, revised and, as required, validated. All personnel preparing, revising, reviewing or approving documents shall be specifically assigned to this work and be given access to appropriate information upon which to base their input. Personnel using documents shall be aware of and use appropriate and correct documents. Records relating to personnel and records that describe the status, configuration and characteristics of items and services, describe the performance of processes and represent objective evidence of quality shall be specified, prepared, reviewed, approved and maintained. All records shall be legible, complete and identifiable. A records system shall be established to provide the identification, collection, indexing, filing, storing, maintenance, retrieval and disposal of records. Retention times of records and associated test materials and specimens shall be established to be consistent with the type of records, materials and specimens involved.

ISO-9001-94	*10CFR50t* Appendix B, *ASME NQA-1-2000* and *DOE/RW-0333P*	*10CFR830.120, 830.122* and *DOE 0.0414*
V. Purchasing Purchaser-Supplied Product: For purchased items that directly impact the quality of the product or service provided, a system must be available to ensure that all requirements are adequately specified and communicated to the supplier. The system also must provide mechanisms for qualifying suppliers and verifying that requirements are met. A system must be in place to verify that all client-supplied products, services or materials are acceptable prior to their use.	**Control of Purchased Material, Parts and Components:** 1. Measures shall be established to ensure that purchased material, equipment and services, whether purchased directly or through contractors and subcontractors, conform to the procurement documents. 2. These measures shall include provisions, as appropriate, for: a. Source evaluation and selection b. Objective evidence of quality furnished by the contractor or subcontractor c. Inspection at the contractor or subcontractor source d. Examination of products upon delivery 3. Documentary evidence that material and equipment conform to the procurement requirements shall be available at the nuclear power plant or fuel-reprocessing plant site prior to installation or use of such material and equipment. 4. a. This documentary evidence shall be retained at the nuclear power plant or fuel-reprocessing plant site. b. It shall be sufficient to identify the particular requirements, such as codes, standards or specifications, met by the purchased material and equipment. 5. The effectiveness of the control of quality by contractors and subcontractors shall be assessed by the applicant or designee at intervals consistent with the importance, complexity and quantity of the product or services.	**Procurement:** Procured items and services shall meet established requirements and perform as specified. Suppliers shall be evaluated and selected on the basis of specified criteria. Requirements necessary to ensure the quality of items and services shall be developed and specified in the procurement documents. Evidence that purchased items and services meet procurement requirements shall be available before they are used. Requirements for reporting deviations from procurement requirements shall be specified in the procurement documents. Inspections and testing of specified items, services and processes shall be conducted using established acceptance and performance criteria. The level of inspection and testing and the degree of independence of personnel shall be established. Administrative controls, such as hold points and status indicators, shall be used to preclude the bypassing of required inspections and tests. Any inadvertent use, installation or operation of items, services and processes that have not passed the required inspections and test shall be prevented.

ISO-9001-94	*10CFR50t* Appendix B, *ASME NQA-1-2000* and *DOE/RW-0333P*	*10CFR830.120, 830.122* and *DOE 0.0414*
VI. Product Identification and Traceability: All products, services and materials must be identified during all stages of production, delivery and installation. When required by clients, records must be maintained for tracing product back to its origins.	**Identification and Control of Materials, Parts and Components:** 1. Measures shall be established for the identification and control of materials, parts and components, including partially fabricated assemblies. 2. These measures shall ensure that identification of the item is maintained by heat number, part number, serial number or other appropriate means, either on the item or on records traceable to the item, as required throughout fabrication, erection, installation and use of the item. 3. These identification and control measures shall be designed to prevent the use of incorrect or defective material, parts and components.	**Inspection and Testing for Acceptance and Control:** Management defines the types of work that require formal inspections. A process is established to specify when and what type of inspection is to be performed for the types of work to be inspected. Inspection types include source, in-process, final, receipt, maintenance and in-service. Administrative controls and status indicators are incorporated into the inspection process. These controls and indicators are used to preclude inadvertent bypassing of required inspections and to prevent inadvertent operation of the item or process. The inspection process may allow inspections to be performed by the organizational unit responsible for the work, another department or an outside agency independent of the responsible organization. Individuals inspect their own work to ensure they have achieved the desired level of quality; however, personnel performing the work do not inspect their own work for acceptance. Personnel responsible for performing acceptance inspections are technically competent. Management ensures that inspectors plan properly. Planning addresses such attributes as item characteristics, work processes, inspection techniques, hold and witness points, acceptance criteria, and the organization or individuals responsible for conducting the inspections. Appropriate tests are conducted to demonstrate that items and processes will perform as intended. The test process includes bench tests, proof tests before installation, preoperational tests, operational tests, post-modification tests and post-maintenance tests.

ISO-9001-94	*10CFR50t* Appendix B, *ASME NQA-1-2000* and *DOE/RW-0333P*	*10CFR830.120, 830.122* and *DOE 0.0414*
VII. Process Control: The company must identify all processes that directly affect the quality of the product or service and ensure that these processes are carried out under controlled conditions, including: • Formal approval of process design and equipment. • Documented work instructions. • Development of quality plans describing how the process is to be monitored. • A suitable working environment. • Documented quality criteria.	**Control of Special Processes:** 1. Measures shall be established to ensure that special processes, including welding, heat treating and nondestructive testing, are: a. Controlled. b. Accomplished by qualified personnel using qualified procedures in accordance with applicable codes, standards, specifications, criteria and other special requirements	

ISO-9001-94	***10CFR50t* Appendix B, *ASME NQA-1-2000* and *DOE/RW-0333P***	***10CFR830.120, 830.122* and *DOE 0.0414***
VIII. Inspection and Testing Inspection and Test Status: The supplier/client must verify that the product conforms with all appropriate specifications throughout the manufacturing process, including receiving, work-in-progress and finished goods. These verifications must be in accordance with documented quality plans and procedures. The supplier/client must establish and maintain a means for clearly identifying the quality status of all product throughout production and installation.	**Inspection and Test Control:** 1. A program for inspection of activities affecting quality shall be: a. Established. b. Executed by or for the organization performing the activity to verify conformance with the documented instructions, procedures and drawings for accomplishing the activity. 2. Such inspection shall be performed by individuals other than those who performed the activity being inspected. 3. Examinations, measurements, or tests of material or products processed shall be performed for each work operation where necessary to ensure quality. 4. If inspection of processed material or products is impossible or disadvantageous, indirect control by monitoring processing methods, equipment and personnel shall be provided. 5. Both inspection and process monitoring shall be provided when control is inadequate without both. 6. If mandatory inspection hold points, which require witnessing or inspecting by the applicant's designated representative and beyond which work shall not proceed without the consent of its designated representative, the specific hold points shall be indicated as inappropriate. 7. A testing program shall be established to ensure that all testing required to demonstrate that SSC will perform satisfactorily in service is: a. Identified. b. Performed in accordance with written test procedures, which incorporate the requirements and acceptance limits contained in applicable design documents. 8. The test program shall include, as appropriate: a. Proof tests prior to installation. b. Preoperational tests. c. Operational tests during nuclear power plant or fuel-reprocessing plant operation of SSC. 9. Test procedures shall include provisions for ensuring that: a. All prerequisites for the given test have been met. b. Adequate test instrumentation is available and used. c. The test is performed under suitable environmental conditions. 10. Test results shall be: a. Documented. b. Evaluated to ensure that test requirements have been satisfied.	

ISO-9001-94	*10CFR50t* Appendix B, *ASME NQA-1-2000* and *DOE/RW-0333P*	*10CFR830.120, 830.122* and *DOE 0.0414*
IX. Inspection Measuring and Test Equipment: A system must be established to control, calibrate and maintain all inspection, measuring and test equipment that is used to demonstrate that products conform to specified requirements.	**Control and Measuring of Test Equipment:** 1. Measures shall be established to assure that tools, gauges, instruments, and other measuring and testing devices used in activities affecting quality are properly: a. Controlled. b. Calibrated, and adjusted at specified periods to maintain accuracy within necessary limits.	
X. Control of Nonconforming Product: The company must establish a system to ensure that any product, component, or raw material that does not fully conform to requirements is prevented from further processing or shipment. The nonconforming material control system must specifically address the identification, documentation, evaluation, segregation and disposition of nonconforming material as well as the notification of all involved functions.	**Nonconforming Materials, Parts or Components:** 1. Measures shall be established to control materials, parts or components that do not conform to requirements in order to prevent their inadvertent use or installation. 2. These measures shall include, as appropriate, procedures for: a. Identification. b. Documentation. c. Segregation. d. Disposition. e. Notification to affected organizations. 3. Nonconforming items shall be: a. Reviewed and accepted. b. Rejected. c. Repaired or reworked in accordance with documented procedures.	
XI. Corrective Action: The company must establish and maintain a process for investigating the root causes of nonconforming product and for taking corrective actions that will prevent future occurrences. This process should specifically verify that the corrective actions taken have been effective.	**Corrective Action:** 1. Measures shall be established to ensure that conditions adverse to quality, such as failures, malfunctions, deficiencies, deviations, defective material and equipment, and nonconformances are promptly identified and corrected. 2. In the case of significant conditions that affect quality, the measures shall ensure that the cause of the condition is determined, and corrective action taken to preclude repetition. 3. The identification of the significant condition adverse to quality, the cause of the condition and the corrective action shall be taken, and then: a. Documented. b. Reported to appropriate levels of management.	**Nonconformance Control and Corrective Actions:** Management fosters a "no-blame" culture to encourage employees to identify nonconforming items, services and processes. Management ensures that adequate resources are provided for identifying, resolving and preventing nonconforming conditions. Management is also involved in the resolution of difficult issues and provides a process for resolving professional differences of views and opinions. Line managers establish and implement measures promptly to identify, document, classify, analyze, correct, eliminate and follow up on activities, items, services or processes that do not meet established requirements and goals, or do not result in the anticipated quality.

All personnel have the opportunity to identify nonconforming items, services and processes. All personnel are encouraged to identify and suggest improvements via the management system.

Personnel responsible for classifying and analyzing nonconformances have an adequate understanding of the area in which they are working and access to pertinent background information concerning the nonconformance. They are also independent from cost and schedule considerations.

Determination of the root cause requires a through investigation by technically qualified, experienced personnel The investigation required includes the participation of the personnel involved, such as craftsmen, operators and those identifying the deficiency, in order to gain complete understanding of the problem. The managers responsible for the determination of the root cause assign sufficient resources to the task.

Nonconforming items are properly controlled to prevent their inadvertent test, installation or use. They are reviewed and either accepted, rejected, repaired or reworked.

Reworked and repaired and replacement items are inspected and tested in accordance with the original inspection and test requirements or specified alternatives.

Performance data and other quality-related information are analyzed to identify quality trends that adversely affect the achievement of quality and to identify opportunities to improve items, services and processes. In order to identify commonalities, this analysis considers information from external sources and is not limited to one activity, one site or one supplier.

The QA program requires that all processes be analyzed to identify trends that may adversely affect item, service and process quality. This analysis includes information performance data, internal and external failure costs and prevention costs.

The QA program also requires that levels of performance be established. These levels are directly related to the item or service provided by an organization and based on the goals and objectives set by management. Once the levels have been established, performance is measured against them. These measurements are monitored on a regular basis to ascertain whether or not improvements in item, service or process quality are necessitated. Management plays an integral role in this activity.

ISO-9001-94	*10CFR50t* Appendix B, *ASME NQA-1-2000* and *DOE/RW-0333P*	*10CFR830.120,* *830.122* and *DOE 0.0414*
XII. Handling, Storage, Packaging and Delivery: The company must have a policy to prevent damage during handling and storage. The company also must have an effective system for controlling the movement of critical commodities in and out of inventory.	**Handling Storage and Shipping:** 1. Measures shall be established to control the handling, storage, shipping, cleaning and preservation of material and equipment in accordance with work and inspection instructions to prevent damage or deterioration. 2. When necessary for particular products, special protective environments, such as inert gas atmosphere, specific moisture content levels and temperature levels, shall be: a. Specified. b. Provided.	
XIII. Quality Records: The company should retain all records demonstrating that the quality system elements are being adequately addressed. The company also must specify the appropriate retention period for such records. These records must be legible and stored in such a way as to prevent damage. In addition, they must be easily retrieved.	**Records:** 1. Sufficient records shall be maintained to furnish evidence of activities affecting quality. 2. The records shall include at least the following: a. Operating logs. b. Results of reviews. c. Inspections. d. Tests. e. Audits. f. Monitoring of work performance. g. Materials analyses. 3. The records shall also include closely related data, such as qualifications of personnel, procedures and equipment. 4. Inspection and test records shall, as a minimum: a. Identify the inspector or data recorder. b. The type of observation. c. The results. d. The acceptability. e. The action taken in connection with any deficiencies noted. 5. Records shall be identified and retrievable. 6. Consistent with applicable regulatory requirements, the applicant shall establish requirements concerning record retention, such as: a. Duration. b. Location. c. Assigned responsibility.	

ISO-9001-94	*10CFR50t* **Appendix B,** *ASME NQA-1-2000* **and** *DOE/RW-0333P*	*10CFR830.120, 830.122* **and** *DOE 0.0414*
XIV. Internal Quality Audits: The company must have an internal system for auditing the effectiveness and compliance of all quality system elements. The company also must establish a formal schedule for conducting these audits as well as maintaining records of each audit and the resulting actions.	**Audits:** 1. A comprehensive system of planned and periodic audits shall be carried out to: a. Verify compliance with all aspects of the QA program. b. Determine the effectiveness of the program. 2. The audits shall be performed: a. In accordance with written procedures or checklists. b. By appropriately trained personnel not having direct responsibilities in the areas being audited. 3. Audit results shall be: a. Documented and reviewed. b. Management having responsibility in the area audited must conduct the above. 4. Follow-up action, including re-auditing of deficient areas, shall be taken where indicated.	**Management Self-Assessment:** The thrust of management self-assessment is to identify, correct and prevent management problems that hinder the achievement of the organization's objectives. This code establishes the requirement for a routine and continuing assessment of the management system by the organization's managers. This self-assessment methodology is in addition to the traditional audit/appraisal that determines the adequacy and extent of the QA program development, documentation and implementation in accordance with specified requirements. This basic requirement improves on the standard stipulation in many QA programs that requires management to regularly assess the adequacy of the portion of the program for which it is responsible and ensure its effective implementation. This standard requirement is typically achieved, on an annual basis, by an independent consultant or group of consultants on behalf of management, and it addresses compliance issues rather than broad categories of management issues. Management self-assessment goes beyond such matters as conformance to regulations, item standards or established procedures. These areas are addressed by basic requirement 10. An effective management self-assessment evaluates issues such as: • Mission of the organization • Whether employees • Understand the mission • What is expected of the organization • Whether the expectations are being met • Opportunities for improving quality and enhancing safety • How to make better use of human resources. The results of the management self-assessment are documented. Decisions and related actions resulting from the recommendations are promptly followed up to evaluate their effectiveness. The assessment process involves all levels of management, but overall responsibility for management self-assessments is retained by senior management. It is essential that senior management directly participate in this process.

ISO-9001-94	*10CFR50t* Appendix B, *ASME NQA-1-2000* and *DOE/RW-0333P*	*10CFR830.120, 830.122* and *DOE 0.0414*
XV.		**Independent Assessment:** Management establishes and implements a process of independent assessments. These assessments are conducted by an organizational unit independent of the work and focus on improving the overall performance of items, services and processes. A system of planned and documented internal and external audits is carried out to assess the adequacy and effectiveness of the QA program. Assessment personnel operate as an arm of, and as an advisor to, senior management. The assessments focus on the performance of work and actions, and include the review and evaluation of QA documents. Assessment personnel view the organization being assessed as the "customer," so as to give meaningful feedback on the organization's performance. Assessments are conducted by peers who are technically competent to review and evaluate the work and processes being assessed. These individuals do not have direct responsibility in the area being assessed. The allocation of resources for assessments is determined. Assessment schedules are flexible enough to allow additional attention to areas of questionable performance or critical or complex work. Assessment results are tracked and resolved by management having responsibility for the area assessed. The assessment organization schedules a follow-up review of deficient areas as appropriate. Assessment responses are documented and include, as applicable, actions to correct the deficiency, identification of causes and actions to prevent recurrence, lessons learned and actions to be taken for improvement.

ISO-9001-94	*10CFR50t* Appendix B, *ASME NQA-1-2000* and *DOE/RW-0333P*	*10CFR830.120, 830.122* and *DOE 0.0414*
XVI. Servicing: A system must be established to define and monitor the quality of services provided after sale.		
XVII. Statistical Techniques: The company should be able to demonstrate that appropriate statistical techniques are used to ensure the quality of the product or service provided.		
XVIII. Contract Review: A system must be developed to ensure that customer needs are understood and that the appropriate resources (e.g., personnel, methods, material, equipment) are available to produce a product or provide a service that meets those needs.		

Appendix 11B: Nuclear Procurement Issues Committee Joint Audit Program

This description is intended to provide an understanding of the Nuclear Procurement Issues Committee (NUPIC) and the joint audit program facilitated by the NUPIC organization.

NUPIC was formed in 1989 as a result of the merger between the utility members of the Coordinating Agency for supplier evaluation (CASE) nuclear section and nuclear supplier quality assurance committee (NSQAC). The membership represents all of the operating nuclear power plants in the United States and several non-U.S. nuclear utilities. As a unified industry organization, NUPIC has combined the most effective elements of previous audit in the U.S. organizations to provide a cost-effective and quality-effective program for the performance of supplier audits by its members.

The joint audit program is based on the utilization of an industry-wide standardized approach to the performance of supplier audits. The audits are conducted with an approved standardized checklist using the performance-based methodology consistent with *EPRI NP 6630,* "Guidelines for Performance-Based Supplier Audits."

The audit program is administered in accordance with the following general provisions: Audits are performed by joint utility audit teams typically comprising four team members, one of whom is a technical specialist. These audits are conducted and documented in accordance with the QA program of the utility, which provides the audit team leader and other standardized guidance. NUPIC members have access to all audit data, including the audit report, audit checklist and all identified findings/observations and follow-up documentation. Each member evaluates the results of the audit to determine appropriate action in accordance with their own QA program. In Figure 11.B.1 is a typical audit summary sheet.

It should be noted that an NUPIC joint audit does not preclude a member from performing its own audit if deemed necessary.

As a result of the joint audit program, significant benefits have been realized.

BENEFITS TO NUPIC MEMBERS

- Consistent basis for evaluation of suppliers
- Audit focus upon product quality and performance-related activities
- Audit scope more in-depth with larger audit teams
- Audit frequency provides more contact with suppliers
- Increased influence for resolution of supplier problems
- Cost-effective use of limited resources
- Reduction in multiple/redundant audits

BENEFITS TO SUPPLIERS

- Consistency in audit scope and baseline requirements
- Increased knowledge of industry needs/expectations and regulations
- Industry interface for resolution of suppliers problems

SUMMARY SHEET

SUPPLIER INFORMATION		
SUPPLIER:	NUPIC Supplier #:	
ADDRESS:		
CITY:	STATE:	ZIP CODE:
TELEPHONE NO:	FAX NO.:	
PRODUCT/SERVICE:		
CODE STAMP AND AUTHORIZATIONS:		

AUDIT SCOPE							
ANSI N45.2	☐	ANSI N45.2.13	☐	IEEE 323	☐	SNT-TC-1A	☐
ANSI N45.2.2	☐	ANSI N45.2.23	☐	IEEE 344	☐	OTHER:	
ANSI N45.2.6	☐	ANSI N101.4	☐	ASME NCA 3800	☐		
ANSI N45.2.9	☐	10CFR50/Appendix B	☐	ASME NCA4000	☐		
ANSI N45.2.11	☐	NUREG 0040 (Y/N)		ASME SECT XI	☐		
ANSI N45.2.12	☐	IEEE 383	☐	ANSI/ASME NQA-1	☐		

SUPPLIER CONTACTS		
SENIOR COMPANY OFFICER:	TITLE:	PHONE:
SENIOR QA OFFICER:	TITLE:	PHONE: EMAIL:

AUDIT INFORMATION	
AUDIT ID NO.:	AUDIT DATES:

AUDIT TEAM INFORMATION

AUDIT TEAM/UTILITY	NAME	PHONE	EMAIL:	CHECKLIST SECTIONS AUDITED
TEAM LEADER				
TEAM MEMBER				
TEAM MEMBER				
TEAM MEMBER				
TEAM MEMBER				
TECHNICAL SPECIALIST				
(SPECIFY DISCIPLINE)				

Audit Team Leader: Date:

NUPIC Representative: Date:

FIGURE 11.B.1. TYPICAL AUDIT SUMMARY SHEET

SUMMARY SHEET

Supplier QA Manual: Revision: Date:

Audit Section	Section Description	C	E	Implementation Status	Comments/Findings
I	Order Entry	√	√		
II	Design		√		
III	Software Quality Assurance	√	√		
IV	Procurement	√	√		
V	Fabrication/Assembly Activities, Material Control and Handling, Storage and Shipping	√			
VI	Special Processes				
VII	Tests and Inspections	√			
VIII	Calibration	√			
IX	Document Control/Adequacy	√	√		
X-A	Organization/Program	√	√		
X-B	Nonconforming Items/Part 21	√	√		
X-C	Internal Audit	√	√		
X-D	Corrective Action	√	√		
X-E	Training/Certification	√	√		
X-F	Field Services	√	√		
X-G	Records	√	√		

IMPLEMENTATION KEY

S – SATISFACTORY U - UNSATISFACTORY N/A - NOT APPLICABLE

C = Recommended for Calibration Suppliers

E = Recommended for Engineering Services Suppliers

FIGURE 11.B.1. (Continued)

Please note that a NUPIC Joint Audit does not constitute nor imply any industry-wide endorsement. Approval or disapproval of a supplier's QA program should not be used in supplier advertising materials.

Suppliers of products and services to the NUPIC member utilities are invited to participate in the process of improving the performance and effectiveness of joint audits. The following mechanisms solicit pertinent input by audited companies and welcome supplier involvement for NUPIC improvements:

a. Completion and return of a NUPIC audit/survey feedback questionnaire (provided by the audit team leader on conclusion of the audit). This questionnaire provides a mechanism to enhance this process by identifying areas that may be improved in future audits.
b. Attendance at the annual NUPIC Vendor Interface Committee (VIC) meeting, normally led in June, which features presentations and discussions on various issues of importance to the nuclear power industry. Efforts are made for suppliers of nuclear equipment and services to interact with nuclear customers, participate in training programs and suggest means for enhancing procurement quality services.
c. Publication of articles of general interest in the NUPIC newsletter. The newsletter is distributed to the NUPIC member utilities and those suppliers subject to NUPIC joint audits and commercial grade surveys as a vehicle for exchanging factual, unbiased information.

Additionally, NUPIC maintains an Internet website, which provides general information about the NUPIC organization.

NUPIC welcomes active supplier involvement and any recommendations designed to improve the NUPIC joint audit process. The NUPIC VIC chairman can be contacted to report concerns, request information of the next NUPIC vendor meeting, offer suggestions and submit newsletter articles, etc. You are encouraged to contact the VIC chairman at the address listed on the NUPIC website: http://www.NUPIC.com.

Hopefully this information has provided a better understanding of the NUPIC joint audit program. This program provides the nuclear industry with a meaningful and cost-effective approach to supplier qualification.

Appendix 11C: ASME Quality Assurance Committee Review of *NQA-1* and *ISO-9001*

COMPARISON OF
ASME NQA-1 AND ISO-9001

Basic Requirement 1 **Supplement 1S-1** **Organization**	**4.1.2.1 Responsibility and Authority** **4.1.2.3 Management Representative**
Requires that a quality organization be established with a responsible level of management identified.	Requires that a quality organization be established with a responsible level of management identified.

COMPARISON

NQA-1 and ISO-9001 have equivalent requirements

CONCLUSION

A quality program that meets the requirements of either standard would meet the requirements of both standards

Basic Requirement 2 **Quality Assurance Program**	**4.2 Quality System**
Requires that a quality program be developed, documented and implemented.	Requires that a quality program be developed, documented and implemented.

COMPARISON

NQA-1 and ISO-9001 have equivalent requirements

CONCLUSION

A quality program that meets the requirements of either standard would meet the requirements of both standards

(continues)

COMPARISON OF
ASME NQA-1 AND ISO-9001

Basic Requirement 6
Supplement 6S-1
Document Control

Requires that documents specifying quality requirements be identified, reviewed, approved and controlled.

4.5 Document Control

Requires that documents specifying quality requirements be identified, reviewed, approved and controlled.

COMPARISON

NQA-1 and ISO-9001 have equivalent requirements except that NQA-1 allows minor revisions to be made without the same review and approval required for the original document or major revisions. ISO-9001 requires that a master list, or equivalent, is established to identify current revision of the documents and it recommends identification of the nature of the changes to the documents.

CONCLUSION

A quality program that meets the requirements of ISO-9001 will meet the requirements of NQA-1. Conversely, a quality program that meets the requirements of NQA-1 may not meet the requirements of ISO-9001.

Basic Requirement 7
Supplement 7S-1
Control of Purchased Items and Services

Requires that the purchase of items and services is controlled. Provides for such evaluation, evaluation of objective evidence, source inspection, audit and examination

4.6.1 General
4.6.2 Assessment of Subcontractors
4.6.4 Verification of Purchased Product
4.7 Purchaser Supplied Product

Requires that the purchase of items and services is controlled. Provides for such evaluation, evaluation of objective evidence, source inspection, audit and examination.

COMPARISON

NQA-1 contains additional requirements for external audits, control of nonconformance, procurement planning and commercial-grade items. Commercial-grade items are not applicable to ISO-9001. NQA-1 provides specific requirements for the control of purchased items and services.

CONCLUSION

A quality program that meets the requirements of ISO-9001 will meet the requirements of NQA-1. Conversely, a quality program that meets the requirements of NQA-1 may not meet the requirements of ISO-9001.

(continues)

COMPARISON OF
ASME NQA-1 AND ISO-9001

Basic Requirement 8 **Supplement 8S-1** **Identification and Control of Items**	**4.10 Inspection and Testing** **4.8 Product Identification and Traceability**
Requires that items be identified and controlled to prevent the use of nonconforming items.	Requires items be identified and controlled to prevent the use of nonconforming items.

COMPARISON

NQA-1 contains additional requirements for the control of limited life items. The other requirements for identification and control of items are equivalent.

CONCLUSION

A quality program that meets the requirements of NAQ-1 will meet the requirements of ISO-9001. Conversely, a quality program that meets the requirements of ISO-9001 may not meet the requirements of NQA-1.

Basic Requirement 9 **Supplement 9S-1**	**4.9 Process Control**
Requires that all processes affecting be controlled. Special processes must be performed in accordance with written procedures by qualified individuals using qualified equipment	Requires that all process affecting be controlled. Special processes must be performed in accordance with written procedures by qualified individuals using qualified equipment.

COMPARISON

NQA-1 and ISO-9001 have equivalent requirements.

CONCLUSION

A quality program that meets the requirements of either standard would meet the requirements of both standards.

(continues)

COMPARISON OF
ASME NQA-1 AND ISO-9001

Basic Requirement 10
Supplement 10S-1
Inspection

Requires that inspections performed to verify conformance of items to specified criteria be controlled. Items may not be used until after any required inspection has been successfully accomplished.

4.10 Inspection and Testing

Requires that inspections performed to verify conformance of items to specified criteria be controlled. Items may not be used until after any required inspection has been successfully accomplished.

COMPARISON

NQA-1 contains additional requirements for the qualification of inspection personnel. NQA-1 contains more specific requirements for records.

CONCLUSION

A quality program that meets the requirements of NQA-1 will meet the requirements of ISO-9001 conversely, a quality program that meets the requirements of ISO-9001 may not meet the requirements of NQA-1.

Basic Requirement 11
Supplement 11S-1 and 11S-2
Test Control

Requires that tests performed to verify conformance of items to specified criteria be controlled. Items may be used until any required test has been successfully accomplished.

4.10 Inspection and Testing

Requires that tests performed to verify conformance of items to specified criteria be controlled. Items may be used until after any required test has been successfully accomplished.

COMPARISON

NQA-1 contains additional requirements for the evaluation of test results by personnel that have specific authority. NQA-1 contains more specific requirements for records.

CONCLUSION

A quality program that meets the requirements of ISO-9001 will meet the requirements of NQA-1. Conversely, a quality program that meets the requirements of NQA-1 may not meet the requirements of ISO-9001.

(continues)

COMPARISON OF
ASME NQA-1 AND ISO-9001

Basic Requirement 12
Supplement 12S-1
Control of Measuring and Test Equipment

Requires measuring and test equipment be controlled and calibrated.

4.11 Inspection, Measuring and Test Equipment

Requires that measuring and test equipment Be controlled and calibrated.

COMPARISON

ISO-9001 contains a more comprehensive and detailed definition of the requirements for the control of inspection, measuring and test equipment. It also includes specific provisions for checking the capability of test hardware (such as jigs, fixtures, templates, patters) and test software. While the intent of NQA-1 is the same, it does not contain similar explicit requirements.

CONCLUSION

A quality program that meets the requirements of ISO-9001 will meet the requirements of NQA-1. (Basic Requirement 12 and Supplement 12-S). Conversely, a quality program that meets the requirements of NQA-1 may not meet the requirements of ISO-9001.

Basic Requirement 13
Supplement 13S-1
Handling, Storage and Shipping

Requires that product be controlled to prevent damage, loss or deterioration.

4.15 Handling, Storage, Packaging and Delivery

Requires that product be controlled to Prevent damage, loss or deterioration.

COMPARISON

ISO-9001 requirements are more programmatic in nature, NQA-1 requirements more specific. ISO-9001 includes a requirement for detecting deterioration of product(s) through periodic assessment of the condition of products in storage.

CONCLUSION

A program meeting NQA-1 may not completely meet the requirements of ISO-9001. Similarly, a program meeting ISO-9001 may not completely meet the requirements of NQA-1.

(continues)

COMPARISON OF
ASME NQA-1 AND ISO-9001

Basic Requirement 14
Inspection, Test and Operating Status

Requires that items be identified as to their conformance to specified requirements to prevent the use of nonconforming items. Also, the authority to identify the status of the items must be documented.

4.12 Inspection and Test Status

Requires that items be identified as to their conformance to specified requirements to prevent the use of nonconforming items. Also, the authority to identify the status of the items must be documented.

COMPARISON

NQA-1 and ISO-9001 have equivalent requirements.

CONCLUSION

A quality program that meets the requirements of either standard would met the requirements of both standards. One exception is that ISO-9001 specifically addresses test software while NQA-1 does not.

Basic Requirement 15
Supplement 15S-1
Control of Nonconforming Items
And Supplement 7S-1, Section 9
Control of Supplier Nonconformances

Requires that nonconforming items be identified and controlled until final disposition.

4.13 Control of Nonconforming Product

Requires that nonconforming items be identified and controlled until final disposition.

COMPARISON

NQA-1 and ISO-9001 have equivalent requirements. However, NQA-1, Supplement 15S-1 contains more detailed requirements concerning identification, segregation and dispositioning of nonconforming items.

CONCLUSION

A quality program that meets the requirements of NQA-1 will meet the requirements of ISO-9001. Conversely, a quality program that meets the requirements of ISO-9001 may not meet the requirements of NQA-1.

(continues)

COMPARISON OF
ASME NQA-1 AND ISO-9001

Basic Requirement 16
Corrective Action

Requires that conditions adverse to quality be identified and corrected. For significant conditions adverse to quality, corrective actions must be documented and reported to management, and the cause of the condition must be determined and corrected.

4.14 Corrective Action

Requires that conditions adverse to quality be identified and corrected. For significant conditions adverse to quality, corrective actions must be documented and reported to management, and the cause of the condition must be determined and corrected.

COMPARISON

Both standards take a programmatic approach to the identification and elimination of conditions that lead to nonconformance. ISO-9001 is more comprehensive in the identification of sources of information that are to be used for detection and elimination of causes of nonconforming product. NQA-1 is more specific in requirements for reporting to management, and for corrective action to be applied to any condition, adverse to quality of services, operations, etc., as well as product.

CONCLUSION

The programmatic and reporting requirements of the two standards are somewhat different. A quality program that meets the requirements of one standard may not meet the requirements of the other standard.

Basic Requirement 17
Supplement 17S-1
Quality Assurance Records

Requires that a document system be implemented to generate, validate, store, retrieve and dispose of records.

4.13 Quality Records

Requires that a document system be implemented to generate, validate, store, retrieve and dispose of records.

COMPARISON

NQA-1 contains more specific requirements for storage facilities and for indexing, distribution, identification, classification and storage of records. NQA-1 provides requirements for generation, validation and receipt of records not included in ISO-9001.

CONCLUSION

A quality program that meets the requirements of NQA-1 will meet the requirements of ISO-9001. Conversely, a quality program that meets the requirements of ISO-9001 may not meet the requirements of NQA-1.

(continues)

COMPARISON OF
ASME NQA-1 AND ISO-9001

Basic Requirement 18
Supplement 18-S1
Audits

Requires that a system of planned and documented internal and external audits be conducted.

4.1.2.2 Verification Resources and Personnel
4.17 Internal Quality Audits
4.18 Training

Required that a system of planned and documented internal and external audits be conducted.

COMPARISON

NQA-1 contains more additional requirements for external audits, audit plan and selection of the audit team. NQA-1 provides more specific requirements for scheduling, personnel, performance and documentation.

CONCLUSION

The programmatic and reporting requirements of NQA-1 will meet the requirements of ISO-9001. Conversely, a quality program that meets the requirements of ISO-9001 may not meet the requirements of NQA-1.

ISO-9001 Requirements Without Specific NQA-1 Counterparts

ISO-9001 also contains the following requirements. These requirements are addressed by NQA-1 throughout the document as part of the quality program but are not treated explicitly as separate sections.

4.3 Contract Review

Requires that each contract be reviewed to determine if the organization can adequately fulfill its contractual responsibility.

4.19 Servicing

Requires that procedures be established for servicing products.

4.20 Statistical Techniques

Allows for statistical techniques to be used for verification (inspection and test).

Appendix 11D: IAEA Review of *ISO-9001* Quality Assurance for Nuclear Projects

Internationally the IAEA code is used as the basis for the quality assurance (QA) arrangements for operating nuclear power plants. Nearly all utilities and regulators, particularly those outside the area of use, permit the use of *ISO-9001* as an acceptable standard for suppliers, with any additional requirements for safety issues forming part of the contractual arrangements or technical specification.

The British Nuclear Industries Forum (which comprises regulators, utilities and suppliers) has outlined the intention and application of QA documents in the nuclear arena as shown in Figure 11.C.1. In the U.K., all suppliers of safety-related spares or equipment are *ISO-9000*-certified and nothing else. A few are certified to ASME for the U.S. market.

There is no logical argument for staying with *10CFR50* Appendix B or *NQA-1*. Since ISO addresses all the management system requirements and is only short on design controls for safety-related equipment, which is overcome by placing additional management system requirements in the contracts.

British Nuclear Industries Forum
Working Group on Quality

Briefing Note **September 1996**

International Quality Code for Nuclear Installations

- The IAEA nuclear safety series code *50-C-Q* (QA for safety in nuclear power plants and other nuclear installations) is due to be issued in September 1996 supported by a suite of guides. This will replace the existing code *50-C-QA*.
- The U.K. nuclear industry provided a major input to the development of this suite of documents.
- The U.K. nuclear industry cites that licensees that intend to adopt intend to adopt *50-C-Q* in place of *British Standard 5882*.
- The code comprises a set of principles of QA for safety in nuclear installations.
- The code is focused on ensuring nuclear safety operations at a nuclear installation, and not on product quality; it is not intended for use by suppliers to the industry.
- The code has been designed for use between the regulatory body and the nuclear site licensee.
- The NII were key players in the development of and review process of the principles embodied in the documents and promote their use. They view the code as an appropriate alternative to *BS 5882*.
- The code is aligned with current total quality management (TQM) and business excellence objectives. It is consistent with the EFQM model and continuous improvement in that it is performance-oriented.

FIGURE 11.C.1. SUMMARY OF BRITISH NUCLEAR INDUSTRIES WORKING GROUP ON QUALITY RECOMMENDATIONS. (Figure continues)

British Nuclear Industries Forum
Working Group on Quality

Briefing Note **September 1996**

International Quality Code for Nuclear Installations

- Adoption of the code in place of *BS 5882* is unlikely to initiate significant change to establish QA arrangements, or to working practices requiring only minor adjustments to high-level documents.
- The code encourages the use of other related international standards for the licensee/supplier interface.
- Site licensees may wish to adopt *ISO-9000* in addition to *50-C-Q* for all or part of their organization as required by their customers.
- Application of Quality standards/code across nuclear industry interfaces in the U.K.

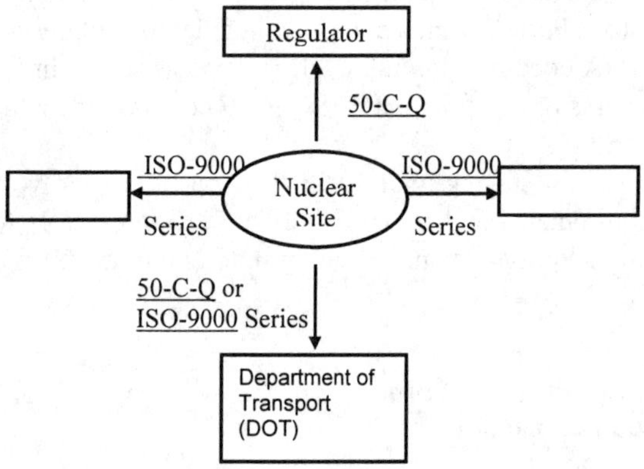

- There are 14 guides that support the code. These guides provide just one approach to satisfy the principles of the code and reflect best practice.
- There is no need or desire for third-party certification to *50-C-Q.*
- It is not appropriate for licensees to specify *50-C-Q* as a requirement to its suppliers.

FIGURE 11.C.1. (Continued)

Appendix 11E: Construction-Lessons Learned/Transition to Operations' Practical Quality Assurance—A Better Approach to Nuclear Construction and Operations, 1989

INTERNATIONAL CONFERENCE ON QUALITY IN THE NUCLEAR POWER INDUSTRY

CONSTRUCTION-LESSONS LEARNED/TRANSITION TO OPERATIONS

PRACTICAL QUALITY ASSURANCE—A BETTER APPROACH TO NUCLEAR CONSTRUCTION AND OPERATIONS

By Roger F. Reedy, President

Reedy Associates, Inc.
15951 Los Gatos Blvd., Suite 1
Los Gatos, CA 95032

ABSTRACT

Nuclear power plant construction is dormant in the United States because of increased costs. A large portion of these costs are associated with inappropriate application of quality assurance (QA) rules. QA, as practiced in the U.S., has focused more on documentation than on important equipment characteristics. Fortunately, there is a way to meet the QA/QC (quality control) requirements of codes, standards and regulations; reduce costs by billions of dollars; *and* have better nuclear plants. The necessary changes are consistent with common sense and good engineering practices. With these changes, we can have affordable electricity and safe plants.

INTRODUCTION

Construction of commercial nuclear power plants in the United States started in the early 1950s. The construction practices were practical and efficient, and the outlook for the power industry was very optimistic for low-cost electricity. In the mid-1960s the construction boom seemed almost unlimited. This continued into the 1970s, and then conditions started to change. This paper reviews the reasons for the change, many of which are familiar to all, and suggests a course for the future, both for construction and operation.

A BETTER MANAGEMENT TOOL

In 1970, quality assurance (QA) was introduced to the industry through *10CFR50,* Appendix B. The time was right for better management control of quality because of the rapid growth and complexity of nuclear construction. QA is needed as a set of management controls for complicated projects that may affect public health and safety. *10CFR50,* Appendix B was definitely a step in the right direction. The principles of Appendix B had been used and were proved to be quite effective in the nuclear navy, NASA and military applications.

Proper QA helps get the job done in a practical and efficient manner and, consequently, helps save money. Personal experience demonstrated this was always the case. My employer in the 1960s and early 1970s built the first three field-erected nuclear reactors without any required weld repairs. The crew was very disappointed when a small indication showed in one weld on the fourth vessel, but knew that nothing is perfect. For the same vessels, all machined surfaces (field-machined) met all specified tolerances without any correction. The result was an improved schedule and lower costs. The company's nuclear QA program in the shop was considered so effective, it was applied to all standard work in that shop with the same results. The emphasis was on "zero defects" in the products being fabricated.

What happened to change this approach to QA? Somehow, it seems the industry lost sight of "zero defect" quality assurance and placed the emphasis on proving quality by documentation. In addition, the industry started to concentrate on meeting the "exact" wording of codes and standards. This caused results never envisioned by engineers writing the codes and standards.

INDEPENDENCE OF QUALITY ASSURANCE GROUPS

Quality can best be built into a product by teamwork. The quality control (QC) group should be part of the company team reporting to production management, but Appendix B has often been interpreted as requiring complete independence of both the QA and QC groups. When the QC group is not part of the production team, it becomes an outsider and is almost forced to act as a police group. When this occurs, there is often resentment on the part of workers, which results in conflict between inspectors and craftsmen. This fact is pointed out in the NRC report, *NUREG 1055,* "Improving Quality and the Assurance of Quality in the Design and Construction of Nuclear Power Plants."

In the mid-1970s, such conflicts did occur and some inspectors claimed they were being harassed and intimidated. The result was the NRC issued a rule against harassment and intimidation of inspectors. Unfortunately, this rule has increased the gap between the inspectors and the workers, and has sometimes made it very difficult to correct an inspector who is performing incorrectly. This situation has in some cases caused reverse intimidation against management. Some inspectors have used the intimidation rule as a means of ensuring they are not overruled. This type of action could go counter to good engineering and management practices.

When the QA and QC are considered independent from cost and schedule considerations, work may be delayed for trivial considerations. Unfortunately, the pursuit of trivia became widespread in the industry and the effect was to almost double construction costs and schedules as well as stop construction of new plants because of cost considerations.

NUREG 1055 also recommends that QC activities and personnel be part of the production group. This type of organizational setup was used by the company that built the defect-free nuclear reactors. That company has always had the QC group as part of the production group. This means the QC group was always part of "the team." This setup improves quality and eliminates potential conflicts.

CONCENTRATION ON DOCUMENTATION

Another problem responsible for extreme cost escalation was based on the rather simple definition in QA standard *N45.2* (*NQA-1,* "Quality Assurance Program Requirements for Nuclear Facilities"). It is obvious from reviewing military QA specifications, upon which Appendix B is based, that only an *item of hardware* can be nonconforming. This makes sense, because if the hardware conforms to specifications, codes standards and design requirements, it is a quality item. The crucial change to the definition in *N45.2 (NQA-1)* was to add a few simple words, which, over time, changed practical QA for the hardware into a huge documentation review. We have all seen the photographs of large valves, with the documentation stacked next to it. The volume of the paper is greater than the volume of the valve.

The *NQA-1* definition is given below. The italicized words show how emphasis was placed on documentation. If the documentation was not right, the work stopped or was considered defective. This was bad because there was no definition that identified when the documentation was adequate, which meant the inspectors could make their own definitions. QA standards identified the titles of records to be kept, but no definition for the contents of records was identified.

> "Nonconformance: A deficiency in characteristic, *documentation or procedure* that renders the quality of an item *or activity* unacceptable or *indeterminate.*"

With no definition of "adequate" documentation, anyone could claim that the documentation was deficient. A study of nonconformance reports (NCRs) at several plant sites showed that a very large majority of the NCRs were written on documentation, not equipment.

Two other definitions that compounded the problem are:

> "Acceptance Criteria: Specified limits placed on characteristics of an item, *process or service* defined in codes, standards or other requirement documents."

> "Characteristic: Any property or attribute of an item, *process or service* that is distinct, describable and measurable."

The italicized words in these three definitions completely changed QA practices in the United States, making them rely heavily on documentation reviews. This practice was criticized by the NRC in *NUREG 1055,* but codes and standards have not been revised to correct the problem even though *NUREG 1055* was written more than five years ago.

Once the QA practices started concentrating heavily on documentation, the door was open for highly inflated material costs. The cost of documentation had to be included in the price structure. In order to obtain the proper documentation, it is not uncommon to pay astronomical prices for small items. Three recent cases fully illustrate the scope of the problem.

1. A small pipe coupling, which could be bought for $1 at the local hardware store, cost $500. This is worse than the famous $400 hammer in military contracts, but is justified by the supplier because of the need for traceability and documentation.
2. A stainless-steel pipe elbow which could be purchased through the suppliers catalogue for $5,000, cost $100,000 in order to obtain the documentation.
3. After $60,000 of piping was delivered to a nuclear plant site, it was determined that QA documentation was required. The cost for obtaining the documentation requested was $18,000. This is a 30% markup on material that was adequate as delivered, except for the

associated paper. The increase in price provided nothing to improve the quality of the material.

This increase in material costs can be partly attributed to programmatic and documentation requirements imposed upon the material manufacturers and suppliers. In order to ensure that the proper material is installed and used in the plant, Section III of the ASME code requires that material manufacturers and suppliers be surveyed and audited to determine that QA programs are implemented. Both the NRC and the *NQA-1* standard strongly imply the same approach should be used. The Section III code has provisions for ASME to accredit material manufacturers and suppliers by use of a survey as one means of ensuring an acceptable QA program. The surveys and audits concentrate on implementation of the QA criteria and on documentation. This system of surveys and audits helps inflate the price of the material as detailed above, but it does not and cannot ensure that the proper material will be delivered. Both fraud and incompetence on the part of material manufacturers and suppliers, neither of which can be determined by surveys and audits, can cause the wrong material to be delivered to the site or manufacturing facility.

Section III of the ASME code allows material to be "upgraded" from one material specification to another. Using this process, it is possible to supply material of lower strength specifications as higher strength material. If done in a fraudulent or incompetent manner, inferior material can be supplied as good material. QA surveys and audits cannot prevent this. NRC *Bulletin 88-05* showed that this situation currently exists and may affect both plants under construction and those in operation. The surveys and audits cost utilities millions of dollars each year, but have been proved to be ineffective for detecting these problems.

The only practical and effective means to ensure the proper material is received is to test the material upon receipt. Receipt of the proper material was never a problem in the pressure vessel and piping industry before the advent of formal QA programs. The accent on documentation escalated prices and provided incentives for new suppliers to "upgrade" material. Prior to 1970 there was no incentive or code provisions for material suppliers to "upgrade" material.

The QA surveys and audits of material manufacturers and suppliers should be stopped now. A better approach can be based on a receipt-testing program as allowed by Appendix B.

The issue of documentation is also very expensive in other ways. The cost of collection and maintenance of documentation at a nuclear plant exceeds millions of dollars annually, even after the plant is in operation. Some utilities may be saving as many as 1,000,000 documents a year, and most of these documents are many pages long. Effort is now being made to reduce the problem by identifying the essential information to save, rather than saving all documents generated, but it may be several years before revised standards are published. As noted above, in the past, the NRC documents and QA standards identified the records to save by name, but nowhere did they identify the information to be included in the records. The result of course is that records at different plants often contain different information. Requirements are met when *all* information is kept, but storage, cataloging and retrieval is difficult and time-consuming. It is possible that the needed information isn't included in the correct record because requirements for record contents was never identified. If a utility could stop the practice of retaining "all documents," and keep only "essential data," cost savings would amount to more than $1,000,000 annually.

CLASSIFICATION AS SAFETY-RELATED

Everyone understands that "safety-related" items are more important than non-safety-related items. However, the distinction between the two is often blurred because of where the line can be drawn when there is an indirect relationship. For example, consider a prestressed concrete containment vessel. The concrete shell and the liner material should be considered very important. The concrete for strength, and the liner for leak-tightness. The prestressing tendons are also important for strength. Is the tube as important or of a secondary concern? It provides no strength or leak-tightness. Is the protective grease in the tube as important? It also provides no strength or leak-tightness. Is the grease cap at the top of the tube as important? Because each of these are connected to each other, sometimes all are classified the same, and the cost of the grease and the grease cap may be 10 to 100 times the cost of commercial items. The fact is that neither the grease nor the grease cap have any *direct* relationship to safety, and adequate grease and grease fittings can be purchased commercially without imposing a nuclear QA program.

PURSUIT OF TRIVIA

Similar to the example of the grease and the grease cap, there are many examples of similar pursuit of trivia. All these examples and the one presented earlier are based on actual occurrences, all increased cost and delayed schedules, and none had anything to do with plant safety nor increased quality. If repairs were made because of any of these actions, the quality of the item may have actually suffered, even though all requirements were met and adequate safety margins were still provided.

At one plant, 500 weld repairs were made in stainless-steel welds. The final radiographs showed the welds had no defects after repair. The inspector however refused to accept the welds because an in-process examination was not documented. He insisted that a code case be requested from ASME or that the repairs had to be cut out. A code case was published to resolve the issue, which required an ultrasonic examination of the weld in lieu of documentation of the in-process examination. The problem with the solution was that it was extremely difficult at that time to use ultrasonic testing to obtain meaningful results in stainless-steel welds. The inspector understood this, but now he had an official way in which he could accept the repairs. The welds were shown to be good, not by the code case, but by the radiographs taken after the repairs. The in-process documentation is meaningless after the radiographs show the welds to be adequate.

The following cases are highlights of more trivial concerns that had no effect on safety, but are examples of issues that cost the nuclear industry billions of dollars.

- An NRC inspector found some fillet welds unacceptable because some small areas of these welds were 1/64″ under the specified size. This was trivial because fillet welds are designed by engineers to the nearest 1/16″. Using standard engineering practices, measurements of these welds should be to the nearest 1/32″. Four pages of this paper are thicker than 1/64″, so it is easy to see the insignificance of this amount of undersize. Since then, codes were changed to accept rather large areas of fillet welds that were even greater than the 1/16″ undersized. Unfortunately, many slightly undersized welds were rejected and repaired before the situation was corrected in 1984.
- Some components were placed on "hold" because the documentation was signed in blue ink rather than black ink as required by procedures. At the time, documentation in blue ink didn't reproduce well. However, the "hold" cost, time and money caused nonconformance reports to be written, and was completely unrelated to safety. Because the procedures weren't followed, inspectors felt the QA program had been violated.
- A welder's qualifications are usually valid for three- to six-month intervals, depending on his welding activities. If the welder doesn't weld with the procedure within the appropriate period of time, the qualification expires. QA inspectors sometimes required good welds to be removed because the welders welded within a day or two after their qualification expired. This happened because the QC inspector was calculating time using 30 days as a month. Therefore he considered the three-month requirement to be 90 days and six-month requirement to be 180 days. He was not willing to accept 31-day months. The welds were good, but because of a difference in definition regarding timing, the welds were replaced. As pointed out in *NUREG 1055,* situations like this put the craftsman welder in conflict with the inspectors.
- At a number of sites, piping material was required to be cut out of piping systems because when installation was complete, some documentation was lost. The fact that the documentation was reviewed several times during installation of work had no bearing on the case. Replacement was difficult, costly and most likely created higher residual stresses in the system. Replacement of the material certainly did not improve quality.

- At one plant, an NCR was generated to address a small dent in a 5/8″ tube for instrument piping. The first recommended disposition was "use-as-is." In order to resolve the issue, the NCR became 29 pages long and had 60 signatures of approval. The time to clear the NCR was nine months; during that period no work could be performed on the tubing. The thickness of the tube was more than three times that required for the highest pressure in any system in the plant.
- *NRC-I&E Bulletin 79-02* concerned base plates. It recommended that base-plate flexibility be taken into account for seismic analysis of piping systems. This action cost untold millions of dollars and shocked professional structural engineers. Base-plate flexibility was not an issue and was neglected as insignificant before the bulletin was issued. Before computers it couldn't be calculated. It is still considered unnecessary by many professional engineers today. The cost of performing the analysis was staggering, but changes as a result of the calculations were minimal.
- *NRC-I&E Bulletin 79-14* occurred after TMI and instigated the practice of checking to ensure that drawings show all dimensions *exactly* as measured in the field. During the field walk-down, tolerances were often ignored. After the "as-built" dimensions were recorded, new stress analysis were performed using the recorded dimensions. Costs associated with the stress analysis often exceeded $15,000,000. Although it seldom caused any plant modifications, it delayed plant construction and had no meaning with regard to increased plant safety.

COST OF NEEDLESS NCRS

Several years ago I studied the administrative cost of unnecessary NCRs for nuclear plant construction. Based on this study, presented at a conference in 1986, the cost was estimated as $500,000,000 for a nuclear plant. This was only the administrative cost and did not include the cost of resolution of the NCRs. The $500,000,000 was based on the study, which showed 95% of the NCRs at a plant construction site were either related to documentation issues not associated with the equipment or were resolved "use-as-is," which meant that the equipment was adequate. Further, the number of NCRs was about 800 per month when all organizations at the site were considered, and the administrative cost was estimated at $5500/NCR, based on the estimates by utilities. If a 10-year construction time is assumed, $500,000,000 is the result. Although the estimate was published two years ago, the only criticism is that the number was underestimated by at least a factor of two.

If all the unnecessary engineering analysis and unnecessary repairs and cost of delays are added together, it is easy to see that the unnecessary work may have cost more than $1 billion per plant site of the most recent vintage. If steps could be taken now to eliminate a repetition of this unnecessary work, the cost of nuclear plant construction could be reduced to a practical level.

The codes and standards were often changed to help eliminate the concerns raised regarding minor issues. In 1969, the code for nuclear piping was about 70 pages long. In 1980, the code requirements associated with piping in a nuclear plant were more than 1,200 pages long. Physically, piping systems in nuclear plants built to the 1969 requirements are almost identical to the piping systems built to the 1980 requirements. There is no significant increase in safety due to all the changes and increased number of pages, but the big detrimental change is that acceptance of engineering judgment has been eliminated from the wording in the code.

Let's examine some of the issues the code tried to address. A good illustration of a trivial issue is socket-welding requirements.

In order to illustrate the detail this issue caused by incorrect interpretations of code requirements, let's review how the ASME Section III code kept revising requirements for socket welds. For the most part, this issue of socket welds concerns 1″-diameter (NPS 1) piping and less. Because piping of this size is generally two to three times the thickness required for pressure, and the consequences of failure are minor compared to 20″-diameter pipe breaks, one would probably not consider the issue to be of high priority, but because the number of socket welds is a large percentage of the number of total pipe welds, the issue is important with regard to plant cost. At one plant, the cost of installation of piping and tubing of 1″-diameter and less changed from an original estimate of about $6,000,000 to about $60,000,000 because of similar issues.

The 1967 *ANSI B31.1* piping code has a sketch showing the use of socket welds. The sketch shows a gap between the tube and the coupling, which is specified as "1/16″ minimum before welding." In the installation of millions of socket welds in fossil plants, compliance with the sketch was never a problem. The figure illustrates the expected gap before welding, with no requirements for verification or documentation after welding. When that detail was used at nuclear plants, the ASME code had to be revised to state, "1/16″ approximately before welding." This change was made because QC inspectors were rejecting socket welds that had gaps slightly exceeding 1/16″, or when they couldn't verify that the gap was 1/16″ before welding or they could not verify that a 1/16″ gap existed after welding. The intent of this code change was to show that the dimension was only an approximate dimension, and that an exact measurement was not required.

The purpose of providing the 1/16″ gap was only to provide a guideline of good workmanship for the welders. Without a gap, there may be a problem that the fillet weld would crack due to the natural shrinkage, which occurs after welding. If the fillet weld is made and no crack occurs, there is no problem and the size of the gap is immaterial. However, with legalistic interpretations, QC inspectors determined that the gap had to be verified both before and after welding.

In order to verify the gap after welding, some QC inspectors determined that the fitting had to be radiographed. If the radiograph showed the gap was 1/16″, the fitting may have been accepted. If it were less than 1/16″, the good weld had to be cut out and rewelded. In cutting out the weld, the length of the socket fitting was sometimes reduced and the inspectors then sometimes insisted that, because the fitting was now reduced in length and no longer met the dimensions of the appropriate standard for socket coupling, it therefore had to be replaced. This meant that two fillet welds had to be replaced and a new fitting procured and installed because the "approximate gap" was not present after welding. There is no requirement of any kind for these radiographs to have been made.

After radiographing to determine the gap (even the radiograph showed an adequate gap), often the radiograph would show the weld was not completely fused at the root or base of the fitting. The QC inspectors then rejected the weld because the code stated, "In making fillet welds, the weld metal shall be deposited in such a way that adequate penetration into the base metal at the root of the weld is secured." Note that "adequate penetration" was never defined by the code, so this left the inspector to decide what was "adequate penetration." The code does not require that socket welds be radiographed to verify "adequate penetration" and does not require replacing socket welds that do not show full fusion at the base of the socket. The only required examination is on the surface of the weld. To solve this problem, the ASME Section III code was revised in 1983 to delete the sentence regarding "adequate penetration." Then in 1984, the code was again revised by adding a sentence to state, "The gap need not be present or verified after welding." All of these changes occurred without any change in design philosophy, but all were necessary to attempt to stop incorrect interpretations regarding a very minor issue.

A related problem occurred regarding the size of the socket weld. Originally, the minimum size was stated in the code as being 1 to 1/4 times the thickness of the pipe or tube. For most socket fittings, this was possible, but in some sizes, there was not enough material to obtain a weld of this size. The fitting thickness was only 1.09 times the thickness of the mating tube. Therefore the figure in the code was revised to state that the fillet weld need only be 1.09 times the thickness of the tube. This change from 1 to 1/4 to 1.09 was interpreted to mean that socket welds now had to be measured to an accuracy of 0.01 inch. This was not the intention, and action is now being considered to change the dimension from 1.09 times the thickness of the tube to one times the thickness. In addition, a further change was required to indicate the 1.09 was based on nominal pipe-wall thickness, because QC inspectors were interpreting that the size required was based on the actual measured thickness of each pipe at the socket weld.

The cost ramifications on these small socket welds can be illustrated by the fact that QC inspectors often required "hold points" for socket-welded fittings. This means that at least two "hold points" were often required for each fitting, one before welding and one after. The time of making one socket-weld connection has increased from approximately 15 minutes to about four hours. When it is understood that there are hundreds of thousands of small socket welds needed for each nuclear plant, the associated cost is staggering. In 1967, no verifications of socket welding were required to be made or documented.

Another illustration of impractical interpretations occurred several years ago when one contractor replaced another at a nuclear site. The first contractor reviewed radiographs of completed storage tanks and rejected 75 radiographs (out of 400) as not meeting code requirements, even though the manufacturer and authorized nuclear inspector accepted the radiographs as being in full compliance. When the second contractor took over the site work activities, he also reviewed the radiographs and rejected 75 as not meeting code standards. The problem, however, was that only 25 of the two groups of 75 radiographs were the same. Also, none of the radiographs showed actual flaws. The rejections were primarily based on film density, penetrameter image or interpretation of questionable indications. The installation of the tanks was held up for almost a year, and there were no confirmed flaws in the welds.

These examples of actual occurrences illustrate how inspections changed from practical engineering, common sense and good teamwork to the philosophy of finding fault no matter how trivial the concern. All of these examples caused tremendous increases of cost and provided no extra assurance of quality or safety. In fact, any associated repairs probably increased residual stresses and therefore caused more harm than good. The results of the efforts of these QA groups were counter to the objectives of *10CFR50,* Appendix B, and the cost consequences have turned many rate payers against nuclear power. As stated by the cartoon character Pogo many years ago, "We have found the enemy, and they are us!"

CHANGES FOR THE FUTURE

Is there any way to learn from these examples? I believe there is. We can operate within the confines of Appendix B and construct quality nuclear plants at less than half the price of the latest vintage plants. Many of the same changes in QA practices can drastically reduce the cost of operating nuclear plants. My own experiences point out ways to reduce operating costs by many millions of dollars annually. To do so requires management cooperation and a minor budget allocated to cost-savings ideas.

Some industry changes that can be made immediately are simple and outlined below:

- Change the ASME *NQA-1* standard to redefine a "nonconformance" as only applying to an item.
- Nuclear plants should establish one group of senior professional engineers as being responsible for interpreting design requirements and codes and standards. All design questions, code interpretations and questions regarding acceptance of an item should be made by this group, and if necessary, they can obtain the official interpretations from the Codes and Standards Committees.
- Engineering tolerances must be established for all items.
- The NRC must modify its rule on intimidation, so that "reverse intimidation" does not cause safety problems. The engineer must be able to overrule the inspector regarding acceptability of an engineered item. The NRC and nuclear plant management must ensure that the responsible engineers are not overruled by inspectors. Equipment installed at nuclear sites are usually engineered items, and therefore professional engineers must make the final decision regarding acceptability and suitability for service. To allow professional engineers to be overruled by inspectors could cause serious safety issues. At the same time, the engineer must be held responsible for his work.
- Designate responsible individuals to take responsibility. When everyone is responsible, no one is responsible. Limit approval signatures on any document to one person and hold him responsible.
- Follow the recommendations of *NUREG-1055* and keep the QC inspectors as part of the production or operations team.
- Stop inspecting code items and documentation for adequacy when they have already been certified as meeting ASME code requirements.
- Identify the "essential data" to be maintained at nuclear sites and use "verified" summaries to reduce the proliferation of documentation.

- Rewrite the QA standards to emphasize quality results and eliminate prescriptive details as to how activities are to be accomplished.
- Instead of performing surveys and audits of material manufacturers and suppliers, initiate material testing programs, which will verify the material to be installed in a nuclear plant and track the performance of suppliers. The testing program for steel can be based on both strength and chemistry and should allow sampling as appropriate. For other materials, other appropriate characteristics would be measured.

CONCLUSION

We can simplify designs and standardize nuclear plants, but the resulting cost savings will amount to nothing if we can't stop the impractical and inappropriate approach to QA that has occurred in the last 15 years. If the implementation of the QA program doesn't save money, it is probably a faulty program looking for the wrong issues. The workers of this country are craftsmen. They are proud of quality work. We must establish programs that encourage quality work. Make inspectors part of the team and use QA programs as management tools to understand construction and operating activities as they occur.

Implementing changes such as these will improve quality and make nuclear power affordable. Then, perhaps, the issue of "prudency" will fade away, because nuclear power will be cost-effective. Let us learn from our past experiences. There are better ways to do things, but we must change now. It is possible to reduce plant operation, maintenance and construction costs immediately, and at the same time have better assurance of quality and reliability. Quality and economics are not oxymoronic terms.

Appendix 11 F: Summary of a Typical Approach to a Graded QA Program Manual

FORWARD

This appendix is meant to summarize current philosophy associated with graded quality assurance (QA) as it applies to QA program manual. It is not intended to be part of the QA program manual.

The graded approach to QA considered in this manual can be divided into three levels:

- Level 1: Typically employed in construction and procurement of nuclear safety-related products, materials and services in high-hazard facilities (PC-5 and PC-4).
- Level 2: Typically employed in construction and procurement of safety-related and for mission-essential products, materials and services in moderate-hazard facilities (PC-3).
- Level 3: Typically employed in non-safety-related, commercial grade products, materials and services (PC-2 and PC-1).

Levels 1 and 2 QA is always a formal program and are presented in this manual. Level 3 is an informal program consisting of a series of project standard operating procedures and is not included in this manual.

The three levels include:

Level 1

- Development of a detailed QA program manual, which regulates all aspects of the project QA program.
- Formation of a project QA organization, which is intended to assist the line organization in meeting its QA responsibilities and commitments.
- Establishes that QA is the responsibility of the project organization.
- Provides for assessment (management and independent) of all aspects of project QA.
- Provides for formal procedures for resolution of design changes and nonconformances identified between "as-designed" and "as-built" construction.
- Provide QA requirements associated with design review.
- Provides for audits of lower-tier vendors of products, materials and services.

Level 2

- Development of a QA manual, which regulates the QA program and includes reference to applicable regulatory, corporate or client policy procedures, specifications and contract requirements.
- Does not usually have a project QA organization attached to the project.

- Provides for management and independent assessment by QA personnel not attached to the project.
- Provides for the requirement of documented resolution of nonconformances identified between "as-designed" and "as-built" construction.

Level 3

- No formal manual is required. The QA effort is provided by reference to applicable regulatory, corporate or contract specifications.
- QA program is the responsibility of and is administered by the project line organization.
- No routine or scheduled audit function.
- Resolution of design changes and nonconformances is by field engineers who are competent to evaluate the intent of the design basis and who are allowed to use judgment in the evaluation of the significance of the nonconformance.

Professional services QA manuals are typically organized using the *10CFR830.122,* IAEA Safety Series 50/SG-Q format while QA program manuals for suppliers or vendors of products and services are organized using the *10CFR50* Appendix B, *ASME NQA-1* and *ISO-9001-74* formats.

Generally, there are two types of Quality Assurance programs, one applied to professional services organizations such as lawyers, medical doctors and professional engineers, and the other applied to suppliers and vendors of safety-related products and services.

- *10CFR50,* Appendix B, QA Criteria for Nuclear Power Plants
- *ASME NQA-1-2000*, QA Program for Nuclear Power Plants
- IAEA Safety Series, No. 50-C/SG-Q, QA for Safety in Nuclear Power Plants and other Nuclear Installations.
- *10CFR830*.120 and DOE Order 420.1A, QA
- *ISO-9001-94*, QA Criteria and Requirement

INDEX